Recent Technologies for Enhancing Performance and Reducing Emissions in Diesel Engines

J. Sadhik Basha
International Maritime College Oman, Sohar, Oman

R.B. Anand
National Institute of Technology, Tiruchirappalli, India

IGI Global
DISSEMINATOR OF KNOWLEDGE

A volume in the Advances in Mechatronics and Mechanical Engineering (AMME) Book Series

Published in the United States of America by
IGI Global
Engineering Science Reference (an imprint of IGI Global)
701 E. Chocolate Avenue
Hershey PA, USA 17033
Tel: 717-533-8845
Fax: 717-533-8661
E-mail: cust@igi-global.com
Web site: http://www.igi-global.com

Library of Congress Cataloging-in-Publication Data

Names: Basha, J. Sadhik, 1977- editor. | Anand, R. B., 1976- editor.
Title: Recent technologies for enhancing performance and reducing emissions in diesel engines / J. Sadhik Basha and R.B. Anand, editors.
Description: Hershey, PA : Engineering Science Reference, an imprint of IGI Global, [2020] | Includes bibliographical references and index. | Summary: "This book provides research on technical and environmental enhancements to the emission and combustion characteristics of diesel engines. While highlighting topics such as biodiesel emulsions, nanoparticle additives, and mathematical modeling, this book explores the potential additives that have been incorporated into the performance of diesel engines in order to positively affect the environment"-- Provided by publisher.
Identifiers: LCCN 2019042201 (print) | LCCN 2019042202 (ebook) | ISBN 9781799825395 (hardcover) | ISBN 9781799825401 (paperback) | ISBN 9781799825418 (ebook)
Subjects: LCSH: Diesel motor--Alternative fuels. | Diesel motor--Combustion. | Motor vehicles--Motors (Diesel)--Exhaust gas.
Classification: LCC TP343 .R37 2020 (print) | LCC TP343 (ebook) | DDC 621.43/6--dc23
LC record available at https://lccn.loc.gov/2019042201
LC ebook record available at https://lccn.loc.gov/2019042202

This book is published in the IGI Global book series Advances in Mechatronics and Mechanical Engineering (AMME) (ISSN: 2328-8205; eISSN: 2328-823X)

British Cataloguing in Publication Data
A Cataloguing in Publication record for this book is available from the British Library.

For electronic access to this publication, please contact: eresources@igi-global.com.

Advances in Mechatronics and Mechanical Engineering (AMME) Book Series

J. Paulo Davim
University of Aveiro, Portugal

ISSN:2328-8205
EISSN:2328-823X

Mission

With its aid in the creation of smartphones, cars, medical imaging devices, and manufacturing tools, the mechatronics engineering field is in high demand. Mechatronics aims to combine the principles of mechanical, computer, and electrical engineering together to bridge the gap of communication between the different disciplines.

The **Advances in Mechatronics and Mechanical Engineering (AMME) Book Series** provides innovative research and practical developments in the field of mechatronics and mechanical engineering. This series covers a wide variety of application areas in electrical engineering, mechanical engineering, computer and software engineering; essential for academics, practitioners, researchers, and industry leaders.

Coverage

- Medical robotics
- Control Systems Modelling and Analysis
- Biologically Inspired Robotics
- Mechanisms and machines
- Bioengineering Materials
- Computational Mechanics
- Control Methodologies
- Design and Manufacture
- Intelligent Sensing
- Computer-Based Manufacturing

IGI Global is currently accepting manuscripts for publication within this series. To submit a proposal for a volume in this series, please contact our Acquisition Editors at Acquisitions@igi-global.com or visit: http://www.igi-global.com/publish/.

The Advances in Mechatronics and Mechanical Engineering (AMME) Book Series (ISSN 2328-8205) is published by IGI Global, 701 E. Chocolate Avenue, Hershey, PA 17033-1240, USA, www.igi-global.com. This series is composed of titles available for purchase individually; each title is edited to be contextually exclusive from any other title within the series. For pricing and ordering information please visit http://www.igi-global.com/book-series/advances-mechatronics-mechanical-engineering/73808. Postmaster: Send all address changes to above address.

Titles in this Series

For a list of additional titles in this series, please visit: www.igi-global.com/book-series

Diverse Applications of Organic-Inorganic Nanocomposites Emerging Research and Opportunities
Gabriele Clarizia (Institute on Membrane Technology, National Research Council, Italy) and Paola Bernardo (Institute on Membrane Technology, National Research Counci, Italy)
Engineering Science Reference • © 2020 • 235pp • H/C (ISBN: 9781799815303) • US $195.00

Handbook of Research on Artificial Intelligence Applications in the Aviation and Aerospace Industries
Tetiana Shmelova (National Aviation University, Ukraine) Yuliya Sikirda (Flight Academy of National Aviation University, Ukraine) and Arnold Sterenharz (EXOLAUNCH GmbH,Germany)
Engineering Science Reference • © 2020 • 517pp • H/C (ISBN: 9781799814153) • US $295.00

Design and Optimization of Sensors and Antennas for Wearable Devices Emerging Research and Opportunities
Vinod Kumar Singh (S. R. Group of Institutions Jhansi, India) Ratnesh Tiwari (Bhilai Institute of Technology, India) Vikas Dubey (Bhilai Institute of Technology, India) Zakir Ali (IET Bundelkhand University, India) and Ashutosh Kumar Singh (Indian Institute of Information Technology, India)
Engineering Science Reference • © 2020 • 196pp • H/C (ISBN: 9781522596837) • US $215.00

Global Advancements in Connected and Intelligent Mobility Emerging Research and Opportunities
Fatma Outay (Zayed University, UAE) Ansar-Ul-Haque Yasar (Hasselt University, Belgium) and Elhadi Shakshuki (Acadia University, Canada)
Engineering Science Reference • © 2020 • 278pp • H/C (ISBN: 9781522590194) • US $195.00

Automated Systems in the Aviation and Aerospace Industries
Tetiana Shmelova (National Aviation University, Ukraine) Yuliya Sikirda (Kirovograd Flight Academy of the National Aviation University, Ukraine) Nina Rizun (Gdansk University of Technology, Poland) Dmytro Kucherov (National Aviation University, Ukraine) and Konstantin Dergachov (National Aerospace University – Kharkiv Aviation Institute, Ukraine)
Engineering Science Reference • © 2019 • 486pp • H/C (ISBN: 9781522577096) • US $265.00

Design, Development, and Optimization of Bio-Mechatronic Engineering Products
Kaushik Kumar (Birla Institute of Technology, India) and J. Paulo Davim (University of Aveiro, Portugal)
Engineering Science Reference • © 2019 • 312pp • H/C (ISBN: 9781522582359) • US $225.00

Nonlinear Equations for Problem Solving in Mechanical Engineering
Seyed Amir Hossein Kiaeian Moosavi (Babol Noshirvani University of Technology, Iran) and Davood Domairry Ganji (Babol Noshirvani University of Technology, Iran)

IGI Global
DISSEMINATOR OF KNOWLEDGE

701 East Chocolate Avenue, Hershey, PA 17033, USA
Tel: 717-533-8845 x100 • Fax: 717-533-8661
E-Mail: cust@igi-global.com • www.igi-global.com

Table of Contents

Detailed Table of Contents

Diesel-water emulsion has been used in diesel engine combustion for a long time with encouraging results, but the point of efficiency and NOx trade-off represent a highly challenging task for diesel engines. A new approach was used in this study. The new blends which were obtained by mixing diesel-neem oil biodiesel blend (70:30 by volume) with water (5% by volume), span-80 surfactant (1% by volume), and cetane enhancing additive of Di-tertiary butyl peroxide (0.5% by volume). The blend is designated as B3. This chapter investigates performance and emission characteristics of a single cylinder diesel engine running on B3 fuel. Performance and emission of the engine fueled by B3 fuel results were compared with diesel (D), diesel-biodiesel blend (B1), and diesel-biodiesel with water emulsion through surfactant (B2). B3 fuel had better performance and improved emissions than B1 fuel and diesel fuel, with NOx emission especially reduced by up to 35%.

Feed stock cost and NOX emission are the major barriers for commercialization of biodiesel. Waste cooking oil is well identified as one of the cheapest feed stocks for biodiesel production. This chapter reduces NOX emission of waste cooking oil biodiesel. Test fuel blends are prepared by mixing diesel (20 to 50 v/v%), butanol (5 v/v%), and waste cooking oil biodiesel (45 to 75 v/v%). Fuel properties of waste cooking oil biodiesel are enhanced due to addition of diesel and butanol. Brake specific energy consumption of the blends is higher than diesel fuel. Harmful emissions like carbon monoxide, nitrous oxide, and smoke opacity are lower for blends than diesel fuel. Increasing biodiesel concentration in blend also reduces hydrocarbon emission to a significant extent. The obtained results justify the suitability of proposed cheap blends for diesel engine emission reduction.

An experimental investigation was conducted to disclose the outcomes of oxygenate mixture as additives in Jatropha biodiesel on the performance, combustion, and emission characteristics of a direct injection compression ignition engine. The experiments were conducted in an instrumented single-cylinder, air-cooled, four-stroke, direct-injection diesel engine, equipped with data acquisition system, AC alternator, and an electric loading device. Four oxygenate additives, namely, Ethylene Glycol (C2H6O2), Di methyl Carbonate (C3H6O3), 2-Butoxyethanol (C6H14O2), & Propylene Glycol (C3H8O2) were selected and nine different combinational oxygenate test fuels were prepared attaining ratios of 1, 2, and 4% volume of oxygenates with biodiesel. A significant reduction of emissions such as CO by 60%, Unburned HC by 11%, and smoke emissions by 27% were observed. Substantial improvement in brake thermal efficiency by 6% was observed, while NO emission increased marginally by 4%.

Preface

This book presents the recent technologies which have been adopted in diesel engines to reduce its harmful emissions and to enhance its performance. Today, the world is primarily confronted with two critical issues viz., fossil fuel depletion and environmental pollution. The problems related to environmental pollution are predominantly generated from various power generation sectors such as diesel engine power plants, gas turbine power plants, automotive sector, transportation sector, etc. Researchers, environmentalists, scientists, and automobile professionals have put their strenuous efforts to reduce the intensity of the harmful pollutants such as HC, NO_x, CO, CO_2, etc. Owing to the extortionately harmful pollutants emitted by these sectors, the global environment has of late been worsening leading to severe climatic changes viz., global warming, acid rain, the ozone layer depletion, greenhouse effect. Viewed from this perspective, with the objective of protecting the ecological environment, the current technical community has attempted with copious techniques to solve the above global issues on espousing three potential methods from the power generation sectors, particularly on diesel engine power plants namely hardware modification of engine, exhaust gas treatment, and improvisation of fuel properties.

The chapters of this book emphasizes the recent techniques which are being adopted to enhance the performance of the diesel engines and to reduce the level of harmful pollutants from them. The book consisting of 12 chapters, begins with a comprehensive review of recent biodiesel technologies which will be beneficial for the researchers and environmentalists exploiting research in the area of alternative fuels. In chapter 2, the utilization of plant biomass for the production of renewable and sustainable biofuels is discussed in accordance with the latest emission legislations. The recent techniques on biofuel upgradation and emission aspects are described in-depth in chapter 3. The fourth chapter aims at a detailed review of fuel-additives that help to improve the properties of diesel/biodiesel fuels. The incorporation of fuel-additives with diesel/biodiesel and its effects on performance and emission attributes of diesel engines are presented in chapter 5. Chapter 6 depicts the operating characteristics of the diesel engine operated on using biogas with the objective of reducing the intensity of NO_x and smoke.

The experimental results of a diesel engine fueled with nano-additives blended tamarind biodiesel fuels are discussed in chapter 7. The eighth chapter deals with the neem biodiesel utilized as a fuel in a VCR engine, and eventually, the experimental results. In chapter 9, the power generation from a diesel engine using the waste feedstock (non-edible utilized cooking oil) blended with butanol additives are analyzed. Chapter 10 presents the consequences of the performance and emission characteristics based on the experimental studies using various oxygenated additives blended with Jatropha biodiesel in a diesel engine. The eleventh chapter reports an innovative investigation in a diesel engine modified Port Charged Compression-Ignition (PCCI) on utilizing various biodiesel fuels with regard to performance and emission features. Finally, the chapter 12 presents an optimization technique of alternative fuels in a detailed manner. All the above chapters are streamlined and organized with due respect to the contribution of the experts, who are acknowledged below.

We sincerely thank our heartfelt gratitude to Dr. Hilal Ali Azzan Al Hadhrami, Dean, IMCO, Oman, Dr. Amer Hamood Amer Al Habsi, DDAA, IMCO, Oman, Mr. Abdul Aziz Salim Abdullah Al Hinaai, DDAF, IMCO, Oman and Dr. Mini Shaji Thomas, Director, National Institute of Technology, Tiruchirappalli, India for their moral support. We also owe our gratitude to the authors of different chapters who shared with us some of their latest research findings in the area of IC Engines and alternative fuels. The editors are extremely thankful to IGI Global Publishers, for providing us an opportunity to execute this arduous task in a short period of time. The editors would like to express their heartfelt thanks to Katelyn Hoover, Assistant Acquisitions Editor, IGI Global, and Jan Travers, Director of Intellectual Property and Contracts, IGI Global, for their support and guidance right from the proposal phase. We appreciate Maria Rohde, Assistant Development Editor, IGI Global for her full support and confidence in any situation and thank her for instilling in us a sense of confidence throughout this project and quick feedback.

We candidly acknowledge the reviewers who have dedicated their time and shared their valuable expertise in reviewing the chapters. It is our earnest hope that this book will be beneficial to the young researchers, academicians and industry experts working in the area of IC Engines and alternative fuels. We would also like to express thanks to our colleagues and friends who provided us with valuable suggestions and moral support during the preparation of this book. Finally, we express our sincere thanks to the Almighty and our parents, and all those who supported us with their commitment to the completion of this project.

J. Sadhik Basha

R. B. Anand
Editors

Chapter 1
Biodiesel Production: Processes and Technologies

Avinash Alagumalai
https://orcid.org/0000-0002-6024-2760
GMR Institute of Technology, India

ABSTRACT

Biodiesel, the best-suited replacement for petroleum diesel, is now drawing attention of researchers owing to its advantages and potential for environmental conservation. In this perspective, this chapter explores the need for biodiesel in present-day scenario by highlighting its properties, advantages, and disadvantages. The chapter presents an overview of different techniques proposed by researchers to produce biodiesel. Among different approaches, the emphasis is on catalytic transesterification, non-catalytic transesterification, microwave heating, and ultrasound assisted processes. The chapter also briefly notes the effects of experimental factors on final product recovery.

INTRODUCTION

Over the past few decades, fossil fuels from underground carbon resources such as coal, oil, and gas continue to play a crucial role in global energy systems. Fossil energy is an underlying driver of the commercial revolution and thus of technological, social, economic and developmental progress. Energy has played a hugely positive role in global change. However, these non-renewable energy sources (fossil fuels) even have a negative impact as they negatively affect human health, produce emissions like carbon monoxide, carbon dioxide, unburned hydrocarbons, oxides of nitrogen and other ozone-depleting substances. The earth should thus reconcile the role of energy in social and economic development with the need to decarbonize, reduce our dependence on fossil fuels and move towards lower-carbon energy sources (Hannah & Max, 2018).

The fossil fuel production and use began with coal. Its original exploitation dates back to 4000 B.C. in China, where burning occurred out of brown coal (one of the few types of coal). Be that as it may, considerable coal ignition was usually associated with the beginning of the industrial revolution (Golas & Needham, 1999). Figure 1 shows the global use of non-renewable energy sources like coal, oil, and

DOI: 10.4018/978-1-7998-2539-5.ch001

gas from 1800 onwards. By and large, one could see that the worldwide use of fossil energy has expanded more than 1300-overlap. From Figure 1 it is observed that coal was the only fossil source until 1860s. The use of crude oil and natural gas production began two or three decades later. Also, Figure 1 shows that twentieth century saw an improvement in the use of fossil fuels, with coal falling from 96 percent of overall production in 1900 to about 30 percent in 2000. After 2000, crude oil is the most significant source of energy with around 39 percent of fossil fuels, followed by coal and natural gas with 33 and 28 percent, respectively.

Crude oil is a blend of hydrocarbons formed from prehistoric plants and animals buried in the primal mud of swamps, lakes, and oceans. Crude oil is a non-renewable energy source, and it exists in liquid form underground pools or reservoirs, in tiny spaces within sedimentary rocks and near the surface in tar (or oil) sands (World Health Organization, 1989). About 100 countries worldwide produce crude oil. According to 2017 statistics (U.S. EIA, 2018a), 48 percent of the world's total crude oil production came from the top five countries (Russia-13%, Saudi Arabia-13%, United States-12%, Iraq-6% and Iran-5%) as shown in Figure 2.

The crude oil production and import scenarios reported in BP Energy Outlook 2030 (2012) are as follows:

- The U.S. imports about 68 percent of its crude oil needs from other countries, while India imports 79 percent of its crude oil needs.
- The Middle East and African generation do not fully meet Asia's energy needs, but the re-balancing of global energy exchange due to the improved net position in America is also a key factor.
- India will increasingly depend on imports of oil, coal and natural gas to meet its growing energy needs.
- European net imports and imports as an offer of use are increasing mainly due to declining local oil and gas production and increasing gas use. The development of net imports is primarily attributable to natural gas.
- Import reliance, estimated as the share of demand met by net imports, is increasing for most major energy importers except the U.S.

Figure 1. Global primary energy consumption by fossil fuel, measured in terawatt-hours (TWh)
Source: Vaclav, 2017

Figure 2. Top five crude oil producing countries, 1980-2017 (millions barrels per day)
Source: U.S. EIA, 2018a

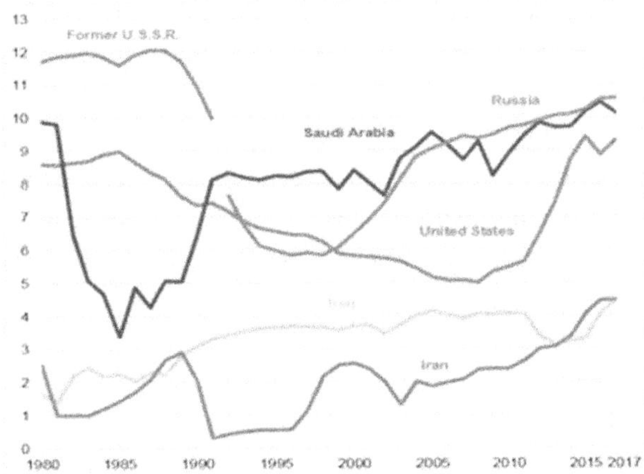

- In the year 2030, current energy importers are expected to import 40 percent more energy than today, with deficits in Europe and the Asia-Pacific region being offset by supply developments in the Middle East, the former Soviet Union, Africa, South and Central America.

As detailed in the U.S. EIA (2018b), global crude oil prices are determined by supply and demand. Economic development is the main factor influencing demand. Petroleum products from crude oil and other liquid hydrocarbons account for about one-third of global energy consumption.

On the other hand, seasonal changes in demand for petroleum products can control the balance of supply and demand for crude oil and its market costs. For instance, crude oil markets tend to be more grounded in the last quarter of the year when global interest in heating oil is supported by both cold climate and inventory build-up, and is weaker in early spring period as interest in heating oil declines in a hotter climate. Furthermore, the Organization of the Petroleum Exporting Countries (OPEC) can have a unique power on prices by setting production targets for its members.

According to the U.S. EIA (2018b), global crude oil prices will average \$73 a barrel in 2018 and \$74/barrel in 2019. The report also states that the average price of crude oil will rise to \$85.70 per barrel by 2025 (in 2017 dollars, which removes the effect of inflation). By 2030 global demand will push the price of oil to \$92.82/barrel. By 2040, costs will be \$106.08/barrel (again in 2017 dollars). By this time, the oil wells have been depleted, making it more expensive to extract oil. By 2050, oil prices will be \$113.56/barrel.

Based on the above actualities, there is a need to find an alternative fuel for alleviating world energy and economic crises.

Biodiesel: A Prudent Step

There is no doubt that the depletion of readily available fossil fuels and their increasing costs are presently the most important problems to be addressed. In this perspective, several alternative fuels are being considered as substitutes for conventional fuels in the transport sector. They consist of liquid and gaseous fuels. Among the liquid fuels, biofuels like alcohol and various vegetable oils receive more attention due to their renewable nature.

Biofuels from different biomasses have therefore demonstrated to be an invaluable source of renewable energy (Agarwal, 2007). Many studies have been carried out in recent days on the conversion of biomass and waste into biodiesel (Akarte, 2004). Biodiesel is generally described as monoalkyl esters of long-chain fatty acids produced by chemically reacting lipids (Alenezi et al., 2013). Although there are some notable advantages of biodiesel, the manufacturers still face various challenges in producing high-quality biodiesel with consistent fuel properties, irrespective of feedstock used (Banga & Varshney, 2010). The potential feedstocks available in each country for biodiesel production are listed in Table 1.

Properties of Biodiesel

For commercial use, the biodiesel produced should be assessed to meet the international fuel standards (ASTM D 6751 and EN 14214 are most commonly used). The specifications and test methods set by ASTM D 6751 and EN 14214 compared to petroleum diesel are compared in Table 2. Generally, the properties specified in the standards are broadly classified into two groups. One reflects the chemical composition of biodiesel (such as cetane number, density, viscosity, etc.) and the other the quality of fuel produced. Typically, the biodiesel has to meet both the properties in pure as well as in blended form (NBB 2007).

To ensure safe operation of biodiesel in CI engines, it should be free from the unreacted catalyst, methanol, and free glycerol. When the transesterification reaction is not complete, all these unreacted compounds end up in the final product. The final biodiesel product should contain less than 0.24% of free glycerol as set by ASTM standard. This is because excessive free glycerol content in biodiesel may block the fuel filters and also can promote combustion problems. On the other hand, the presence of residual methanol (even up to 1%) will decrease the flash point of produced fuel (even as less then 40°C). According to fuel standards, the amount of methanol present in the final product should be very low (0.2% max). Although there is no limit prescribed for residual catalyst in the standards, it is limited by the sulfated ash specification level (up to 0.02% by mass) (Harrington, 1986).

Advantages and Disadvantages of Using Biodiesel

Biodiesel is a sustainable and renewable biofuel which is an excellent surrogate to diesel and it possesses some significant advantages such as non-toxicity, biodegradability, enhanced lubricity, lower vapour pressure, higher flash point, no PAH compounds, and low or no sulfur content. On the other hand, higher density and viscosity, slight decrease in fuel economy on energy basics, more prone to oxidation, unfavorable cold flow properties and high production cost are some of the disadvantages. The notable advantages and disadvantages of using biodiesel are depicted in Figure 3.

BIODIESEL PRODUCTION METHODS

Biodiesel is usually produced by the transesterification reaction of vegetable oils or animal fats in the presence of catalyst and alcohol under typical stirring with batch or continuous processes. Due to economic reasons, the selection of the most efficient method for biodiesel production has gained importance in recent

Table 1. Potential biodiesel feedstocks (Avinash et al. 2014)

Country	Feedstocks
Argentina	Soybean
	Sunflower
	Crambe abyssinica
	Jatropha macrocarpa
Australia	Physic nut (Jatropha curcas)
	Pongam (Milletia Pinnata)
	Indian mustard (Brassica juncea)
	Calophyllum inophyllum
Bangladesh	Pongamia Pinnata
	Rubber seed
Brazil	Soybean
	Sugarcane
	Palm
Canada	Canola
	Sunflower
	Soybean
Chile	Residual wood (from Radiata pine and Eucalyptus)
	Camelina
	Rapeseed
China	Rapeseed
	Micro algae
Cuba	Jatropha curcas
	Neem
	Moringa
Greece	Sunflower
	Cotton seed
	Rapeseed
	Tomato seed
	Tobacco seed
	Pumpkin seed
India	Jatropha curcas
Iran	Palm
	Jatropha curcas
	Castor
	Algae
	Fish oil
Ireland	Animal tallow
	Used cooking oil
	Silage grass
Italy	Rapeseed
	Soybean
	Sunflower
Malaysia	Palm
	Jatropha Curcas
	Calophyllum inophyllum
Mali	Jatropha curcas
Mexico	Palm
	Tallow
Mozambique	Jatropha curcas
Norway	Animal fat
	Fish residues
Pakistan	Jatropha curcas
Peru	Palm
	Jatropha curcas
Tanzania	Jatropha curcas
Thailand	Palm
	Jatropha curcas
Turkey	Sunflower
	Rapeseed
United states	Soybean
Zimbabwe	Jatropha curcas

continues in next column

Table 1. Continued

Table 2. Specifications and test methods of biodiesel compared to petroleum diesel

Biodiesel standards		Biodiesel Europe EN 14214:2003	Biodiesel USA ASTM D 6751-07b	Petroleum Diesel EN 590:1999
Property	**Unit**			
Density 15°C	g/cm^3	0.86-0.90		0.82-0.845
Viscosity 40°C	mm^2/s	3.5-5.0	1.9-6.0	2.0-4.5
Distillation	% @ °C		90%, 360°C	85%, 350°C - 95%, 360°C
Flashpoint	°C	120 min	93 min	55 min
Sulfur	mg/kg	10 max	15 max	350 max
Water	mg/kg	500 max	500 max	200 max
Oxidation stability	hrs;110°C	6 hours min	3 hours min	N/A (25 g/m^3)
Cetane number		51 min	47 min	51 min
Acid value	mg KOH /g	0.5 max	0.5 max	
Methanol	% mass	0.20 max	0.2 max or Fp <130°C	
Ester content	% mass	96.5 min		
Monoglyceride	% mass	0.8 max		
Diglyceride	% mass	0.2 max		
Triglyceride	% mass	0.2 max		
Free glycerol	% mass	0.02 max	0.02 max	
Total glycerol	% mass	0.25 max	0.24 max	
Iodine value		120 max		
Phosphorus	mg/kg	10 max	10 max	
Sodium and potassium	mg/kg	5 max	5 max	
Calcium and Magnesium	mg/kg	5 max	5 max	

Source: Biofuel Systems Group Limited, https://www.biofuelsystems.com/about.htm

years. In this point of view, the researchers have investigated various techniques such as non-catalytic transesterification or supercritical processes, microwave assisted and ultrasound assisted processes. It has been discovered that these techniques have several differences compared to conventional methods.

Catalytic Transesterification

The most commonly used technique of biodiesel production is catalytic transesterification (Demirbas, 2009). In this process, biodiesel (monoalkyl esters of long-chain fatty acids) is produced by the reaction of triglycerides with simpler alcohols such as methanol or ethanol. The process is also carried out in the presence of a catalyst, such as an acid catalyst, an alkali catalyst or heterogeneous catalyst or an enzyme catalyst (Dubé et al., 2007). The chemical structure of the transesterification reaction is given in Equation 1 (Patel et al. 2015).

Figure 3. Advantages and disadvantages of biodiesel
Source: Copyright © 2018, by Avinash A and Murugesan A, ROC: L-76839/2018

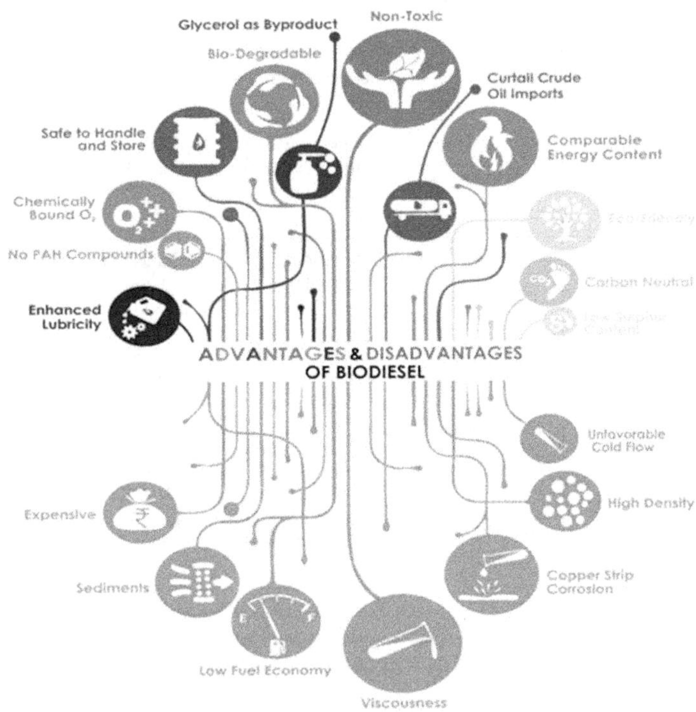

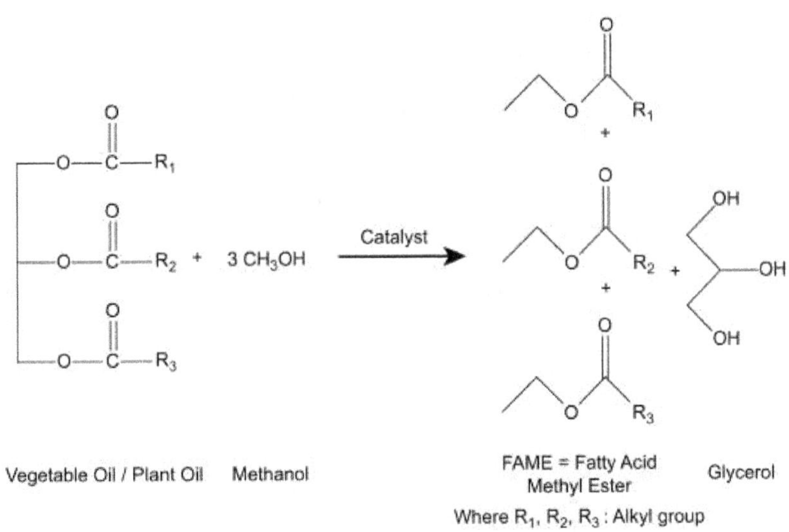

Several studies have reported that catalytic transesterification method can produce high-quality biodiesel. Also, this method is economical and suitable for large-scale biodiesel production (Wang et al., 2006). The influence of different catalysts on biodiesel conversion in catalytic transesterification is portrayed in Table 3.

Table 3. Biodiesel production studies in catalytic transesterification

Catalyst type	Catalyst	Feedstock	Reaction conditions	Yield %	Ref.
Acid catalyst	12-Tungstophosphoric Acid	Canola oil	Temperature of 200 °C, 9:1 methanol to oil molar ratio, 3 wt. % of catalyst loading	90%	Kulkarni et al. (2006)
	Trifluoroacetic acid	Soybean oil	Temperature of 120 °C, 20:1 methanol to oil molar ratio, 2.0 M of catalyst loading, and a reaction time of 5 h.	98.4%	Miao et al. (2009)
	p-toluenesulfonic acid	Corn oil	Temperature of 80 °C, 6:1 methanol to oil molar ratio (dimethyl ether as co solvent), 4 wt. % of catalyst loading, and a reaction time of 2 h.	100%	Guan et al. (2009)
	H_2SO_4	Mixed oil (50% sunflower and 50% soybean oil)	Temperature of 60 °C, 6:1 methanol to oil molar ratio, 2.5% of catalyst loading, and a stirring speed of 300 rpm	96.6%	Farag et al. (2011)
	Acetic acid	Soybean oil	Temperature of 250 °C, 30:1 methanol to oil molar ratio, stirring speed of 300 rpm and a reaction time of 1 h	95%	Go et al. (2014)
Base catalyst	NaOH	Jatropha oil	Temperature of 65 °C, 5:01 methanol to oil molar ratio, 0.8% of catalyst loading, and a reaction time of 1 h.	95.5%	Ojolo et al. (2011)
		Soybean oil	Temperature of 60±1 °C, 6:01 methanol to oil molar ratio, 1% of catalyst loading, and a reaction time of 1 h.	90%	Keera et al. (2011)
		Cottonseed oil	Temperature of 60±1 °C, 6:01 methanol to oil molar ratio, 1% of catalyst loading, and a reaction time of 1 h.	98.50%	Keera et al. (2011)
		Waste frying oil	Temperature of 50 °C, 7.5:1 methanol to oil molar ratio, 0.5% of catalyst loading, and a reaction time of 30 min.	96%	Uzun et al. (2012)
		Waste cooking oil	Temperature of 65 °C, 9:1 methanol to oil molar ratio, 0.75 wt. % of catalyst loading, and a stirring speed of 500 rpm for 1 h.	95.05%	Avinash & Murugesan (2018)
	KOH	Pongamia oil	Temperature of 65 °C, 12:1 methanol to oil molar ratio, 1% of catalyst loading, and a stirring speed of 360 rpm for 1 h.	98%	Meher et al. (2006)
		Duck tallow	Temperature of 65 °C, 6:01 methanol to oil molar ratio, 1% of catalyst loading, and a reaction time of 180 min.	97%	Chung et al. (2009)
		Fish oil	Temperature of 32 °C, 6:01 methanol to oil molar ratio, 0.5 wt. % of catalyst loading, and a reaction time of 1 h.	96%	Fadhil & Ali (2013)

continues on following page

Table 3. Continued

Catalyst type	Catalyst	Feedstock	Reaction conditions	Yield %	Ref.
Hetero-geneous catalysts	Calcined AlMg hydrotalcite	Soybean oil	Temperature of 180 °C, 2.0 g of oil, 0.88 g of methanol, 0.1 g of catalyst, and a reaction time of 60 min.	92%	Di Serio et al. (2006)
	$LiNO_3/\gamma Al_2O_3$	Palm oil	Temperature of 60 °C, 65:1 methanol to oil molar ratio, 10 wt.% of catalyst, and a reaction time of 3 h.	93.4%	Semwal et al. (2011)
	$NaNO_3/\gamma Al_2O3$			95.1%	
	$KNO_3/\gamma Al_2O_3$			94.7%	
	$Mg(NO_3)_2/\gamma Al_2O_3$			10.4%	
	$Ca(NO_3)_2/\gamma Al_2O_3$			94.3%	
	$S–ZrO_2$	Soybean	Temperature of 120 °C, 20:1 methanol to oil molar ratio, 5 wt.% of catalyst, and a reaction time of 1 h.	98.6%	Yang & Xie (2007)
	CaO/Al_2O_3	Palm oil	Temperature of 60 °C, 15:1 MeOH to oil molar ratio, 9 wt% of catalyst, and a reaction time of 3 h.	86.38 ±0.44%	Uprety et al. (2016)
	Birch bark ash		Temperature of 60 °C, 12:1 MeOH to oil molar ratio, 3 wt% of catalyst, and a reaction time of 3 h.	69.7 ±0.83%	
	Flyash from wood pellet		Temperature of 60 °C, 12:1 MeOH to oil molar ratio, 9 wt% of catalyst, and a reaction time of 3 h.	87.76 ±1.14%	
	Egg shell ash	Waste vegetable oil	Temperature of 65± 5 °C, 22.5:1 methanol to oil molar ratio, 3.5 wt% of catalyst, and a reaction time of 5 h 30 min.	91%	Tshizanga et al. (2017)
	Tucumã peels	Soybean oil	Temperature of 80 °C, 15:1 methanol to oil molar ratio, 1 m/m of catalyst, and a reaction time of 4 h.	97%	Mendonça et al. (2019)
Enzymes	Lipozyme TL	Soybean oil	Temperature of 40 °C, 1:1 alcohol to oil molar ratio, 0.04 wt. % of enzyme, and a stirring time of 150 rpm.	66%	Du et al. (2005)
	Silica gel- lipozyme TL	Soybean oil	Temperature of 40 °C, 1:1 alcohol to oil molar ratio, 0.06 wt. % of enzyme, and a stirring time of 150 rpm.	90%	Du et al. (2005)
	Lipozyme TL IM and Novozym 435	Rapeseed oil	Temperature of 35 °C, 4:1 alcohol to oil molar ratio, 3% Lipozyme TL IM and 1% Novozym 435, and a stirring time of 130 rpm for 12 h.	95%	Li et al. (2006)
	Pseudomonas cepacia	Jatropha oil	Temperature of 40 °C, 4:1 ethanol to oil molar ratio, 0.05 g enzyme, and a stirring time of 200 rpm for 24 h.	92%	Shah et al. (2007)
	Novozym 435	Soybean oil	Temperature of 37 °C, 3:1 alcohol to oil molar ratio, 5 g of enzyme with 100 g of oil.	90%	Cerveró et al. (2014)
	MAS1	Waste cooking oil	Temperature of 30 °C, 3:1 methanol to oil molar ratio, 5 wt.% enzyme and a reaction time of 24 h	95.5%	Wang et al. (2017)

Alkaline catalyzed transesterification (suitable for free fatty acids less than 2.5 wt%) is the process used to produce most of the biodiesel produced today (Leung et al. 2010). As reported by Carlos et al. (2011), alkaline catalyzed transesterification offers high biodiesel yield with low reaction time, temperature and catalyst concentration when compared to acid and heterogeneous catalysts. In the alkaline transesterification reaction, the common vegetable oils or fat called triglycerides, are reacted with an alcohol in the presence of alkaline catalyst. The alcohols that can be used for the transesterification reaction include methanol, ethanol, propanol, butanol, and amyl alcohol. Among these alcohols, methanol and ethanol are most commonly used in the transesterification reaction because they are the simplest form of alcohols and are readily available. Between methanol and ethanol, methanol is widely used in biodiesel production because of its low cost, physical and chemical advantages (Subramaniam et al. 2013). As reported by Ma & Hanna (1999), methanol reacts rapidly with triglycerides and easily dissolves the alkaline catalyst. Sodium hydroxide and potassium hydroxide are the two main alkaline catalysts used in the transesterification reaction.

Non-Catalytic Transesterification or Supercritical Process

The supercritical method is the catalyst free process for biodiesel production. Biodiesel production can be effortlessly accomplished by the supercritical process in the absence of a catalyst. A supercritical liquid is any substance at a temperature and pressure above its critical point. It can diffuse through solids like a gas, and break up materials like a liquid. Some of these liquids are environmentally friendly and cost-effective. Usually, water, carbon dioxide, and alcohol are used as supercritical liquids (Özçimen & Yücel, 2011). Supercritical liquids have different application domains. One of these applications is the biodiesel production which was first reported by Saka & Kusdiana (2001). Besides, numerous studies have been carried out on biodiesel production under supercritical conditions since 2001.

Demirbas (2005) reported that supercritical methanol transesterification of different vegetable oils could produce more than 95% yield within the first 10 min of the reaction. In addition to commercial methanol, crude bio-methanol prepared through wood gasification can produce more than 97% biodiesel yield through the supercritical process as reported by Isayama & Saka (2008). Apart from methanol, researchers have used supercritical ethanol (Demirbaş, 2008; Gui et al., 2009) and added enzymes with supercritical alcohols to increase the yield (Varma et al., 2010). Researchers have also used carbon dioxide and acetic acid together with supercritical alcohols to accelerate the biodiesel conversion (Wei et al., 2013). The carbon dioxide was added to appreciably increase the breakdown of fats by decreasing the reaction temperature. On the other hand, acetic acid was added to decrease the glycerol formation and increase the hydrolysis of fats. Some researchers have used solvents like methyl acetate and dimethyl carbonate in the place of supercritical alcohols for high product recovery (Ilham & Saka, 2010; Tan et al., 2010). Overall, the environmental impact assessment study of supercritical biodiesel production process reveals that low energy requirements with high product yield and better glycerol decomposition are attainable (Lee & Saka, 2010; Marulanda, 2012; Okoro et al., 2018). The effects of different supercritical fluids on biodiesel yield reported in literature are illustrated in Table 4.

As seen in table 3, supercritical methanol produces higher biodiesel yield within a short reaction time. However, it can be observed that the performance reduces with supercritical ethanol. Thus, it is suggested to go for mixed method (supercritical alcohol+enzyme) to improve the yield. Also, the use of solvents instead of supercritical alcohols improves product recovery.

Table 4. Biodiesel production studies in non catalytic transesterification

Supercritical fluid	Feedstock	Reaction conditions	Yield %	Ref.
Supercritical methanol	Rapeseed oil	Temperature of 350°C, pressure of 19 Mpa, 42:1 alcohol to oil molar ratio, and a reaction time of 4 min.	95%	Saka &Kusdiana (2001)
	Soybean oil	Temperature of 280°C, pressure of 12.8 Mpa, 24:1 alcohol to oil molar ratio, and a reaction time of 10 min.	98%	Cao et al. (2005)
	Cottonseed oil	Temperature of 273°C, 41:1 alcohol to oil molar ratio, and a reaction time of 8 min.	98%	Demirbaş (2008)
	Chicken fat	Temperature of 400°C, pressure of 41.1 Mpa, 6:1 alcohol to oil molar ratio, and a reaction time of 6 min.	88%	Marulanda et al. (2010)
Supercritical ethanol	Cottonseed oil	Temperature of 230°C, 41:1 alcohol to oil molar ratio, and a reaction time of 8 min.	70%	Demirbaş (2008)
	Refined palm oil	Temperature of 349°C, pressure > 6.3 Mpa, 33:1 alcohol to oil molar ratio, and a reaction time of 30 min.	79.2%	Gui et al. (2009)
Supercritical alcohol+ enzyme (Novozyme 435)	Sesame oil	Temperature of 350°C, pressure of 200 bar, 40:1 alcohol to oil molar ratio, and a reaction time of 40 min.	90% (Supercritical methanol) 100% (Supercritical ethanol)	Varma et al. (2010)
Solvent (supercritical methyl acetate)	Palm oil	Temperature of 399°C, 30:1 solvent to oil molar ratio, and a reaction time of 59 min.	97.6%	Tan et al. (2010)

Microwave-Assisted Process

In the production of biodiesel, heating coils are usually used to heat the raw materials. The production can also be done by an alternative treatment technique known as the microwave-assisted method (Figure 4). The microwave method of biodiesel production has become increasingly important in recent years (Motasemi and Ani, 2012; Wahidin et al., 2014; Suryanto et al., 2018). In this method, microwave radiations influence molecular movements such as ion migration or dipole rotations but do not change the molecular structure (Refaat et al., 2008). In general, microwave frequencies range from 300 MHz to 30 GHz, but a frequency of 2.45 GHz is preferred in laboratory applications (Taylor et al., 2005). In this method of biodiesel production, the lipid, alcohol and base catalyst contain both polar and ionic sites. Microwaves stimulate the smallest degree of variation of polar molecules and ions, directing to molecular friction and thus the chemical reaction is initiated (Taylor et al., 2005).

Azcan and Danisman (2008) attempted to produce biodiesel from rapeseed oil by microwave heating (67% of 1200 W). In their work, it has been reported that microwave irradiation achieves around 94% biodiesel yield within a short period (5 min). It was also observed that the quality of biodiesel was not affected by the sort of catalyst used. However, the higher biodiesel yields, energy savings and rapid conversion can be improved by using different catalysts. Chen et al. (2012) reported that higher biodiesel yield can be achieved with sodium methoxide (0.75 wt.%) as a catalyst in microwave heating process.

Figure 4. Microwave-assisted process for biodiesel production (Lin et al., 2015)

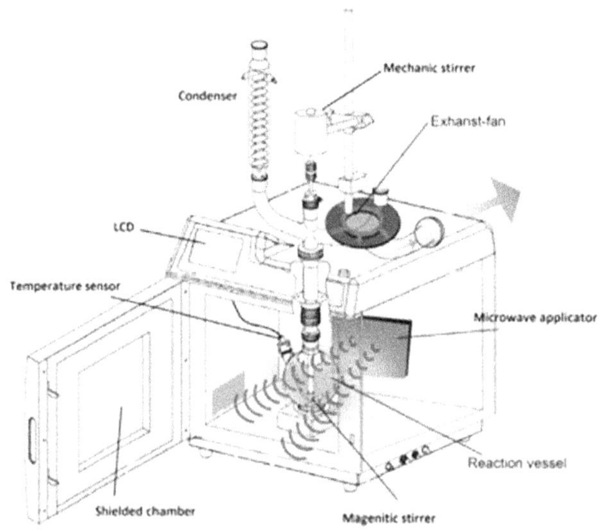

Several research studies suggest the use of ionic liquid catalysts to improve the biodiesel yield. Lin et al. (2015) reported that more than 98% methyl ester yield is possible with the ionic liquid catalyst (1 wt.% of ionic liquid with 0.75wt. % of NaOH) in transesterification of jatropha oil by microwave irradiation technique. A similar observation was reported by Handayani and Hadiyanto (2017) in biodiesel production from *Calophyllum inophyllum* oil by microwave heating using ionic liquids.

In addition to different types of catalysts, one step and two step biodiesel production by microwave heating can also significantly influence the biodiesel conversion as reported by researchers (Cheng et al. 2013). Chee Loong and Idris (2016) reported that one step biodiesel production by simultaneous cooling and microwave heating can boost the conversion by 5 times compared to the conventional method. Studies by Bakar et al. (2014) disclosed that the 14% increase in palm biodiesel yield is achievable with two step (esterification and transesterification) microwave heating compared to conventional heating. On the whole, the impact of microwave heating on biodiesel yield of different feedstocks reported in the literature is exemplified in Table 5.

Ultrasound Assisted Method

In the ultrasound-based method of biodiesel production, ultrasonic waves are used for better agitation and effective surface contact between oil and alcohol molecules (Özçimen & Yücel, 2011). Ultrasonic waves are sound waves with a frequency beyond the frequencies to which the human ear can respond. The normal hearing range is between 16 Hz and about 18 kHz, while ultrasound is usually considered to be in the order of 20 kHz to 100 MHz. According to researchers, ultrasonic irradiation has the most critical effect called cavitation. The formation and collapse of microbubbles during cavitation are responsible for most of the significant chemical reactions (Kumar et al. 2010a; Kojima & Takai, 2019). This cavitation also provides the activation energy required for initiating transesterification reaction. Also, ultrasonication has the potential to reduce the mean droplet distribution by more than 40% compared to standard impellers (Wu et al. 2007).

Table 5. Biodiesel production studies in microwave assisted transesterification

Feedstock	Reaction conditions	Yield %	Ref.
Frying oil	Temperature of 60°C, microwave irradiation of 750 W, 6:1 alcohol to oil molar ratio, 0.5 wt.% of catalyst loading and a reaction time of 4 min.	87%	Saifuddin & Chua (2004)
Cotton seed oil	Temperature of 60°C, microwave irradiation of 1200 W (21%), 6:1 alcohol to oil molar ratio, 1.5 wt.% of catalyst loading and a reaction time of 7 min.	92.4%	Azcan& Danisman, (2007)
Safflower seed oil	Temperature of 60°C, microwave irradiation of 300 W, 10:1 alcohol to oil molar ratio, 1 wt.% of catalyst loading and a reaction time of 6 min.	98.4%	Düz et al.(2011)
microalgae	Temperature of 65°C, microwave irradiation with 70% power, and a reaction time of 10 min.	86.4%	Wahidin et al. (2014)
Jatropha oil	Temperature of 70°C, 9:1 alcohol to oil molar ratio, 1 wt.% of ionic liquid with 0.75wt. % of NaOH and a reaction time of 6 min.	98.5%	Lin et al. (2015)
Palm oil	Temperature of 70°C, microwave irradiation of 400 W, 12:1 alcohol to oil molar ratio, 1 wt.% of catalyst loading and a resistance time of 1.75 min.	99.4%	Choedkiatsakul et al. (2015)
microalgae	Temperature of 65°C, microwave irradiation of 800 W, and a reaction time of 10 min.	75%	Chee Loong et al. (2017)
Waste oil	Temperature of 70°C, microwave irradiation of 325 W, 1 g/g of oil of catalyst loading and a irradiation time of 200 s.	97%	Thirugnanasambandham et al. (2017)
Cotton seed oil	Microwave irradiation of 400 W, 12:1 alcohol to oil molar ratio, 0.5 wt.% of catalyst loading and a reaction time of 5 min.	99%	Suryanto et al. (2018)

Figure 5. Ultrasound-assisted process for biodiesel production (Kojima & Takai, 2019)

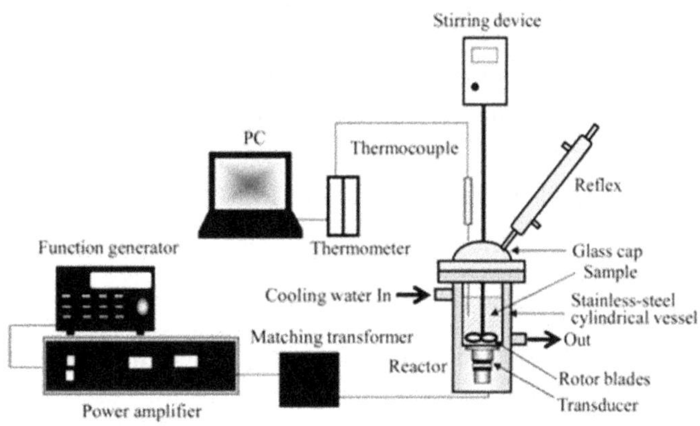

Gao et al. (2013a, 2013b, 2013c) attempted to produce biodiesel from three different feedstocks such as palm oil, soybean oil and waste cooking oil by ultrasonication technique. It was observed in their work that more than 95% biodiesel yield with palm & soybean oils, and over 90% yield with waste cooking oil can be achieved within a shorter reaction period of 30 min. Investigations by Koutsouki et al. (2015) and Kumar et al. (2010a, 2010b) proved that more than 97% biodiesel conversion is achievable with shorter reaction time (<30 min) by ultrasonic method of biodiesel production from Cynara cardunculus

L. seed oil, coconut oil and Jatropha oil. Findings by Singh et al. (2007) disclosed that a higher yield of over 97% within a shorter reaction of 5 min is also possible by maintaining the input energy between 125 kJ to 215 kJ.

In addition to commonly available feedstocks, researchers have also derived ultrasound assisted biodiesel from microorganisms and wastewater treatment plant sludge. Zang et al. (2014, 2016) investigated the potential of microbial and sludge lipids by ultrasonication aided in-situ transesterification technique. It was reported in their work that ultrasonication aided in-situ transesterification technique is a potential alternative to two-step transesterification process and this technique can produce higher biodiesel yield within shorter reaction time (23 h less compared to without ultrasonic assistance).

Researchers have also proposed techniques to quantity the conversion of triglycerides into fatty acid esters during ultrasoniction process. Reyman et al. (2014) proposed a new technique to monitor the reaction and quantify the fatty acid esters produced during the reaction. According to this technique, it is very easy to measure methyl ester using a linear correlation between the intensity of Fourier transform infrared peak (1,437 cm^{-1}) and the % conversion. The biodiesel production by ultrasonic assisted transesterification reported in the literature is illustrated in table 6.

MAIN FACTORS AFFECTING THE YIELD OF BIODIESEL

Molar Ratio of Alcohol to Oil

Several research studies have reported that the alcohol quantity is one of the most critical factors influencing biodiesel yield (Zhang et al., 2003 and Freedman et al., 1986). Theoretically, 1 mole of triglyceride requires 3 moles of alcohol to produce 3 moles of fatty acid ester or biodiesel and

1 mol of glycerol. However, in actual practice, excess alcohol is used to ensure the complete conversion of free fatty acids to fatty acid esters. This excess alcohol will also ensure faster ester conversion. However, increasing the alcohol beyond the level will not have a positive effect on biodiesel yield and will increase the cost of recovering alcohol. Thus, alcohol to oil molar ratio of 6:1 is generally preferred in the literature using an alkaline catalyst. However, oils with a high content of free fatty acids may require a high molar ratio of up to 15:1 (Sprules & Price, 1950).

Concentration of Catalyst

The catalyst concentration is the second most important factor that affects the biodiesel yield during the transesterification process. For industrial-scale production, alkaline catalysts are preferred over acid catalysts, because alkaline catalysts are less corrosive for industrial plants. The lower concentration of catalyst does not promote the complete conversion of triglycerides into esters. Thus, the increase in catalyst concentration increases the biodiesel yield. As reported in literature, sodium hydroxide is widely used alkaline catalyst. It has been found that the optimum catalyst concentration for sodium hydroxide is up to 1.5 wt/wt% of the oil. The biodiesel yield begins to drop beyond this optimum value because the addition of more catalyst increases the alkalinity of triglycerides to produce more soap (Eevera et al., 2009).

Table 6. Biodiesel production studies in ultrasonic assisted transesterification

Feedstock	Reaction conditions	Yield %	Ref.
Soybean oil	Temperature of 40°C, 20 kHz amplitude level, 6:1 methanol to oil molar ratio, catalyst concentration of 1.5 wt.%, and a reaction time of 15 min.	99%	Colucci et al. (2005)
Soybean oil	Temperature of 45°C, 19.7 kHz amplitude level, 6:1 methanol to oil molar ratio, and a reaction time of 30 min.	100%	Ji et al. (2006)
Soybean oil	Temperature of 89°C, 100% amplitude, and a reaction time of 5 min.	99%	Singh et al. (2007)
Fish oil	Temperature of 20-60°C, 25-35 kHz amplitude level, 6:1 methanol to oil molar ratio, catalyst concentration of 0.8 wt.%, and a reaction time of >30 min.	98%	Armenta et al. (2007)
Jatropha oil	50% amplitude, 3 wt.% of catalyst loading, 9:1 methanol to oil molar ratio and a reaction time of 15 min.	98.53%	Kumar et al. (2010a)
Waste cooking oil	Temperature of 20-25°C, 20 Hz amplitude level, 1.5:1 alcohol to oil ratio, catalyst concentration of 0.3%, and a reaction time of 20 min	81%	Thanh et al. (2010)
Soybean oil	Temperature of 40°C, 40 kHz amplitude level, 6:1 alcohol to oil ratio, catalyst concentration of 6%, and a reaction time of 4 h.	96%	Yu et al.(2010)
Palm oil	Temperature of 65°C, 30 kHz amplitude level, 9:1 alcohol to oil ratio, catalyst concentration of 3%, and a reaction time of 1 h.	95%	Mootabadi et al., (2010)
Palm oil	Temperature of 65°C, 20 kHz amplitude level, 9:1 alcohol to oil ratio, catalyst concentration of 2.8%, and a reaction time of 50 min.	92%	Salamatinia et al., (2010)
Palm Oil	Temperature of 50-60°C, 40 Hz amplitude level, methanol to oil weight ratio of 1.3, catalyst concentration of 0.5%, and a reaction time of 30 min.	95%	Gao et al. (2013a)
Soybean oil	Temperature of 50°C 45 Hz amplitude level, 6:1 methanol to oil molar ratio, 1 wt.% catalyst concentration, and a reaction time of 30 min.	97%	Gao et al. (2013b)
Waste cooking oil	Temperature of 40-50°C, 30 Hz amplitude level, methanol to oil weight ratio of 1.3, catalyst concentration of 2%, and a reaction time of 30 min.	91%	Gao et al. (2013c)
Silybum marianum oil	Temperature of 60°C, 8:1 methanol to oil molar ratio of 1.3, catalyst concentration of 1.5 wt.%, and a reaction time of 20 min.	95.75%	Takase et al. (2014)
Soybean oil	Temperature of 60°C, 43kHz of ultrasound, 6:1 methanol to oil molar ratio, and a reaction time of 2.5 h.	88%	Kojima et al. (2019)

Reaction Time

The conversion rate of triglycerides to esters increases with increasing reaction time. Initially, the reaction is much slower due to improper mixing and dispersing of the alkoxide solution with triglycerides. As soon as they are mixed and dispersed, the reaction begins and proceeds faster. The reaction time of up to 90 minutes is generally preferred for alkaline catalyst transesterification. The further extension of the reaction time has no significant influence on the biodiesel yield. Excess reaction time, however, reduces product yield and produces more soap by reversing reaction (Eevera et al. 2009 and Ma et al. 1998).

Reaction Temperature

The reaction temperature strongly influences the transesterification reaction rate. In general, a higher reaction temperature is required to decrease the viscosity of the oil. This promotes the reaction rates and shortens the reaction time. However, the reaction temperature must be lower than the boiling point of the alcohol to avoid the escape of alcohol during the reaction stage. The reaction temperature of 50-60°C is commonly suggested for catalyst transesterification (Eevera et al., 2009 and Ma et al., 1999).

Mixing Intensity

The mixing intensity is also an essential parameter in the transesterification process. Once the catalyst-alcohol mixture is added to the oil, the neutralization reaction starts, and further mixing is not necessary. A better mixing, however, ensures an almost complete and faster conversion of free fatty acids in triglycerides to fatty acid esters. As reported by Murugesan et al. (2015), the reaction is incomplete at lower stirring rates and better at optimum value (550 rpm). However, a further increase in the stirrer speed did not lead to a better yield, but to a reduction in the ester yield.

CONCLUDING REMARKS

The cost and demand for fossil fuels are growing day by day and the expected fall in the price will be a tremendous challenge to the countries that rely on them for export earnings, such as many of the Middle East countries that have long looked to oil and gas sales as the foundation of their economy.

From this survey, it is figured out that acid catalyzed transesterification needs higher operation temperature and long reaction time than the base catalyzed transesterification. On the other hand, base catalyzed transesterification is fast, economical and advantageous from the industrial point of view. Coming to heterogeneous catalysts, easy separation and reusability of the catalyst are the biggest advantages. However, long reaction time is needed for the completion of reaction particularly in case of heterogeneous catalysts derived from natural materials. Finally, enzymatic transesterification can be carried even at room temperature and this technique requires very long reaction time (even up to a day) to obtain higher yields.

Although new methods for biodiesel production such as supercritical processes, microwave, and ultrasound assisted processes offer more advantages such as shorter reaction times and more efficient reaction, these processes also pose some challenges. For instance, high energy demand and a large amount of alcohol are the biggest challenges of the supercritical process. On the other hand, microwave process of biodiesel production is still on a pilot scale, and it cannot be carried out on a large-scale in biodiesel production due to the microwave penetration depth. The safety feature is an additional shortcoming of microwave irradiation technique. Finally, ultrasonic biodiesel production may be more favorable for small-scale biodiesel production, but it is not feasible in large-scale extension as several ultrasound probes are required. Based on the above considerations, catalytic transesterification is the most efficient and cost-effective technology for large-scale biodiesel production. However, other methods are more advantageous for pilot scale production. Further exploration of the literature clarifies that there is a need to optimize biodiesel process parameters to obtain high yield.

REFERENCES

Agarwal, A. K. (2007). Biofuels (alcohols and biodiesel) applications as fuels for internal combustion engines. *Progress in Energy and Combustion Science*, *33*(3), 233–271. doi:10.1016/j.pecs.2006.08.003

Akarte, R. R. (2004). An overview of energy and issues of future', In *Proceedings First National Conference on Energy and Fuel Issues of Future, PIET, pp. 150-160,* Pune, India.

Alenezi, R., Santos, R. C. D., Raymahasay, S., & Leeke, G. A. (2013). Improved biodiesel manufacture at low temperature and short reaction time. *Renewable Energy*, *53*, 242–248. doi:10.1016/j.renene.2012.11.019

Armenta, R. E., Vinatoru, M., Burja, A. M., Kralovec, J. A., & Barrow, C. J. (2007). Transesterification of fish oil to produce fatty acid ethyl esters using ultrasonic energy. JAOCS. *Journal of the American Oil Chemists' Society*, *84*(11), 1045–1052. doi:10.100711746-007-1129-2

Avinash, A., & Murugesan, A. (2018). Prediction capabilities of mathematical models in producing a renewable fuel from waste cooking oil for sustainable energy and clean environment. *Fuel*, *216*, 322–329. doi:10.1016/j.fuel.2017.12.029

Avinash, A., & Murugesan, A., (2018), Advantages & disadvantages of biodiesel, Indian copyright, ROC Number: L-76839/2018.

Avinash, A., Subramaniam, D., & Murugesan, A. (2014). Bio-diesel - A global scenario. *Renewable & Sustainable Energy Reviews*, *29*, 517–527. doi:10.1016/j.rser.2013.09.007

Azcan, N., & Danisman, A. (2007). Alkali catalyzed transesterification of cottonseed oil by microwave irradiation. *Fuel*, *86*(17–18), 2639–2644. doi:10.1016/j.fuel.2007.05.021

Azcan, N., & Danisman, A. (2008). Microwave assisted transesterification of rapeseed oil. *Fuel*, *87*(10–11), 1781–1788. doi:10.1016/j.fuel.2007.12.004

Bakar, S. A., Yusup, S., Ahmad, M. M., Quitain, A. T., Sasaki, M., Goto, M., ... Ahmad, J. (2014). Comparative Study on Two-Step Fatty Acid Methyl Ester (FAME) Production from High FFA Crude Palm Oil Using Microwave Technique and Conventional Technique. *Advanced Materials Research*, *917*, 87–95. doi:10.4028/www.scientific.net/AMR.917.87

Banga, S., & Varshney, P. K. (2010). Effect of impurities on performance of biodiesel: A review. *Journal of Scientific and Industrial Research*, *69*(8), 575–579.

Biofuel Systems Group Limited. Beech Croft, 51 Ruff Lane, Ormskirk, Lancashire, England, L39 4UL. Retrieved from https://www.biofuelsystems.com/ about.htm

Cao, W., Han, H., & Zhang, J. (2005). Preparation of biodiesel from soybean oil using supercritical methanol and co-solvent. *Fuel*, *84*(4), 347–351. doi:10.1016/j.fuel.2004.10.001

Carlos, A., Guerrero, F., Guerrero-Romero, A., & Sierra, F. E. (2011). Biodiesel production from waste cooking oil, Biodiesel: Feedstocks and Processing Technologies InTech, pp. 35-36.

Cerveró, J. M., Álvarez, J. R., & Luque, S. (2014). Novozym 435-catalyzed synthesis of fatty acid ethyl esters from soybean oil for biodiesel production. *Biomass and Bioenergy*, *61*, 131–137. doi:10.1016/j.biombioe.2013.12.005

Chee Loong, T., & Idris, A. (2017). One step transesterification of biodiesel production using simultaneous cooling and microwave heating. *Journal of Cleaner Production*, *146*, 57–62. doi:10.1016/j.jclepro.2016.03.155

Chen, K. S., Lin, Y. C., Hsu, K. H., & Wang, H. K. (2012). Improving biodiesel yields from waste cooking oil by using sodium methoxide and a microwave heating system. *Energy*, *38*(1), 151–156. doi:10.1016/j.energy.2011.12.020

Cheng, J., Yu, T., Li, T., Zhou, J., & Cen, K. (2013). Using wet microalgae for direct biodiesel production via microwave irradiation. *Bioresource Technology*, *131*, 531–535. doi:10.1016/j.biortech.2013.01.045 PMID:23403061

Choedkiatsakul, I., Ngaosuwan, K., Assabumrungrat, S., Mantegna, S., & Cravotto, G. (2015). Biodiesel production in a novel continuous flow microwave reactor. *Renewable Energy*, *83*, 25–29. doi:10.1016/j.renene.2015.04.012

Chung, K. H., Kim, J., & Lee, K. Y. (2009). Biodiesel production by transesterification of duck tallow with methanol on alkali catalysts. *Biomass and Bioenergy*, *33*(1), 155–158. doi:10.1016/j.biombioe.2008.04.014

Colucci, J. A., Borrero, E. E., & Alape, F. (2005). Biodiesel from an alkaline transesterification reaction of soybean oil using ultrasonic mixing. *JAOCS. Journal of the American Oil Chemists' Society*, *82*(7), 525–530. doi:10.100711746-005-1104-3

Demirbas, A. (2005). Biodiesel production from vegetable oils by supercritical methanol. *Journal of Scientific and Industrial Research*, *64*(11), 858–865.

Demirbas, A. (2008). Studies on cottonseed oil biodiesel prepared in non-catalytic SCF conditions. *Bioresource Technology*, *99*(5), 1125–1130. doi:10.1016/j.biortech.2007.02.024 PMID:17433669

Demirbas, A. (2009). Biodiesel from waste cooking oil via base-catalytic and supercritical methanol transesterification. *Energy Conversion and Management*, *50*(4), 923–927. doi:10.1016/j.enconman.2008.12.023

Di Serio, M., Ledda, M., Cozzolino, M., Minutillo, G., Tesser, R., & Santacesaria, E. (2006). Transesterification of soybean oil to biodiesel by using heterogeneous basic catalysts. *Industrial & Engineering Chemistry Research*, *45*(9), 3009–3014. doi:10.1021/ie051402o

Du, W., Xu, Y. Y., Liu, D. H., & Li, Z. B. (2005). Study on acyl migration in immobilized lipozyme TL-catalyzed transesterification of soybean oil for biodiesel production. *Journal of Molecular Catalysis. B, Enzymatic*, *37*(1–6), 68–71. doi:10.1016/j.molcatb.2005.09.008

Dubé, M. A., Tremblay, A. Y., & Liu, J. (2007). Biodiesel production using a membrane reactor. *Bioresource Technology*, *98*(3), 639–647. doi:10.1016/j.biortech.2006.02.019 PMID:16580197

Duz, M. Z., Saydut, A., & Ozturk, G. (2011). Alkali catalyzed transesterification of safflower seed oil assisted by microwave irradiation. *Fuel Processing Technology*, *92*(3), 308–313. doi:10.1016/j.fuproc.2010.09.020

Eevera, T., Rajendran, K., & Saradha, S. (2009). Biodiesel production process optimization and characterization to assess the suitability of the product for varied environmental conditions. *Renewable Energy, 34*(3), 762–765. doi:10.1016/j.renene.2008.04.006

Energy Outlook, B. P. 2030, London, (2012, January). Retrieved from https://www.bp.com/ content/dam/bp/en/corporate/pdf/energy-economics/energy-outlook/bp-energy-outlook-2012.pdf

Fadhil, A. B., & Ali, L. H. (2013). Alkaline-catalyzed transesterification of Silurus triostegus Heckel fish oil: Optimization of transesterification parameters. *Renewable Energy, 60*, 481–488. doi:10.1016/j.renene.2013.04.018

Farag, H. A., El-Maghraby, A., & Taha, N. A. (2011). Optimization of factors affecting esterification of mixed oil with high percentage of free fatty acid. *Fuel Processing Technology, 92*(3), 507–510. doi:10.1016/j.fuproc.2010.11.004

Freedman, B., Butterfield, R. O., & Pryde, E. H. (1986). Transesterification kinetics of soybean oil 1. *Journal of the American Oil Chemists' Society, 63*(10), 1375–1380. doi:10.1007/BF02679606

Gao, M., & Gao, Q. G. (2013b). Base-Catalyzed Transesterification of Soybean Oil by Ultrasonication. *Advanced Materials Research, 724–725*, 356–359. doi:10.4028/www.scientific.net/AMR.724-725.356

Gao, M., Guo, C. B., & Lin, X. (2013c). Alkali-Catalyzed Transesterification of Waste Cooking Oil to Prepare Biodiesel by Ultrasonication. *Advanced Materials Research, 803*, 17–20. doi:10.4028/www.scientific.net/AMR.803.17

Gao, M., Yan, Y. Q., & Lin, X. (2013). Alkali-Catalyzed Transesterification of Palm Oil by Ultrasonication. *Advanced Materials Research, 803*, 13–16. doi:10.4028/www.scientific.net/AMR.803.13

Go, A. W., Sutanto, S., Nguyenthi, B. T., Cabatingan, L. K., Ismadji, S., & Ju, Y. H. (2014). Transesterification of soybean oil with methanol and acetic acid at lower reaction severity under subcritical conditions. *Energy Conversion and Management, 88*, 1159–1166. doi:10.1016/j.enconman.2014.03.014

Golas, P. J., & Needham, J. (1999). Science and Civilisation in China, Cambridge University Press, pp. 186-91.

Guan, G., Kusakabe, K., Sakurai, N., & Moriyama, K. (2009). Transesterification of vegetable oil to biodiesel fuel using acid catalysts in the presence of dimethyl ether. *Fuel, 88*(1), 81–86. doi:10.1016/j.fuel.2008.07.021

Gui, M. M., Lee, K. T., & Bhatia, S. (2009). Supercritical ethanol technology for the production of biodiesel: Process optimization studies. *The Journal of Supercritical Fluids, 49*(2), 286–292. doi:10.1016/j.supflu.2008.12.014

Handayani, P. A., Abdullah, A., & Hadiyanto, H. (2017). Biodiesel production from Nyamplung (Calophyllum inophyllum) oil using ionic liquid as a catalyst and microwave heating system. *Bulletin of Chemical Reaction Engineering & Catalysis, 12*(2), 293–298. doi:10.9767/bcrec.12.2.807.293-298

Hannah, R., & Max, R. (2018). Fossil Fuels. *Published online at OurWorldInData.org,* Retrieved from https://ourworldindata.org/fossil-fuels.

Harrington, K. J. (1986). Chemical and physical properties of vegetable oil esters and their effect on diesel fuel performance. *Biomass*, *9*(1), 1–17. doi:10.1016/0144-4565(86)90008-9

Ilham, Z., & Saka, S. (2010). Two-step supercritical dimethyl carbonate method for biodiesel production from Jatropha curcas oil. *Bioresource Technology*, *101*(8), 2735–2740. doi:10.1016/j.biortech.2009.10.053 PMID:19932022

Isayama, Y., & Saka, S. (2008). Biodiesel production by supercritical process with crude bio-methanol prepared by wood gasification. *Bioresource Technology*, *99*(11), 4775–4779. doi:10.1016/j.biortech.2007.09.056 PMID:17977720

Ji, J., Wang, J., Li, Y., Yu, Y., & Xu, Z. (2006). Preparation of biodiesel with the help of ultrasonic and hydrodynamic cavitation. *Ultrasonics*, *44*(SUPPL.), e411–e414. doi:10.1016/j.ultras.2006.05.020 PMID:16797656

Keera, S. T., El Sabagh, S. M., & Taman, A. R. (2011). Transesterification of vegetable oil to biodiesel fuel using alkaline catalyst. *Fuel*, *90*(1), 42–47. doi:10.1016/j.fuel.2010.07.046

Kojima, Y., & Takai, S. (2019). Transesterification of vegetable oil with methanol using solid base catalyst of calcium oxide under ultrasonication. *Chemical Engineering and Processing - Process Intensification*, 101–106. doi:10.1016/j.cep.2018.12.007

Koutsouki, A. A., Tegou, E., Kontakos, S., Kontominas, M. G., Pomonis, P. J., & Manos, G. (2015). In situ transesterification of Cynara cardunculus L. seed oil via direct ultrasonication for the production of biodiesel. *Fuel Processing Technology*, *134*, 122–129. doi:10.1016/j.fuproc.2015.01.024

Kulkarni, M. G., Gopinath, R., Meher, L. C., & Dalai, A. K. (2006). Solid acid catalyzed biodiesel production by simultaneous esterification and transesterification. *Green Chemistry*, *8*(12), 1056–1062. doi:10.1039/b605713f

Kumar, D., Kumar, G., Poonam, & Singh, C. P. (2010a). Ultrasonic-assisted transesterification of Jatropha curcus oil using solid catalyst, Na/SiO2. *Ultrasonics Sonochemistry*, *17*(5), 839–844. doi:10.1016/j.ultsonch.2010.03.001 PMID:20359932

Kumar, D., Kumar, G., Poonam, & Singh, C. P. (2010b). Fast, easy ethanolysis of coconut oil for biodiesel production assisted by ultrasonication. *Ultrasonics Sonochemistry*, *17*(3), 555–559. doi:10.1016/j.ultsonch.2009.10.018 PMID:19945331

Lee, J. S., & Saka, S. (2010). Biodiesel production by heterogeneous catalysts and supercritical technologies. *Bioresource Technology*, *101*(19), 7191–7200. doi:10.1016/j.biortech.2010.04.071 PMID:20570143

Leung, D. Y. C., Wu, X., & Leung, M. K. H. (2010). A review on biodiesel production using catalyzed transesterification. *Applied Energy*, *87*(4), 1083–1095. doi:10.1016/j.apenergy.2009.10.006

Li, L., Du, W., Liu, D., Wang, L., & Li, Z. (2006). Lipase-catalyzed transesterification of rapeseed oils for biodiesel production with a novel organic solvent as the reaction medium. *Journal of Molecular Catalysis. B, Enzymatic*, *43*(1–4), 58–62. doi:10.1016/j.molcatb.2006.06.012

Lin, Y. C., Chen, S. C., Wu, T. Y., Yang, P. M., Jhang, S. R., & Lin, J. F. (2015). Energy-saving and rapid transesterification of jatropha oil using a microwave heating system with ionic liquid catalyst. *Journal of the Taiwan Institute of Chemical Engineers*, *49*, 72–78. doi:10.1016/j.jtice.2014.11.014

Ma, F., Clements, L. D., & Hanna, M. A. (1998). The effects of catalyst, free fatty acids, and water on transesterification of beef tallow. *Transactions of the ASAE. American Society of Agricultural Engineers*, *41*(5), 1261–1264. doi:10.13031/2013.17292

Ma, F., & Hanna, M. A. (1999). Biodiesel production: A review. *Bioresource Technology*, *70*(1), 1–15. doi:10.1016/S0960-8524(99)00025-5

Marulanda, V. F. (2012). Biodiesel production by supercritical methanol transesterification: Process simulation and potential environmental impact assessment. *Journal of Cleaner Production*, *33*, 109–116. doi:10.1016/j.jclepro.2012.04.022

Meher, L. C., Dharmagadda, V. S. S., & Naik, S. N. (2006). Optimization of alkali-catalyzed transesterification of Pongamia pinnata oil for production of biodiesel. *Bioresource Technology*, *97*(12), 1392–1397. doi:10.1016/j.biortech.2005.07.003 PMID:16359862

Mendonça, I. M., Paes, O. A. R. L., Maia, P. J. S., Souza, M. P., Almeida, R. A., Silva, C. C., ... de Freitas, F. A. (2019). New heterogeneous catalyst for biodiesel production from waste tucumã peels (Astrocaryum aculeatum Meyer): Parameters optimization study. *Renewable Energy*, *130*, 103–110. doi:10.1016/j.renene.2018.06.059

Miao, X., Li, R., & Yao, H. (2009). Effective acid-catalyzed transesterification for biodiesel production. *Energy Conversion and Management*, *50*(10), 2680–2684. doi:10.1016/j.enconman.2009.06.021

Mootabadi, H., Salamatinia, B., Bhatia, S., & Abdullah, A. Z. (2010). Ultrasonic-assisted biodiesel production process from palm oil using alkaline earth metal oxides as the heterogeneous catalysts. *Fuel*, *89*(8), 1818–1825. doi:10.1016/j.fuel.2009.12.023

Motasemi, F., & Ani, F. N. (2012). A review on microwave-assisted production of biodiesel. *Renewable & Sustainable Energy Reviews*, *16*(7), 4719–4733. doi:10.1016/j.rser.2012.03.069

Murugesan, A., Subramaniam, D., & Avinash, A. (2015). An empirical and statistical analysis of biodiesel production by transesterification process. *Biofuels*, *6*(1–2), 79–86. doi:10.1080/17597269.2015.1050643

NBB. (2007). Biodiesel production and quality. Fuel fact sheets, USA: National Biodiesel Board, Retrieved from http://biodiesel.org/what-is-biodiesel/biodiesel-fact-sheets.

Ojolo, S. J. O. B. S., Adelaja, A. O., & Ogbonnaya, M. (2011). Study of an effective technique for the production of biodiesel from jatropha oil. *Journal of Emerging Trends in Engineering and Applied Sciences*, *2*, 79–86.

Okoro, O. V., Sun, Z., & Birch, J. (2018). Catalyst-free biodiesel production methods: A comparative technical and environmental evaluation. *Sustainability (Switzerland)*, *10*(1). doi:10.3390u10010127

Özçimen, D., & Yücel, S. (2011). *Novel Methods in Biodiesel Production. Biofuel's Engineering Process Technology Marco Aurélio dos Santos Bernardes*. Intech Open; doi:10.5772/18750

Patel, N. K., & Shah, S. N. (2015). Biodiesel from Plant Oils. In Food, Energy, and Water: The Chemistry Connection (pp. 277–307). doi:10.1016/B978-0-12-800211-7.00011-9

Refaat, A. A., El Sheltawy, S. T., & Sadek, K. U. (2008). Optimum reaction time, performance and exhaust emissions of biodiesel produced by microwave irradiation. *International Journal of Environmental Science and Technology*, *5*(3), 315–322. doi:10.1007/BF03326026

Reyman, D., Saiz Bermejo, A., Ramirez Uceda, I., & Rodriguez Gamero, M. (2014). A new FTIR method to monitor transesterification in biodiesel production by ultrasonication. *Environmental Chemistry Letters*, *12*(1), 235–240. doi:10.100710311-013-0440-4

Saifuddin, N., & Chua, K. H. (2004). Production of Ethyl Ester (Biodiesel) From Used Frying Oil: Optimization of Transesterification Process using Microwave Irradiation. *Malaysian Journal of Chemistry*, *6*(1), 77–82.

Saka, S., & Kusdiana, D. (2001). Biodiesel fuel from rapeseed oil as prepared in supercritical methanol. *Fuel*, *80*(2), 225–231. doi:10.1016/S0016-2361(00)00083-1

Salamatinia, B., Mootabadi, H., Bhatia, S., & Abdullah, A. Z. (2010). Optimization of ultrasonic-assisted heterogeneous biodiesel production from palm oil: A response surface methodology approach. *Fuel Processing Technology*, *91*(5), 441–448. doi:10.1016/j.fuproc.2009.12.002

Semwal, S., Arora, A. K., Badoni, R. P., & Tuli, D. K. (2011). Biodiesel production using heterogeneous catalysts. *Bioresource Technology*, *102*(3), 2151–2161. doi:10.1016/j.biortech.2010.10.080 PMID:21106371

Shah, S., & Gupta, M. N. (2007). Lipase catalyzed preparation of biodiesel from Jatropha oil in a solvent free system. *Process Biochemistry*, *42*(3), 409–414. doi:10.1016/j.procbio.2006.09.024

Singh, A. K., Fernando, S. D., & Hernandez, R. (2007). Base-catalyzed fast transesterification of soybean oil using ultrasonication. *Energy & Fuels*, *21*(2), 1161–1164. doi:10.1021/ef060507g

Sprules, F. J., & Price, D. (1950). Production of fatty esters. United States: NOPCO CHEM CO. Retrieved from http://www.freepatentsonline.com/ 2494366.html

Subramaniam, D., Murugesan, A., Avinash, A., & Kumaravel, A. (2013). Bio-diesel production and its engine characteristics - An expatiate view. *Renewable & Sustainable Energy Reviews*. doi:10.1016/j.rser.2013.02.002

Suryanto, A., Sabara, Z., Artiningsih, A., & Ismail, H. (2018). Microwave Assisted Trans-esterification of Cotton Seed Oil for Biodiesel Production. *Reaktor*, *18*, 27–30. doi:10.14710/reaktor.18.1.27-30

Takase, M., Feng, W., Wang, W., Gu, X., Zhu, Y., Li, T., ... Wu, X. (2014). Silybum marianum oil as a new potential non-edible feedstock for biodiesel: A comparison of its production using conventional and ultrasonic assisted method. *Fuel Processing Technology*, *123*, 19–26. doi:10.1016/j.fuproc.2014.01.032

Tan, K. T., Lee, K. T., & Mohamed, A. R. (2010). A glycerol-free process to produce biodiesel by supercritical methyl acetate technology: An optimization study via Response Surface Methodology. *Bioresource Technology*, *101*(3), 965–969. doi:10.1016/j.biortech.2009.09.004 PMID:19773156

Taylor, M., Atri, S. S., & Minhas, S. (2005). Evalueserve Analysis: Developments in Microwave Chemistry. Retrieved from www.rsc.org/images/ evalueserve_ tcm18-16758.pdf

Thanh, L. T., Okitsu, K., Sadanaga, Y., Takenaka, N., Maeda, Y., & Bandow, H. (2010). A two-step continuous ultrasound assisted production of biodiesel fuel from waste cooking oils: A practical and economical approach to produce high quality biodiesel fuel. *Bioresource Technology*, *101*(14), 5394–5401. doi:10.1016/j.biortech.2010.02.060 PMID:20219362

Thirugnanasambandham, K., Shine, K., Aziz, H. A., & Gimenes, M. L. (2017). Biodiesel synthesis from waste oil using novel microwave technique: Response surface modeling and optimization. *Energy Sources. Part A, Recovery, Utilization, and Environmental Effects*, *39*(7), 636–642. doi:10.1080/15567036.2016.1196270

Tshizanga, N., Aransiola, E. F., & Oyekola, O. (2017). Optimisation of biodiesel production from waste vegetable oil and eggshell ash. *South African Journal of Chemical Engineering*, *23*, 145–156. doi:10.1016/j.sajce.2017.05.003

Uprety, B. K., Chaiwong, W., Ewelike, C., & Rakshit, S. K. (2016). Biodiesel production using heterogeneous catalysts including wood ash and the importance of enhancing byproduct glycerol purity. *Energy Conversion and Management*, *115*, 191–199. doi:10.1016/j.enconman.2016.02.032

U.S. EIA. (2018a). International Energy Statistics. Retrieved from https://www.eia.gov/

U.S. EIA. (2018b). The Short-term Energy Outlook. Retrieved from https://www.eia.gov/outlooks/steo/

Uzun, B. B., Kılıç, M., Özbay, N., Pütün, A. E., & Pütün, E. (2012). Biodiesel production from waste frying oils: Optimization of reaction parameters and determination of fuel properties. *Energy*, *44*(1), 347–351. doi:10.1016/j.energy.2012.06.024

Vaclav, S. (2017). *Energy Transitions: Global and National Perspectives*. BP Statistical Review of World Energy.

Varma, M. N., Deshpande, P. A., & Madras, G. (2010). Synthesis of biodiesel in supercritical alcohols and supercritical carbon dioxide. *Fuel*, *89*(7), 1641–1646. doi:10.1016/j.fuel.2009.08.012

Wahidin, S., Idris, A., & Shaleh, S. R. M. (2014). Rapid biodiesel production using wet microalgae via microwave irradiation. *Energy Conversion and Management*, *84*, 227–233. doi:10.1016/j.enconman.2014.04.034

Wang, X., Qin, X., Li, D., Yang, B., & Wang, Y. (2017). One-step synthesis of high-yield biodiesel from waste cooking oils by a novel and highly methanol-tolerant immobilized lipase. *Bioresource Technology*, *235*, 18–24. doi:10.1016/j.biortech.2017.03.086 PMID:28351728

Wang, Y., Ou, S., Liu, P., Xue, F., & Tang, S. (2006). Comparison of two different processes to synthesize biodiesel by waste cooking oil. *Journal of Molecular Catalysis A Chemical*, *252*(1–2), 107–112. doi:10.1016/j.molcata.2006.02.047

Wei, C. Y., Huang, T. C., & Chen, H. H. (2013). Biodiesel production using supercritical methanol with carbon dioxide and acetic acid. *Journal of Chemistry*, *2013*, 1–6. doi:10.1155/2013/789594

World Health Organization. (1989). Crude Oil. *IARC Monographs on the Evaluation of Carcinogenic Risks to Humans*, *45*, 119–158. PMID:2664246

Wu, P., Yang, Y., Colucci, J. A., & Grulke, E. A. (2007). Effect of ultrasonication on droplet size in biodiesel mixtures. *JAOCS. Journal of the American Oil Chemists' Society*, *84*(9), 877–884. doi:10.100711746-007-1114-9

Yang, Z., & Xie, W. (2007). Soybean oil transesterification over zinc oxide modified with alkali earth metals. *Fuel Processing Technology*, *88*(6), 631–638. doi:10.1016/j.fuproc.2007.02.006

Yu, D., Tian, L., Wu, H., Wang, S., Wang, Y., Ma, D., & Fang, X. (2010). Ultrasonic irradiation with vibration for biodiesel production from soybean oil by Novozym 435. *Process Biochemistry*, *45*(4), 519–525. doi:10.1016/j.procbio.2009.11.012

Zhang, X., Yan, S., Tyagi, R. D., Drogui, P., & Surampalli, R. Y. (2016). Ultrasonication aided biodiesel production from one-step and two-step transesterification of sludge derived lipid. *Energy*, *94*, 401–408. doi:10.1016/j.energy.2015.11.016

Zhang, X., Yan, S., Tyagi, R. D., Surampalli, R. Y., & Valéro, J. R. (2014). Ultrasonication aided in-situ transesterification of microbial lipids to biodiesel. *Bioresource Technology*, *169*, 175–180. doi:10.1016/j.biortech.2014.06.108 PMID:25050978

Zhang, Y., Dubé, M. A., McLean, D. D., & Kates, M. (2003). Biodiesel production from waste cooking oil: 2. Economic assessment and sensitivity analysis. *Bioresource Technology*, *90*(3), 229–240. doi:10.1016/S0960-8524(03)00150-0 PMID:14575945

ADDITIONAL READING

Biodiesel Board. N. (2009). Benefits of Biodiesel. *October*, 01–02. Retrieved from https://biodiesel.org/docs/ffs-basics/benefits-of-biodiesel.pdf?sfvrsn=4

de Blasio, C. (2019). Biodiesel. In Green Energy and Technology (pp. 253–265). doi:10.1007/978-3-030-11599-9_18

Gajendra Babu, M.K., &, K.A. Subramanian, K.A. (2017). Alternative Transportation Fuels: Utilisation in Combustion Engines. ISBN 9781138076662

National Biodiesel Board. USA. (2014). Introduction to Biodiesel and Glossary of Terms. In Bioenergy (pp. 37–40). doi:10.1016/b978-0-12-407909-0.00002-x

KEY TERMS AND DEFINITIONS

Biodiesel: A fuel comprised of mono-alkyl esters of long chain fatty acids derived from vegetable oils or fats.

Catalyst: A substance to increase the reaction rate without getting consumed in the process.

Feedstocks: The raw materials used to make biodiesel.

Microwave Chemistry: The science of applying microwave radiation to chemical reactions.

Supercritical Fluid: Any substance at a temperature and pressure above its critical point.

Transesterification: Alcohol replacement process to separate glycerol from vegetable oils or fats.

Ultrasonication: The process of using sound energy at high frequencies to break apart particle by cavitation.

Chapter 2
Utilization of Plant Biomass for the Production of Renewable and Sustainable Biofuels With Zero Carbon Emission

Sandip Kumar Singh
https://orcid.org/0000-0003-0094-6253
CSIR-National Chemical Laboratory Pune, India

ABSTRACT

Biofuels for use of transport and industrial purposes have been synthesised on a substantial scale since 1970s, using a set of technologies. Today, biofuels are widely available using sugar, grains, starch-based bioethanol, and oil seeds-based biodiesel. For enhancing the anticipations of product portfolio of plant biomass-to-biofuels formation, it is vital to develop effective conversion technologies for upgradation of abundantly available lignocellulosic biomass resources into value-added co-products particularly biofuels and chemicals. In this chapter, brief synthesis processes and utilization of synthesised biofuels such as methanol, ethanol, butanol, gasoline, diesel, and jet fuel have been outlined for their use in transport sectors either as a neat or blended with gasoline. Biofuels' physico-chemical properties, performances, gas emissions, pros, and cons of various synthesised biofuels' neat and blend are compared with non-renewable fuels. Thenceforth, discussion gradually focuses towards the zero-carbon emission upon the utilization of biofuels derived from plant biomass.

INTRODUCTION

In the USA and around the globe, the public deliberations have grown increasingly concerned considering the impacts of global warming, calling on academics and industries for working rational to mitigate the effects of greenhouse gas emission (Cole et al., 1997; von Blottnitz & Curran, 2007). Many experts have expressed their anxiety considering the increasing costs and sustainability of fossil feedstocks reserve, with some expert people revealed that the global oil production has already notched, and the global energy

DOI: 10.4018/978-1-7998-2539-5.ch002

consumption rate is constantly increasing (Hajjari et at., 2017; Nejat et al., 2015). Considering the issues for example, environmental and non-renewable energy with fossil feedstocks in particular with crude oil, a global biofuel technology needs to be implemented (Luque et al., 2008). Currently a considerable amount of agricultural edible-food residues such as corn, starch, sugar, and oil has been out-putted for the production of first-generation biofuels in North America, Brazil and Europe (Ruth, 2008). In Asia and other countries, a significant part of the productive land has been utilized for the cultivation of biofuels plants for example, palm oil. In order to fulfil the rising global requirement for the energy as biofuels and to reduce the reliance on fossil feedstocks and to also avoid the straining the global food supply, an integrated biorefinery must be processed to acquire the increasing amounts of renewable, sustainable and eco-friendly biofuels from plant biomass, including agriculture and forest residues, municipal waste, animal residues and dedicated bioenergy plants (Luque et al., 2008; Shabih et al., 2018).

For enhancing the anticipations of product portfolio of plant biomass-to-biofuels formation processes, it is vital to develop effective conversion technologies for the valorisation of abundant alternate lignocellulosic biomass into value added co-products. Lignocellulosic biomass is highly recalcitrant structure, assorted class of functional groups and decorated with a set of various linkages (Singh & Dhepe, 2016a, 2018). Plant biomass is mainly consisted of cellulose (homopolymer; C6 sugar), hemicellulose (heteropolymer; C5 and C6 sugars) and lignin (heteropolymer; C6 + C3, C4 or C5; phenolic propenoids) (Mosier et al., 2005; Singh, 2019). Several methods for example, chemical, physical, biological and thermal have been implemented for effective conversion of plant biomass into biofuels (Huber et al., 2006). Among all, the enzymatic process is more effective and remarkable in terms of selectivity for desire products, green process mostly avoid using the harmful chemicals, safe to handle compared to toxic, flammable and corrosive chemicals, *etc.* (Payne et al., 2015; Straathof, 2014). The current technologies used to produce the biofuels from edible sources are bioethanol and that is used in spark ignition engines (Alonso et al., 2010; Serrano-Ruiz et al., 2012). Bioethanol has good properties as a fuel, good for ignition engines, eco-friendly and easily producible at large scales (Balat, 2011; Vohra et al., 2014). Biofuels derived from plant biomass are called as a carbon neutral fuel or zero carbon emission (McMillan, 1997). Zero carbon emission means that the carbon dioxide emitted during the utilization of fuel, for example, in transport or electricity generation, *etc.* is reabsorbed by the plants during the photosynthesis (McMillan, 1997).

This chapter surveys the biofuels, liquid alkanes and alkenes production using a set of various methods from biomass either edible (*e.g.,* sugar, cereals, vegetable oils, *etc.*) or non-edible (*e.g.*, lignocellulosic biomass) source. Today's the most favourable terms are capital and operating economics and environment concerns for the developing countries. These terms could be buffered by bioenergy balance and carbon emissions. Various physico-chemical properties (*e.g.,* research octane number, sulphur content, machine octane number, reid vapor pressure, initial boiling temperature, latent heat of vaporization, sulphur content, corrosivity, *etc.*) of the generated biofuels have been included in this chapter. Reid vapor pressure is a common test method, that helps to measure the volatility of crude oil and petroleum refined products by using the ASTM D 323 test method at 37.8 °C. Moreover, the neat and blended (different ratios with gasoline) properties in terms of spark ignition engines performance of biofuels are discussed as well. Finally, the low or zero oxides of carbon (*e.g.*, carbon monoxide and carbon dioxide) emissions upon utilization of biofuels are incorporated that can reduce or mitigate the effect of global warming.

BIOMASS

Various types of species are present on biosphere, and these species are named as plant, animal, micro-organism, *etc.* The plant, animal and microorganism (living or dead) derived species are collectively known as a biomass. Plant derived biomass is commonly agricultural crops waste, forest residue, wood chips, *etc.* (McKendry, 2002). Plant biomass is currently used to generate heat, electricity, chemicals, materials, and polymers (Agbor et al., 2011; Tilman et al., 2006). Although, it is a good renewable and sustainable resource of C, H and O, which can be utilized and converted into biofuels, chemicals and energy *via* a biorefinery concept (Agbor et al., 2011).

PRODUCTION OF BIOFUELS FROM BIOMASS

Conversion of biomass (plant or animal derived and edible or nonedible biomass) into biofuels, chemicals, materials or polymers is an important step to substitute or reduce the dependency on fossil feedstocks and mitigate the effect of global warming as well (Stöcker, 2008). Prior to conversion of biomass, post- and pre-treatment are important steps for maximum conversion of biomass through minimization of side products or maximum selectivity of desire products (Mosier et al., 2005; Singh et al., 2019; Yang & Wyman, 2008; Yuan et al., 2019). "Figure 1" shows a group of assorted pathways for the production of liquid biofuels from biomass. Some of the conversion technologies shown in "Figure 1" have already been in function at commercial scales, some technologies are tested at large scales, and other are tested at lab scales, and need to be optimized and implemented at large scales (Brown & Brown, 2013; Fatih Demirbas, 2009; Huber et al., 2006). Plant biomass is mainly consisted with cellulose, hemicellulose and lignin and these contents are particularly structured with carbon, hydrogen and oxygen through various linkages (Singh, 2019; Singh & Dhepe, 2016a). Conversion of plant biomass into liquid biofuels, the process involves to remove partially or completely oxygen either in form of oxides of carbon (CO or CO_2) or oxide of hydrogen (H_2O) (Stöcker, 2008). To exit with better fuels efficiency for ignition engine performance, the (C+H)/O ratios must be a high number (Turner et al., 2018). The hydrolysis of biomass into sugars (C5 or C6), (Bär et al., 2018) and further processing into aqueous alkanes, ethanol, *etc.* were soundly recognized processes (Alonso et al., 2010; Binder & Raines, 2010). Thermal conversion processes such as pyrolysis and gasification were existed to convert biomass into bio-oils and synthetic gas, and further processed into a number of assorted biofuels (Huber et al., 2006). Some of the technologies are based on the conversion of edible biomass such as vegetable oils, grains into biofuels through different pathways such as transesterification, hydrogenation, enzymatic hydrolysis, *etc.* (Figure 1) (Bhowmick et al., 2019; Huber et al., 2006; Tenenbaum, 2008).

RENEWABLE AND SUSTAINABLE HYDROCARBONS AND BIOFUELS

Renewable and sustainable hydrocarbons biofuels are known as clean and green biofuels and that can be produced from edible (starch $(C_6H_{10}O_5)_n$, oil (mustard oil, coconut oil, corn oil *etc.*) *etc.*), and non-edible (*e.g.*, lignocellulosic plant biomass) resources (Tenenbaum, 2008). In 2016, International Energy Agency (IEA) added the gaseous fuels (*e.g.*, methane, hydrogen, *etc.*) derived from plant biomass into biofuels. Several methods are known to produce biofuels using chemical, biological, thermal processes through

Figure 1. Schematic representation for production of liquid biofuels from biomass

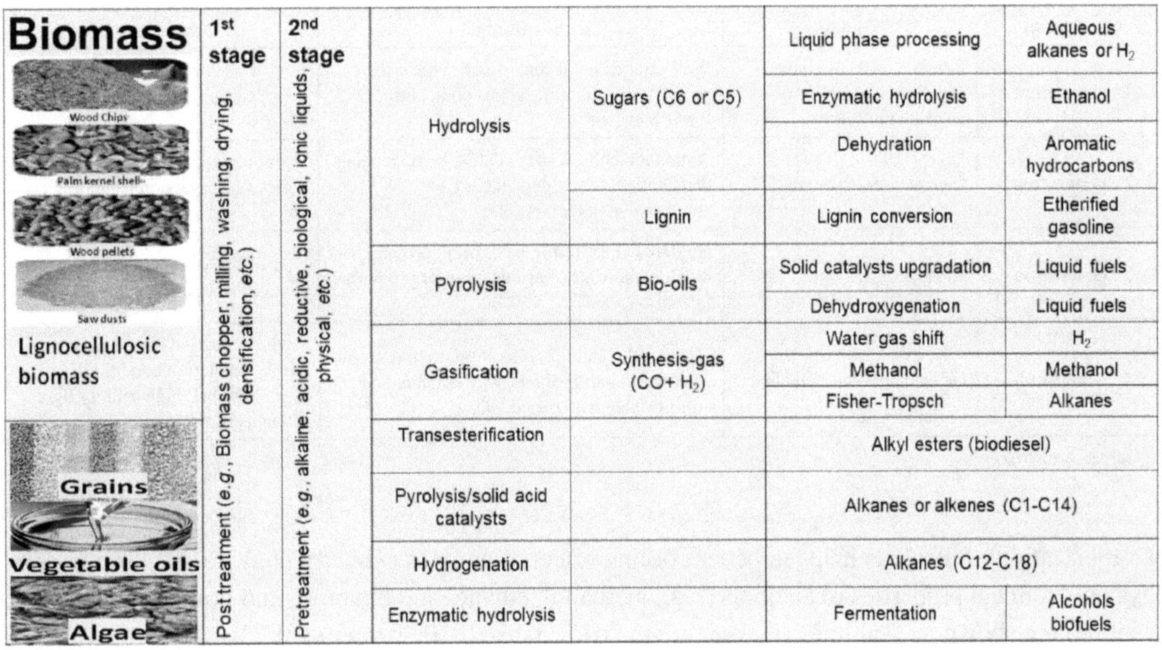

integrated biorefineries pathway. The biofuels product derived from plant biomass through biorefineries way are almost similar to fuels such as gasoline, diesel, or jet fuel derived from crude oil through petroleum refineries way. Petroleum refineries are non-renewable pathways and have a large number of global issues. Interestingly, the biofuels product derived from plant biomass can be used as in liquified fuels vehicles without any engine modification. Moreover, these fuels can be stored and used the same distribution set-up to fill the vehicles tank *via* using the same existing pipelines applied for petroleum derived non-renewable fuels.

In 2011, IEA has reported that 27% global non-renewable transport fuels can be substituted using biofuels by 2050 (https/www.iea.org/renewables2018/transport/ 2018). Therefore, it is important to consider the existing technologies for types of biofuels generation, their advantages and disadvantages as well. The technologies for biofuels generation can be enabled to adopt globally. The types of biofuels generation, their advantages and disadvantages are discussed and shown as in "Table 1" (Aro, 2016).

LIQUID BIOFUELS

Liquid biofuels blend with gasoline or diesel in particular fractions have been associated with characteristic properties, and can be applied to enhance the thermal efficacy, reduce the exhaust emissions, and improved the combustion behaviour of the conventional used diesel or gasoline engine (Karthickeyan et al., 2019). Low viscosity of bioderived alcohols and/or liquid biofuels accelerates the fuel atomization process and enhances the fuel injection process. Moreover, the amount of oxygen contents (up to a certain fraction) in alcohol, boosts the complete combustion of fuel and generates low exhaust emissions. The latent heat of vaporization of alcohols fuel is high compared to diesel and gasoline. Latent heat

Table 1. Types of biofuels generation, used biomass, advantages and disadvantages for biofuels production

Generation	Biomass	Advantage	Disadvantage
1st generation	Edible biomass - grains, seeds, sugars, edible vegetable oils, *etc.*	Well established technologies, large scales production capabilities, cost competitive to fossil fuels, *etc.*	Food verses fuels, low area used efficacy, geographical limitations, *etc.*
2nd generation	Lignocellulosic biomass	Abundant feedstocks available, homogeneous distribution or less dependency on geographical locations, *etc.*	Economical processes, lack of effective conversion technologies, *etc.*
3rd generation	Algae biomass	No demand for fertile land, only carbon dioxide and water, various types of products formation, *etc.*	Cost effective processes, early stage of research, *etc.*
4th generation	Genetically modified algae	Non-arable land for cultivation, high rates of conversion into biofuels, waste, saline and non-portable water used, *etc.*	High production cost, required harsh production conditions (*e.g.*, high pH, high salinity light intensity), *etc.*

Source: Aro, 2016

of vaporization of alcohols helps in better cooling effect of engines (Zaharin et al., 2017). The various physico-chemical properties of alcohols (*e.g.*, methanol, ethanol and butanol), and liquid biofuels (*e.g.*, gasoline) are shown in "Table 2" (Hassan et al., 2015; Saha, 2010; Turner et al., 2018).

Methanol

Methanol is a top ten globally produced chemicals, and it can be synthesised from fossil feedstocks, plant biomass, animal wastes, from carbon dioxide and water or from converting the biomass feedstocks in to synthesis gas over a range of temperature (220-300 °C), pressure (50- 100 bar), and usually in presence of Cu/ZnO based catalyst (Farrauto et al., 2015). Moreover, it can also be produced using an enzymatic or electrolysis process (Obert & Dave, 1999; Steinberg, 1975). As a fuel, methanol has advantages and disadvantages over the convention used fuels such as gasoline or diesel fuel. Methanol has high exhaust gas recirculation and compression ratios in spark ignition engines. Anti nock index (which equates to 99) shows that methanol has 109 research octane number and 89 machine octane number. Moreover, methanol has been projected as a forthcoming biofuel, and as an alternative substitution of H_2 economy. Mobil company in New Zealand between 1981-1984, methanol has been applied to produce gasoline

Table 2. Physico-chemical properties of biofuels and biodiesel

Fuel	Enthalpy of vaporization (ΔH_{vap}, MJ/kg)	Energy density (MJ/L)	Specific energy (MJ/kg)	Air fuel ratio	Cetane number	Latent heat of vaporization (kJ/kg)	Research octane number	Machinery octane number
Methanol	1.20	19.7	15.6	6.5	2	1162.6	136	104
Ethanol	0.92	24.0	30.0	9.0	8	918.4	129	102
Butanol	0.43	29.2	36.6	11.2	17	585.4	96	78
Gasoline	0.36	34.6	46.9	14.6	10-15	317.7	91-99	81-89

Source: Hassan et al., 2015; Saha, 2010; Turner et al., 2018

14,500 bbl/day (Wender, 1996). At kilogram-scale, methanol is primarily synthesised by reactions as shown in equations 1 and 2;

$$CO + 2H_2 \rightarrow CH_3OH \quad (1)$$

$$CO_2 + 3H_2 \rightarrow CH_3OH + H_2O \quad (2)$$

Ethanol

Bioethanol or grain alcohol (EtOH or CH_3CH_2OH) is known as a renewable and sustainable clean, green, advance liquid neutral fuel, and can be used to substitute the non-renewable fossil fuels and to reduce the dependency on foreign fuel feedstocks and global warming as well. It can be produced from different biomass feedstocks (*e.g.*, edible or non-edible biomass). Effective conversion technologies of edible biomass into a liquid ethanol as a biofuel are well known and implemented to produce liquid biofuel at kilogram scales. The general monomer molecular formula of ethanol or ethyl alcohol is CH_3CH_2OH, and it contains one oxygen (34.8%). This oxygen helps to reduce the particulates and NOx emissions, that generates during the bioethanol combustion (Balat et al., 2008). Cellulose and hemicellulose are called as polysaccharides, and the major components of lignocellulosic biomass, and contained two-thirds of dried lignocellulosic biomass. These biomolecules can be processed *via* an enzymatic pathway to produce C6 and C5 sugars, and further transformed into a bioethanol through a fermentation process (Katahira et al., 2006). Theoretical studies showed that the bioethanol (0.51 kg) and carbon dioxide (0.49 kg) as side-products are produced upon the utilization of per kg of glucose and xylose (Hamelinck et al., 2005). Production of ethanol and carbon dioxide from xylose (C5 sugar) and glucose (C6 sugar) is shown as in below reactions (Equations 3 and 4).

$$3C_5H_{10}O_5 \rightarrow 5C_2H_5OH + 5CO_2 \quad (3)$$

$$C_6H_{12}O_6 \rightarrow 2C_2H_5OH + 2CO_2 \quad (4)$$

However, the production technologies of fuels at large scales from nonedible lignocellulosic biomass need to be implemented. Bioethanol has better fuel properties compared to gasoline such as high-octane number, high heat of vaporization, broad flammability limits, and high flame limits. Ethanol fuelled engines worked with low spark compared to gasoline fuelled engines, it is because of self-ignition temperature of ethanol is high than that of gasoline under the same compression ratio (Al-Farayedhi et al., 2004; Costa & Sodré, 2011). Ignition property of ethanol supports that spark ignition engines have high efficacy and shaft power (Al-Farayedhi et al., 2004). High cooling efficacy of fresh cylinder with ethanol engines characterized due to high heat of vaporization. This result suggested that ethanol can be used as a fuel in current engines without any modification (Sadiq Al-Baghdadi, 2000, 2001). Ethanol has better anti-nock property for spark ignition engines, and burned more efficiently compared to gasoline. Therefore, ethanol decreased the peak temperature inside the cylinder, and improved the engines performance (He et al., 2003; Hsieh et al., 2002). Ethanol accumulates both the polar (hydroxyl groups) and nonpolar (alkyl chains) functional groups, therefore, it is soluble in both water (polar) and gasoline (nonpolar) solvents (Gramajo et al., 2004; Rajan & Saniee, 1983).

Various blend ratios of ethanol were characterised with gasoline, and the exhibited properties are shown in "Table 3" (He et al., 2003; Hsieh et al., 2002). Test fuel, "E" means ethanol, and number next to it shows the volume blended in gasoline for example E5 means that 5% ethanol (99.9%) volume/volume (*v/v*) and 95% gasoline (*v/v*) blended in E5. The ASTM stands for American Society for Testing and Materials, used to characterize the fuel properties such as research octane number (RON), reid vapor pressure (RVP), initial boiling temperature, sulphur content, corrosivity, *etc.* "Table 3" shows that as ethanol blend ratio increases from 0 to 30% (*v/v*), RON increases, whereas, RVP value increases and to reach its maximum at E10 and start decreases after that. Most important property like sulphur content decreases as ethanol blend increases from E0 to E30. The effect of methanol blend with gasoline was also studied on the performance of spark ignition engines, and it was characterized that methanol and gasoline blend ratio 15:85 (*v/v*) showed the best engine performance (Abu-Zaid et al., 2004; Turner et al., 2018). Moreover, dual alcohol such as ethanol + methanol or n-butanol + methanol blend with gasoline was also tested with spark ignition engines and results showed that adversely affects were observed of engine performance and exhaust gas emissions compared to neat gasoline or single alcohol blend (El-fasakhany, 2015, 2016; Elfasakhany & Mahrous, 2016).

Table 3. Various blend ratios of ethanol with gasoline and their characteristic properties

Fuel properties	Test fuel					Method
	E0	E5	E10	E20	E30	
RON[a]	95.4	96.7	98.1	100.7	102.4	ASTM D2699
RVP[b] (kPa, 37.8 °C)	53.7	59.3	59.6	58.3	56.8	ASTM D5191
Corrosivity (3 h, 50 °C)	1a	1a	1a	1a	1a	ASTM D130
Lead content (g/L)	<0.0025	<0.0025	<0.0025	<0.0025	<0.0025	ASTM D3237
Washed and unwashed gum (mg/100 mL)	02. and 18.8	0.2 and 18.6	0.2 and 17.4	0.6 and 15	0.2 and 14.4	ASTM D381
Density (kg/L, 15.5 °C)	0.7575	0.7591	0.7608	0.7645	0.7682	ASTM D4052
Distillation temperature (°C)						
IBP[c]	35.5	36.5	37.8	36.7	39.5	ASTM D86
10 (*v/v*)	54.5	49.7	50.8	52.8	54.8	
50 (*v/v*)	94.4	88.0	71.1	70.3	72.4	
90 (*v/v*)	167.3	167.7	166.4	163.0	159.3	
100 (*v/v*) or end point	197.0	202.5	197.5	198.6	198.3	
Heating value (kcal/g)	10.176	9.692	9.511	9.316	8.680	
Carbon (*wt.*%)	86.60	87.70	86.70	87.60	86.00	
Hydrogen (*wt.*%)	13.30	12.20	13.20	12.30	13.90	
Sulphur (*wt.*%)	0.0061	0.0059	0.0055	0.0049	0.0045	ASTM D5453
Residue (v%)	1.70	1.50	1.50	1.50	1.50	

Note: [a] RON- research octane number, [b] RVP- reid vapor pressure, [c] IBP- initial boiling temperature

Source: He et al., 2003; Hsieh et al., 2002

Butanol

Butanol is known as alkyl groups of four carbon and associated of one hydroxyl group at 1-, 2-, tert- or iso- carbon. Butanol can be synthesised in large quantities from renewable resource such as plant biomass and also from non-renewable resources such as crude oils, natural gas and coal, through a chemical, thermal or biological pathway (Ezeji et al., 2007; Ndaba et al., 2015; Trifirò, 2019). Due to its long chain of alkyl groups, it may be applied as a fuel in an internal combustion engine. It is tested and characterised to substitute ethanol up to 100% in gasoline engines, and in diesel and in jet fuel as an additive up to 20% without any engine modification as conventional existing engines. Biobutanol has several advantages, and produced low amounts of carbon monoxide, NOx, SOx and hydrocarbons compared to gasoline. Biobutanol successfully blended with diesel up to 20% (*v/v*) without any spark ignition engines modification. It improves the diesel engine performance in terms of lubricating power of oxidative properties, cold behaviour, low toxic emissions, *etc.* Biobutanol has a high calorific value such as high energy density, high specific energy and low vapor pressure compared to ethanol and has lower calorific values to gasoline as well. The research octane number and machine octane number of n- and t-butanol are 96 & 78, and 105 & 89 respectively. t-butanol can be applied as an additive in gasoline and cannot be used as a neat fuel due to its high melting point (25.5 °C). The energy released per cycle using butanol is high compared to ethanol and methanol, and approximately 10% high in case of gasoline. Butanol has low heat of vaporization than that of ethanol and methanol due to high energy released, it can start engine in smooth and easier ways compared to ethanol or methanol running engine.

Gasoline

Increasing population demands from industries and transports sector, and strict environmental protection policies have motivated and interested to develop gasoline fuel from lignocellulosic biomass (Ruddy et al., 2019). Several methods are known to produce biogasoline from lignocellulosic biomass such as pyrolysis, hydrothermal, (Shamsul et al., 2017) an integrated catalytic transformation process, (Zhang et al., 2015) by coupling of oligomerization and biomass catalytic pyrolysis process, (Wang et al., 2017) catalytic production of gasoline using shape selective catalysts, (Weisz et al. 1979) *etc.* The chemical composition of gasoline is characterized as alkanes (4-8%), alkenes (2-5%), isoalkanes (25-40%), cycloalkanes (3-7%), cycloalkenes (1-4%), aromatic (20-50% including 0.5-2.5% benzene) (Shamsul et al., 2017). Thermal efficacy of biogasoline engine is similar to biodiesel engine, that results encouraging to develop effective conversion technologies for biogasoline production as a production of biodiesel from biomass is already in function. The specific energy (46.9 MJ/kg) and energy density (34.6 MJ/L) of gasoline are high compared to ethanol and methanol ("Table 2") (Hassan et al., 2015). Gasoline derived from plant biomass has similar research octane number (90-110) to gasoline (93) derived from non-renewable crude oil. Biogasoline has similar heating rate (41.0 MJ/kg) to commercial gasoline (43.1 MJ/kg) (Zhang et al., 2007). The high octane rating and low energy content of biogasoline is mainly influenced the spark ignition engines compared to conventional used gasoline. The assorted ethanol: gasoline blends on spark ignition engines characterized that at low scales blend exhibited the improvement in engine torque and brake power, whereas, the increment of engine brake power and torques of E5, E10 and E20 are 2.31%, 2.77% & 4.16, and, 0.29%, 0.59 & 4.77% respectively (Thakur et al., 2017). The gasoline derived from plant biomass contains higher content of oxygen that helps fuels in internal combustion engine in reduction of noxious gases and soot (Shamsul et al., 2017).

Diesel

Biodiesel is a renewable and sustainable liquid biofuel, that either can be used to reduce and/or substitute the dependency on non-renewable liquid fuels as a global energy demand (Corma et al., 2011). Biodiesel can be produced using assorted processes (Ma & Hanna, 1999; Meher et al., 2006). The basic process to produce biodiesel is a transesterification method of fatty acids and further, the process involves as a hydrogenation which saturates the double bonds and generates H_2O or CO_2 as a side product depends on availability of hydrogen (Ma & Hanna, 1999). Two basic reactions such as hydrogenation and decarboxylation occur to produce the renewable biodiesel from triolein (Equations 5 and 6). Hydrogenation process consumes 2.5 times more hydrogen compared to decarboxylation process and produces more carbon in fuel and H_2O as a side product. The hydrogenation and decarboxylation processes of triolein yield 0.862 kg $C_{18}H_{38}$/kg and 0.815 kg $C_{17}H_{36}$/kg ideal liquid fuels respectively. Vegetable oils or animal fat associated with long range of fatty acid chains and yielded a range of n-paraffins through a transesterification process (Karthickeyan, 2019a, 2019b) and resulted the mixture of n-paraffins product suitable for blending application with petroleum-based diesel fuel.

Hydrogenation $C_{57}H_{104}O_6 + 15H_2 \rightarrow 3C_{18}H_{38} + 6H_2O + C_3H_8$ (5)

Decarboxylation $C_{57}H_{104}O_6 + 6H_2 \rightarrow 3C_{17}H_{36} + 3CO_2 + C_3H_8$ (6)

Physico-chemical properties of biodiesel such as cold flow, lubricity, viscosity, cetane number, and oxidative stability agree with the composition of fatty acids. High and low percentages of mono-unsaturated and saturated fatty acids affect the cold flow property of biodiesel respectively. The selection of biodiesel mainly depends on considerable availability, satisfactory performance, low temperature operability and oxidation stability. Moreover, the standard properties of biodiesel also depend on the region specific of countries such as different kinds of climate. United States, European Union, India and Canada are following the different standards as ASTM D6751, EN 14,214, IS 15,607 and CAN/CGSB-3.524 respectively (Husam et al. 2017; Singh et al., 2019). Few of the comparable properties of biodiesel (neat or blended) and non-renewable diesel are shown in "Table 4" (Barabás & Todoruţ, 2011; Husam et al., 2017).

Table 4. Comparable properties of renewable biodiesel (neat and blend) and non-renewable diesel

Properties	B5 blend	B20 blend	B100	Non-renewable diesel	Test method
Flash point (°C)	57 - 74	67 - 82	111 - 168	53.0 - 71.5	ASTM D93 and D7215
Viscosity (mm²/Sec)	2.48 - 4.45	2.74 - 3.75	4.27 - 11.0	2.40 - 4.30	ASTM D445 and D975
Cetane number (min)	51.5	52.2 - 52.3	56.0	50.9	ASTM D613
Calorific value (MJ/kg)	42.13 - 46.00	42.01 - 44.98	37.76 - 37.95	43.15 - 46.35	ASTM D240 and ASTM D4868
Sulphur content (ppm)	-	-	10.94 - 11.69	3.59 - 500	ASTM D5453 and ASTM D2622

Source: Barabás & Todoruţ, 2011; Husam Al-Mashhadani & Fernando, 2017

Jet Fuel

The production of jet fuel from sustainable and renewable bioresources are an auspicious approach for aviation industries to target and reduce the dependency on fossil feedstocks and assists also to mitigate greenhouse gases emission effect. Conventional used jet fuel is produced from crude oil by using a distillation process and temperature ranging from 205 to 260 °C (Liu et al., 2013). The jet fuel is the mixture of hydrocarbons in the range of C8 to C16, and these hydrocarbons are a mixture of several compounds such as alkenes, iso-alkenes, aromatic compounds, naphthenic and derivative of naphthenic (Hileman & Stratton, 2014). The ratio of each compound in jet fuel has a direct relationship for jet fuel properties. The high hydrocarbon contents of alkene have ensured the energy density of jet fuels. Naphthenes have a high capability to decrease the fuel freeze points, that is censorious at high altitude flying. Additional, aromatics and aromatic derivatives assisted to enhance the lubricity of material compatibility and that stopped leakage in some aircrafts. However, the excess amounts of aromatics would be assisted to clean the fuel, therefore, the ratios of aromatic should be controlled in a realistic range (Blakey et al., 2011; Hileman & Stratton, 2014; Liu et al., 2013). The commercial jet fuel has strict specific properties such as a sulphur content, lubricity or viscosity, density, flash point, smoke point, conductivity, naphthalenes, *etc.* Assorted conversion technologies are known to produce jet fuel from biomass-based materials (Bond et al., 2014). The biomass conversion technologies are varied and depended on feedstock types as well. Based on the feedstocks and conversion processes, bio-jet fuel derived from biomass can be exhibited through oil-to-jet fuel, alcohol-to-jet fuel, sugar-to-jet fuel and gas-to-jet fuel processes (Huber et al., 2006; Wei et al., 2019).

ZERO CARBON EMISSIONS

Growing the world population, heterogeneous distribution of fossil feedstocks, fluctuating oil prices, and global warming are the main concerns that have been motivated to develop and transfer the current economy, primarily based on the use of fossil feedstocks, into a biobased economy, based on the use of renewable and sustainable plant biomass. The petro-refinery and bio-refinery can be used to synthesize transport fuels, chemicals, *etc.*, but, the nature of raw material taken as feedstocks are different in both cases. However, due to the fast depletion of fossil feedstocks, plant biomass feedstocks are gaining importance globally through an integrated bio-refinery concept. The utilization of biomass (preferably non-edible lignocellulosic) into valuable biofuels, biochemicals, and energy can be employed through a bio-refinery pathway as shown in "Figure 2". The generated valuable biofuels from plant biomass are basically made up of carbon and hydrogen and in almost negligible concentration of oxygen, upon usage of these fuels release oxides of carbon (*e.g.* CO and CO_2) in atmosphere. The released carbon dioxide is consumed by the plant for their growth *via* photosynthesis process in the presence of water and sunlight. Further, these plants grow and produce edible (starch $(C_6H_{10}O_5)_n$, oil (mustard oil, coconut oil, corn oil *etc.*) *etc.*) and non-edible (lignocellulosic biomass (plant derived) biomass, which is composed of cellulose (40-50 dried *wt.%*), hemicellulose (20-30 dried *wt.%*), lignin (15-25 dried *wt.%*), *etc.*) (Li et al., 2015; Ragauskas et al., 2006; Singh & Dhepe, 2016b). Plant biomass (dead) are again used for bio-refinery concept to gain fuels, chemicals and energy. Consequently, it can be said that the utilization of fuels and chemicals obtained from plant biomass, allows a reutilization of these in a recycle way (carbon neutral way), which is typically reducing the CO_2 levels in the atmosphere through a photosynthesis pathway.

Figure 2. Integrated biorefinery concept for zero carbon emission

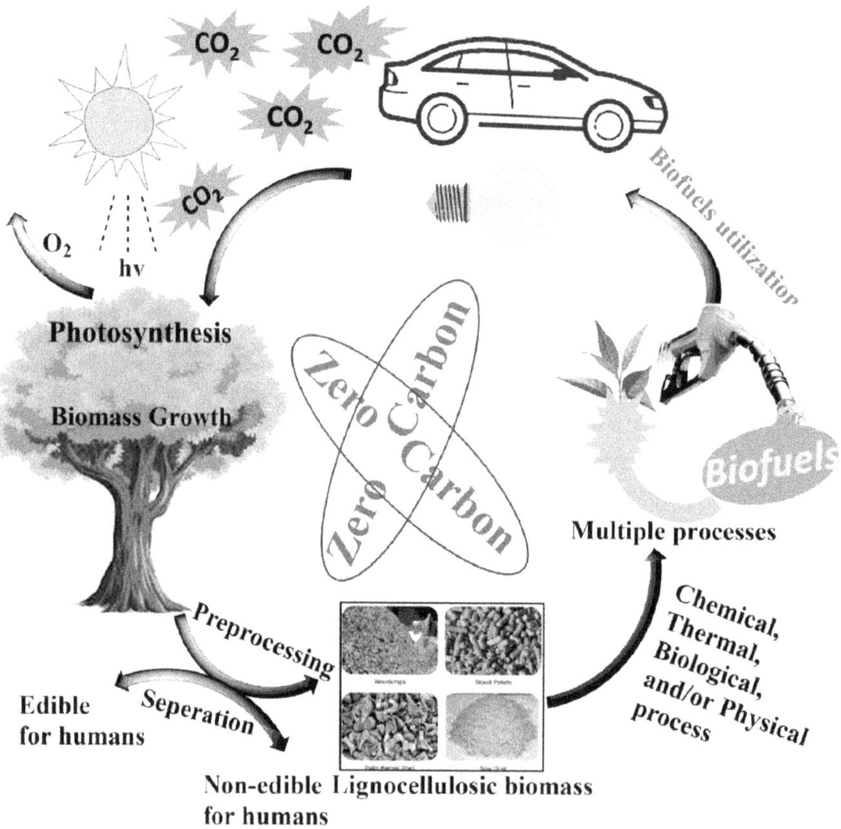

The whole bio-refinery concept is shown in "Figure 2", whereas, in use of non-renewable fossil feedstocks, a million of years is essential for the natural conversion of biomass into fossil feedstocks. In the meanwhile, large amount of CO_2 will be released in the atmosphere, which is the foremost reason for global warming. Hence, it is vital to use biomass in addition to fossil feedstocks which may curtail the problems associated with fossil feedstocks.

CONCLUSION

The strategies goal of plant biomass conversion and utilization need to be effectively developed the operational studies and technologies for feedstocks conversion into economically survival liquids biofuels commodities, such as biomethanol, bioethanol, biobutanol, biodiesel, biogasoline, jet-fuel, *etc*. Primarily, two basic biomass conversion processes such as biochemical and thermochemical are extensively studied and investigated at lab to market scales. However, the biomass conversion into ethanol using biological process is a critical step in 1st - and 2nd- generation ethanol processes. To have better economy of result, researchers are applying several strategies to improve the process economics of the steps involves through an increment of biofuels production efficacy. This chapter discusses the physico-chemical (*e.g.*, research octane number, machine octane number, reid vapor pressure, initial boiling temperature, latent heat of vaporization, sulphur content, corrosivity, *etc.*) characteristics result of biofuels derived from plant biomass that can be used either as a neat fuel or blended with gasoline, diesel, *etc.* in ignition engines. The utilization of liquid biofuels is not just substitute or reduce the dependency of non-renewable fossil fuels, although, assisted to improve the increasingly alarming reports in concerns of global climate strike.

REFERENCES

Abu-Zaid, M., Badran, O., & Yamin, J. (2004). Effect of Methanol Addition on the Performance of Spark Ignition Engines. *Energy & Fuels, 18*(2), 312–315. doi:10.1021/ef030103d

Agbor, V. B., Cicek, N., Sparling, R., Berlin, A., & Levin, D. B. (2011). Biomass pretreatment: Fundamentals toward application. *Biotechnology Advances, 29*(6), 675–685. doi:10.1016/j.biotechadv.2011.05.005 PMID:21624451

Al-Farayedhi, A. A., Al-Dawood, A. M., & Gandhidasan, P. (2004). Experimental Investigation of SI Engine Performance Using Oxygenated Fuel. *Journal of Engineering for Gas Turbines and Power, 126*(1), 178–191. doi:10.1115/1.1615254

Alonso, D. M., Bond, J. Q., & Dumesic, J. A. (2010). Catalytic conversion of biomass to biofuels. *Green Chemistry, 12*(9), 1493–1513. doi:10.1039/c004654j

Aro, E.-M. (2016). From first generation biofuels to advanced solar biofuels. *Ambio, 45*(S1Suppl 1), S24–S31. doi:10.100713280-015-0730-0 PMID:26667057

Balat, M. (2011). Production of bioethanol from lignocellulosic materials via the biochemical pathway: A review. *Energy Conversion and Management, 52*(2), 858–875. doi:10.1016/j.enconman.2010.08.013

Balat, M., Balat, H., & Öz, C. (2008). Progress in bioethanol processing. *Progress in Energy and Combustion Science, 34*(5), 551–573. doi:10.1016/j.pecs.2007.11.001

Bär, J., Phongpreecha, T., Singh, S. K., Kral Yilmaz, M., Foster, C. E., Crowe, J. D., & Hodge, D. B. (2018). Deconstruction of hybrid poplar to monomeric sugars and aromatics using ethanol organosolv fractionation. *Biomass Conversion and Biorefinery, 8*(4), 813–824. doi:10.100713399-018-0330-x

Barabás, I., & Todoruţ, I.-A. (2011). Biodiesel Quality, Standards and Properties. In G. Montero (Ed.), *Biodiesel Quality, Emissions, and By-Products* (pp. 1–28). InTech. doi:10.5772/25370

Bhowmick, P., Jeevanantham, A. K., Ashok, B., Nanthagopal, K., Perumal, D. A., Karthickeyan, V., ... Jain, A. (2019). Effect of fuel injection strategies and EGR on biodiesel blend in a CRDI engine. *Energy, 181*, 1094–1113. doi:10.1016/j.energy.2019.06.014

Binder, J. B., & Raines, R. T. (2010). Fermentable sugars by chemical hydrolysis of biomass. *Proceedings of the National Academy of Sciences of the United States of America, 107*(10), 4516–4521. doi:10.1073/pnas.0912073107 PMID:20194793

Blakey, S., Rye, L., & Wilson, C. W. (2011). Aviation gas turbine alternative fuels: A review. *Proceedings of the Combustion Institute, 33*(2), 2863–2885. doi:10.1016/j.proci.2010.09.011

Bond, J. Q., Upadhye, A. A., Olcay, H., Tompsett, G. A., Jae, J., Xing, R., ... Huber, G. W. (2014). Production of renewable jet fuel range alkanes and commodity chemicals from integrated catalytic processing of biomass. *Energy & Environmental Science*, *7*(4), 1500–1523. doi:10.1039/C3EE43846E

Brown, T. R., & Brown, R. C. (2013). A review of cellulosic biofuel commercial-scale projects in the United States. *Biofuels, Bioproducts, & Biorefining*, *7*(3), 235–245. doi:10.1002/bbb.1387

Cole, C. V., Duxbury, J., Freney, J., Heinemeyer, O., Minami, K., Mosier, A., ... Zhao, Q. (1997). Global estimates of potential mitigation of greenhouse gas emissions by agriculture. *Nutrient Cycling in Agroecosystems*, *49*(1), 221–228. doi:10.1023/A:1009731711346

Corma, A., de la Torre, O., Renz, M., & Villandier, N. (2011). Production of High-Quality Diesel from Biomass Waste Products. *Angewandte Chemie International Edition*, *50*(10), 2375–2378. doi:10.1002/anie.201007508 PMID:21351358

Costa, R. C., & Sodré, J. R. (2011). Compression ratio effects on an ethanol/gasoline fuelled engine performance. *Applied Thermal Engineering*, *31*(2), 278–283. doi:10.1016/j.applthermaleng.2010.09.007

Elfasakhany, A. (2015). Investigations on the effects of ethanol–methanol–gasoline blends in a spark-ignition engine: Performance and emissions analysis. *Engineering Science and Technology, an International Journal, 18*(4), 713-719.

Elfasakhany, A. (2016). Performance and emissions of spark-ignition engine using ethanol–methanol–gasoline, n-butanol–iso-butanol–gasoline and iso-butanol–ethanol–gasoline blends: A comparative study. *Engineering Science and Technology, an International Journal, 19*(4), 2053-2059.

Elfasakhany, A., & Mahrous, A.-F. (2016). Performance and emissions assessment of n-butanol–methanol–gasoline blends as a fuel in spark-ignition engines. *Alexandria Engineering Journal*, *55*(3), 3015–3024. doi:10.1016/j.aej.2016.05.016

Ezeji, T. C., Qureshi, N., & Blaschek, H. P. (2007). Bioproduction of butanol from biomass: From genes to bioreactors. *Current Opinion in Biotechnology*, *18*(3), 220–227. doi:10.1016/j.copbio.2007.04.002 PMID:17462877

Farrauto, R. J., Dorazio, L., & Bartholomew, C. H. (2015). *Introduction to Catalysis and Industrial Catalytic Processes*. John Wiley & Sons.

Fatih Demirbas, M. (2009). Biorefineries for biofuel upgrading: A critical review. *Applied Energy*, *86*, S151–S161. doi:10.1016/j.apenergy.2009.04.043

Gramajo de Doz, M. B., Bonatti, C. M., & Sólimo, H. N. (2004). Water Tolerance and Ethanol Concentration in Ethanol-Gasoline Fuels at Three Temperatures. *Energy & Fuels*, *18*(2), 334–337. doi:10.1021/ef034040a

Hajjari, M., Tabatabaei, M., Aghbashlo, M., & Ghanavati, H. (2017). A review on the prospects of sustainable biodiesel production: A global scenario with an emphasis on waste-oil biodiesel utilization. *Renewable & Sustainable Energy Reviews*, *72*, 445–464. doi:10.1016/j.rser.2017.01.034

Hamelinck, C. N., Hooijdonk, G., & Faaij, A. P. C. (2005). Ethanol from lignocellulosic biomass: Techno-economic performance in short-, middle- and long-term. *Biomass and Bioenergy*, *28*(4), 384–410. doi:10.1016/j.biombioe.2004.09.002

Hassan, S. N., Sani, Y. M., Abdul Aziz, A. R., Sulaiman, N. M. N., & Daud, W. M. A. W. (2015). Biogasoline: An out-of-the-box solution to the food-for-fuel and land-use competitions. *Energy Conversion and Management*, *89*, 349–367. doi:10.1016/j.enconman.2014.09.050

He, B.-Q., Jian-Xin, W., Hao, J.-M., Yan, X.-G., & Xiao, J.-H. (2003). A study on emission characteristics of an EFI engine with ethanol blended gasoline fuels. *Atmospheric Environment*, *37*(7), 949–957. doi:10.1016/S1352-2310(02)00973-1

Hileman, J. I., & Stratton, R. W. (2014). Alternative jet fuel feasibility. *Transport Policy*, *34*, 52–62. doi:10.1016/j.tranpol.2014.02.018

Hsieh, W.-D., Chen, R.-H., Wu, T.-L., & Lin, T.-H. (2002). Engine performance and pollutant emission of an SI engine using ethanol–gasoline blended fuels. *Atmospheric Environment*, *36*(3), 403–410. doi:10.1016/S1352-2310(01)00508-8

Huber, G. W., Iborra, S., & Corma, A. (2006). Synthesis of Transportation Fuels from Biomass: Chemistry, Catalysts, and Engineering. *Chemical Reviews*, *106*(9), 4044–4098. doi:10.1021/cr068360d PMID:16967928

Husam Al-Mashhadani, & Fernando, S. (2017). Properties, performance, and applications of biofuel blends: a review. *AIMS Energy, 5*(4), 735-767.

iea. (2018). Retrieved from www.iea.org/renewables2018/transport/

Karthickeyan, V. (2019a). Effect of cetane enhancer on Moringa oleifera biodiesel in a thermal coated direct injection diesel engine. *Fuel*, *235*, 538–550. doi:10.1016/j.fuel.2018.08.030

Karthickeyan, V. (2019b). Effect of combustion chamber bowl geometry modification on engine performance, combustion and emission characteristics of biodiesel fuelled diesel engine with its energy and exergy analysis. *Energy*, *176*, 830–852. doi:10.1016/j.energy.2019.04.012

Karthickeyan, V., Ashok, B., Nanthagopal, K., Thiyagarajan, S., & Geo, V. E. (2019). Investigation of novel Pistacia khinjuk biodiesel in DI diesel engine with post combustion capture system. *Applied Thermal Engineering*, *159*, 113969. doi:10.1016/j.applthermaleng.2019.113969

Katahira, S., Mizuike, A., Fukuda, H., & Kondo, A. (2006). Ethanol fermentation from lignocellulosic hydrolysate by a recombinant xylose- and cellooligosaccharide-assimilating yeast strain. *Applied Microbiology and Biotechnology*, *72*(6), 1136–1143. doi:10.100700253-006-0402-x PMID:16575564

Li, C., Zhao, X., Wang, A., Huber, G. W., & Zhang, T. (2015). Catalytic Transformation of Lignin for the Production of Chemicals and Fuels. *Chemical Reviews*, *115*(21), 11559–11624. doi:10.1021/acs.chemrev.5b00155 PMID:26479313

Liu, G., Yan, B., & Chen, G. (2013). Technical review on jet fuel production. *Renewable & Sustainable Energy Reviews*, *25*, 59–70. doi:10.1016/j.rser.2013.03.025

Luque, R., Herrero-Davila, L., Campelo, J. M., Clark, J. H., Hidalgo, J. M., Luna, D., ... Romero, A. A. (2008). Biofuels: A technological perspective. *Energy & Environmental Science*, *1*(5), 542–564. doi:10.1039/b807094f

Ma, F., & Hanna, M. A. (1999). Biodiesel production: A review1Journal Series #12109, Agricultural Research Division, Institute of Agriculture and Natural Resources, University of Nebraska–Lincoln.1. *Bioresource Technology*, *70*(1), 1–15. doi:10.1016/S0960-8524(99)00025-5

McKendry, P. (2002). Energy production from biomass (part 1): Overview of biomass. *Bioresource Technology*, *83*(1), 37–46. doi:10.1016/S0960-8524(01)00118-3 PMID:12058829

McMillan, J. D. (1997). Bioethanol production: Status and prospects. *Renewable Energy*, *10*(2), 295–302. doi:10.1016/0960-1481(96)00081-X

Meher, L. C., Vidya Sagar, D., & Naik, S. N. (2006). Technical aspects of biodiesel production by transesterification—A review. *Renewable & Sustainable Energy Reviews*, *10*(3), 248–268. doi:10.1016/j.rser.2004.09.002

Mosier, N., Wyman, C., Dale, B., Elander, R., Lee, Y. Y., Holtzapple, M., & Ladisch, M. (2005). Features of promising technologies for pretreatment of lignocellulosic biomass. *Bioresource Technology*, *96*(6), 673–686. doi:10.1016/j.biortech.2004.06.025 PMID:15588770

Ndaba, B., Chiyanzu, I., & Marx, S. (2015). n-Butanol derived from biochemical and chemical routes: A review. *Biotechnology Reports (Amsterdam, Netherlands)*, *8*, 1–9. doi:10.1016/j.btre.2015.08.001 PMID:28352567

Nejat, P., Jomehzadeh, F., Taheri, M. M., Gohari, M., & Abd. Majid, M. Z. (2015). A global review of energy consumption, CO2 emissions and policy in the residential sector (with an overview of the top ten CO2 emitting countries). *Renewable & Sustainable Energy Reviews*, *43*, 843–862. doi:10.1016/j.rser.2014.11.066

Obert, R., & Dave, B. C. (1999). Enzymatic Conversion of Carbon Dioxide to Methanol: Enhanced Methanol Production in Silica Sol−Gel Matrices. *Journal of the American Chemical Society*, *121*(51), 12192–12193. doi:10.1021/ja991899r

Payne, C. M., Knott, B. C., Mayes, H. B., Hansson, H., Himmel, M. E., Sandgren, M., ... Beckham, G. T. (2015). Fungal Cellulases. *Chemical Reviews*, *115*(3), 1308–1448. doi:10.1021/cr500351c PMID:25629559

Ragauskas, A. J., Williams, C. K., Davison, B. H., Britovsek, G., Cairney, J., Eckert, C. A., ... Tschaplinski, T. (2006). The Path Forward for Biofuels and Biomaterials. *Science*, *311*(5760), 484–489. doi:10.1126cience.1114736 PMID:16439654

Rajan, S., & Saniee, F. F. (1983). Water—ethanol—gasoline blends as spark ignition engine fuels. *Fuel*, *62*(1), 117–121. doi:10.1016/0016-2361(83)90265-X

Ruddy, D. A., Hensley, J. E., Nash, C. P., Tan, E. C. D., Christensen, E., Farberow, C. A., ... Schaidle, J. A. (2019). Methanol to high-octane gasoline within a market-responsive biorefinery concept enabled by catalysis. *Nature Catalysis*, *2*(7), 632–640. doi:10.103841929-019-0319-2

Ruth, L. (2008). Bio or bust? *EMBO Reports*, *9*(2), 130–133. doi:10.1038j.embor.2008.6 PMID:18246105

Sadiq Al-Baghdadi, M. A.-R. (2000). Performance study of a four-stroke spark ignition engine working with both of hydrogen and ethyl alcohol as supplementary fuel. *International Journal of Hydrogen Energy*, *25*(10), 1005–1009. doi:10.1016/S0360-3199(00)00012-4

Sadiq Al-Baghdadi, M. A.-R. (2001). The safe operation zone of the spark ignition engine working with dual renewable supplemented fuels (hydrogen+ethyl alcohol). *Renewable Energy*, *22*(4), 579–583. doi:10.1016/S0960-1481(00)00109-9

Saha, A. (2010). *Vaporization characteristics of pure and blended biofuel droplet injected into hot stream of air. (Master in Science)*. Orlando, FL: University of Central Florida.

Serrano-Ruiz, J. C., Ramos-Fernández, E. V., & Sepúlveda-Escribano, A. (2012). From biodiesel and bioethanol to liquid hydrocarbon fuels: New hydrotreating and advanced microbial technologies. *Energy & Environmental Science*, *5*(2), 5638–5652. doi:10.1039/C1EE02418C

Shabih, F., Amir, H., Muhammad, N., Temoor, A., Muhammad, S., Mohsin, T., ... Romana, T. (2018). Lignocellulosic Biomass: A Sustainable Bioenergy Source for the Future. *Protein and Peptide Letters*, *25*(2), 148–163. doi:10.2174/0929866525666180122144504 PMID:29359659

Shamsul, N. S., Kamarudin, S. K., & Rahman, N. A. (2017). Conversion of bio-oil to bio gasoline via pyrolysis and hydrothermal: A review. *Renewable & Sustainable Energy Reviews*, *80*, 538–549. doi:10.1016/j.rser.2017.05.245

Singh, D., Sharma, D., Soni, S. L., Sharma, S., & Kumari, D. (2019). Chemical compositions, properties, and standards for different generation biodiesels: A review. *Fuel*, *253*, 60–71. doi:10.1016/j.fuel.2019.04.174

Singh, S. K. (2019). Solubility of lignin and chitin in ionic liquids and their biomedical applications. *International Journal of Biological Macromolecules*, *132*, 265–277. doi:10.1016/j.ijbiomac.2019.03.182 PMID:30930266

Singh, S. K., & Dhepe, P. L. (2016a). Isolation of lignin by organosolv process from different varieties of rice husk: Understanding their physical and chemical properties. *Bioresource Technology*, *221*, 310–317. doi:10.1016/j.biortech.2016.09.042 PMID:27648850

Singh, S. K., & Dhepe, P. L. (2016b). Ionic liquids catalyzed lignin liquefaction: Mechanistic studies using TPO-MS, FT-IR, RAMAN and 1D, 2D-HSQC/NOESY NMR. *Green Chemistry*, *18*(14), 4098–4108. doi:10.1039/C6GC00771F

Singh, S. K., & Dhepe, P. L. (2018). Experimental evidences for existence of varying moieties and functional groups in assorted crop waste derived organosolv lignins. *Industrial Crops and Products*, *119*, 144–151. doi:10.1016/j.indcrop.2018.04.002

Singh, S. K., Savoy, A., Yuan, Z., Luo, H., Stahl, S. S., Hegg, E. L., & Hodge, D. B. (2019). Integrated Two-Stage Alkaline-Oxidative Pretreatment of Hybrid Poplar. Part 1: Impact of Alkaline Pre-Extraction Conditions on Process Performance and Lignin Properties. *Industrial & Engineering Chemistry Research*, *58*(35), 15989–15999. doi:10.1021/acs.iecr.9b01124

Steinberg, M. (1975). Electrolytic synthesis of methanol from CO2, US Pat. No. US3959094A.

Stöcker, M. (2008). Biofuels and Biomass-To-Liquid Fuels in the Biorefinery: Catalytic Conversion of Lignocellulosic Biomass using Porous Materials. *Angewandte Chemie International Edition*, *47*(48), 9200–9211. doi:10.1002/anie.200801476 PMID:18937235

Straathof, A. J. J. (2014). Transformation of Biomass into Commodity Chemicals Using Enzymes or Cells. *Chemical Reviews*, *114*(3), 1871–1908. doi:10.1021/cr400309c PMID:23987659

Tenenbaum, D. J. (2008). Food vs. Fuel: Diversion of Crops Could Cause More Hunger. *Environmental Health Perspectives*, *116*(6), 254–257. doi:10.1289/ehp.116-a254 PMID:18560500

Thakur, A. K., Kaviti, A. K., Mehra, R., & Mer, K. K. S. (2017). Progress in performance analysis of ethanol-gasoline blends on SI engine. *Renewable & Sustainable Energy Reviews*, *69*, 324–340. doi:10.1016/j.rser.2016.11.056

Tilman, D., Hill, J., & Lehman, C. (2006). Carbon-Negative Biofuels from Low-Input High-Diversity Grassland Biomass. *Science*, *314*(5805), 1598–1600. doi:10.1126cience.1133306 PMID:17158327

Trifirò, F. (2019). Fuels from Biomass. *TECNICA ITALIANA-Italian Journal of Engineering Science*, *63*(1), 86–93. doi:10.18280/ti-ijes.630112

Turner, J. W., Lewis, A. G., Akehurst, S., Brace, C. J., Verhelst, S., Vancoillie, J., ... Edwards, P. P. (2018). Alcohol fuels for spark-ignition engines: Performance, efficiency and emission effects at mid to high blend rates for binary mixtures and pure components. *Proceedings of the Institution of Mechanical Engineers. Part D, Journal of Automobile Engineering*, *232*(1), 36–56. doi:10.1177/0954407017752832

Vohra, M., Manwar, J., Manmode, R., Padgilwar, S., & Patil, S. (2014). Bioethanol production: Feedstock and current technologies. *Journal of Environmental Chemical Engineering*, *2*(1), 573–584. doi:10.1016/j.jece.2013.10.013

von Blottnitz, H., & Curran, M. A. (2007). A review of assessments conducted on bio-ethanol as a transportation fuel from a net energy, greenhouse gas, and environmental life cycle perspective. *Journal of Cleaner Production*, *15*(7), 607–619. doi:10.1016/j.jclepro.2006.03.002

Wang, T., Zhang, Q., Ding, M., Wang, C., Li, Y., Zhang, Q., & Ma, L. (2017). Bio-gasoline Production by Coupling of Biomass Catalytic Pyrolysis and Oligomerization Process. *Energy Procedia*, *105*, 858–863. doi:10.1016/j.egypro.2017.03.401

Wei, H., Liu, W., Chen, X., Yang, Q., Li, J., & Chen, H. (2019). Renewable bio-jet fuel production for aviation: A review. *Fuel*, *254*. doi:10.1016/j.fuel.2019.06.007

Weisz, P. B., Haag, W. O., & Rodewald, P. G. (1979). Catalytic Production of High-Grade Fuel (Gasoline) from Biomass Compounds by Shape-Selective Catalysis. *Science*, *206*(4414), 57–58. doi:10.1126cience.206.4414.57 PMID:17812451

Wender, I. (1996). Reactions of synthesis gas. *Fuel Processing Technology*, *48*(3), 189–297. doi:10.1016/S0378-3820(96)01048-X

Yang, B., & Wyman, C. E. (2008). Pretreatment: The key to unlocking low-cost cellulosic ethanol. *Biofuels, Bioproducts, & Biorefining*, *2*(1), 26–40. doi:10.1002/bbb.49

Yuan, Z., Singh, S. K., Bals, B. D., Hodge, D. B., & Hegg, E. L. (2019). Integrated Two-Stage Alkaline-Oxidative Pretreatment of Hybrid Poplar. Part 2: Impact of Cu-Catalyzed Alkaline Hydrogen Peroxide Pretreatment Conditions on Process Performance and Economics. *Industrial & Engineering Chemistry Research*, *58*(35), 16000–16008. doi:10.1021/acs.iecr.9b00901

Zaharin, M. S. M., Abdullah, N. R., Najafi, G., Sharudin, H., & Yusaf, T. (2017). Effects of physicochemical properties of biodiesel fuel blends with alcohol on diesel engine performance and exhaust emissions: A review. *Renewable & Sustainable Energy Reviews*, *79*, 475–493. doi:10.1016/j.rser.2017.05.035

Zhang, Q., Chang, J., Wang, T., & Xu, Y. (2007). Review of biomass pyrolysis oil properties and upgrading research. *Energy Conversion and Management*, *48*(1), 87–92. doi:10.1016/j.enconman.2006.05.010

Zhang, Z., Bi, P., Jiang, P., Fan, M., Deng, S., Zhai, Q., & Li, Q. (2015). Production of gasoline fraction from bio-oil under atmospheric conditions by an integrated catalytic transformation process. *Energy*, *90*, 1922–1930. doi:10.1016/j.energy.2015.07.009

Chapter 3
Emission Aspects of Biomass-Based Advanced Second Generation Bio-Fuels in IC Engines

R. Sakthivel
https://orcid.org/0000-0002-0766-0285
Amrita Vishwa Vidyapeetham University, India

Mohanraj T.
Amrita Vishwa Vidyapeetham University, India

Joseph John Marshal S.
Karunya Institute of Technology and Sciences, India

Baranitharan P.
NIT, Tiruchirappalli. India

Tamilvanan A.
https://orcid.org/0000-0001-9406-5005
Kongu Engineering College, India

Gomathi K.
Kongu Engineering College, India

ABSTRACT

Rapid industrialization and growth in population in urban regions augment the pollution levels from transportation sectors, especially from diesel fleets. A wide array of research activities were carried out to satisfy the energy needs as well as reduce the emission levels, which poses a big challenge to the research community. In this situation, biomass-derived fuels provide a ray of hope to the research community to address the emission problem by adapting closed carbon cycle at low cost. This chapter gives an overview to the readers about the present energy scenario, biomass-based fuel, upgradation techniques for biomass fuel, and engine adaptability of biomass-based fuels. This chapter provides a clear glimpse of biomass energy, one of the potential energy resources in the near future.

DOI: 10.4018/978-1-7998-2539-5.ch003

INTRODUCTION

In current scenario, more importance has been given towards the progress of petroleum, coal, and gas-based plants for utilizing the economically feasible feedstock. These feed stocks are generally used for producing various yields like fuel, chemicals, fiber, plastics, pesticides, fertilizers, lubricants, coke, etc. for satisfying the requirements (Bender, 2000; M. F. Demirbas, 2006). These resources are the crucial fuels to produce power and electricity used for engineering as well as domestic applications. Exponential growth of industrialization at worldwide is the most important reason for significant utilization of fossil fuels. However, these fossil resources are considered as non-sustainable resource and produce environmental pollutants or emissions (Nanda, Rana, Sarangi, Dalai, & Kozinski, 2018).

Petroleum based fuels plays a vital role in transportation part. Transport accounts for around 64% fuel utilization, 27% of energy, and 23% of Carbon-dioxide (CO_2) emissions. Burning of fossil fuel in transportation sector is the major reason for increasing the pollution level. CO_2 is one of the most dangerous greenhouse emission from transport sector (Boden, Marland, & Andres, 2017). CO_2 emission from fuel combustion was shown in Figure 1(IEA, 2017). Due to increase in greenhouse gases (GHG), there is a tremendous increase in earth's temperature. This brought upon a change in climatic pattern and environment. 8% of 25,000 species vanished due to these climatic changes. Half the species of plants and animals found on the face of earth are at a risk of extinction. Climatic changes have also affected the frequency, intensity and duration of natural disaster such as hurricane, tornado, tsunami etc. On the other hand, polar ice caps are melting at an alarming rate and polar animals are soon to become endangered species.

Figure 2 demonstrates the global production of petroleum and other fuels. The destruction of fossil fuels has a major contribution to raise the CO_2 level as well as GHG emissions. An unpleasant effect of GHG emission on environment along with decaying petroleum reservations was recognized. Hence,

Figure 1. World CO_2 emissions from fuel combustion from 1971 to 2015 by region (IEA, 2017)

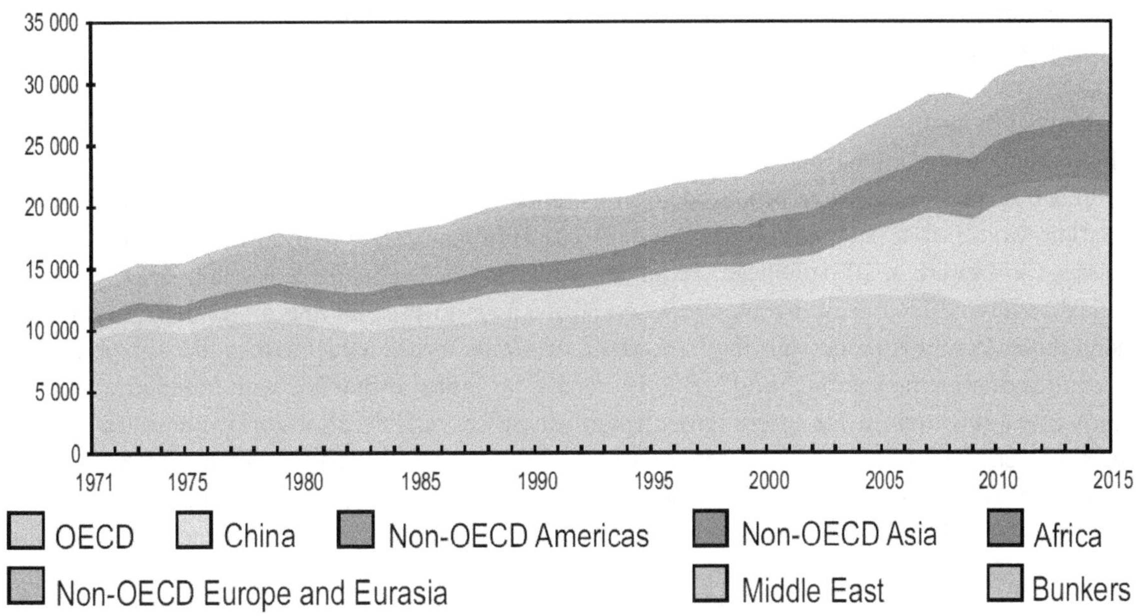

Figure 2. Production of petroleum and other fuels (EIA, 2015)

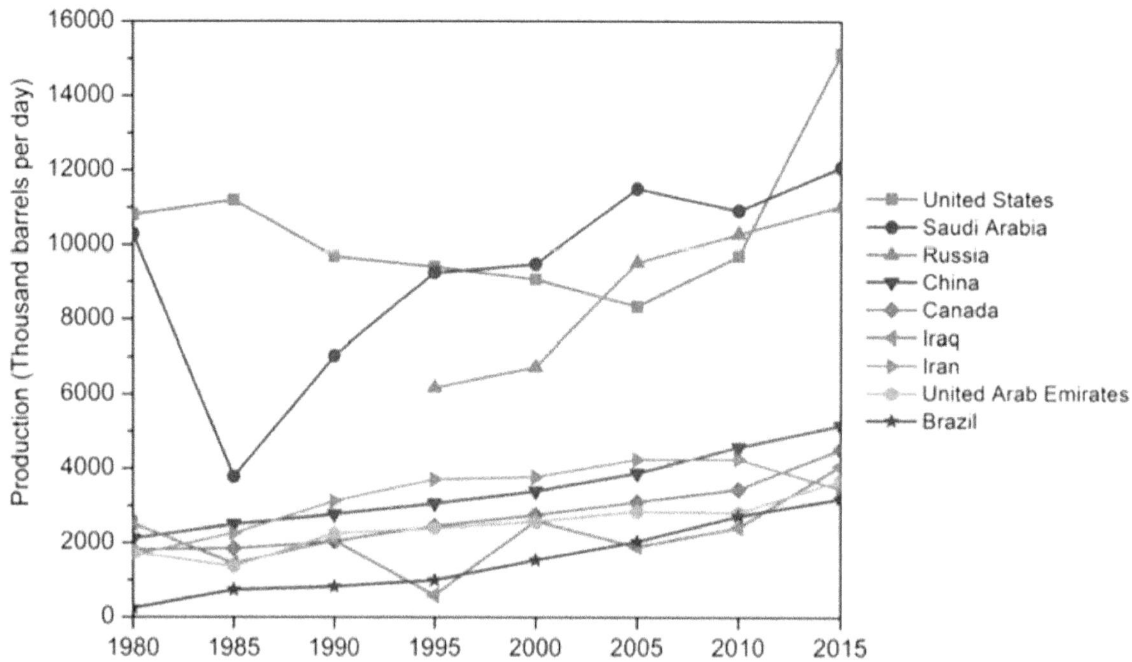

an exploration for environment friendly and sustainable resources of energy for industrial and domestic purposes was initiated in recent years (Sakthivel, Ramesh, Purnachandran, & Shameer, 2018). Accordingly, there is an enhanced interest in extraction and utilization of fuel from plants/organic wastes. As compared to raw diesel, hybrid diesel decreases GHG emissions approximately 24% (Chrysikou, Dagonikou, Dimitriadis, & Bezergianni, 2019). The bio-fuels extracted from renewable and biomaterials has the ability to diminish GHG emissions as well as enhance the overall energy efficiency of the existing fuel systems, and used for different applications (Nanda, Azargohar, Dalai, & Kozinski, 2015). An extensive focus on bio-fuel extraction may perhaps substitute the exploit of petroleum and other fossil fuels in future.

The second generation bio-fuels like bio-alcohols, vegetable oil, biodiesel, bio-DMF, bio-hydrogen, bio-Fischer-Tropsch diesel were produced from the crops not used for food, wheat, corn and similar vegetation, wood, waste through fermentation, esterification, gasification, pyrolysis, thermo-chemical processes (A. Demirbas, 2011; Dinesh, Tamilvanan, Vaishnavi, Gopinath, & Mohan, 2019). Thermo-chemical route and biochemical route are essential methods to produce the bio-fuels from biomass. Among these, Pyrolysis process paying more attention owing to its characteristics like simple process and requires minimum operating pressure. This process has higher efficiency with minimal waste. The plastic waste was converted into energy through pyrolysis process and pyrolysis oil have elevated calorific value which can be utilized as an alternative fuel (Sharuddin, Abnisa, Daud, & Aroua, 2018).

The emission performance of biodiesel with water was investigated. Fuel was prepared with various blends of water with biodiesel, 95% biodiesel + 5% water (BD95W5) and 90% biodiesel + 10% water (BD90W10). The water blend reduces the CO and HC emissions considerably. From the experimental results, it was found that NOx and smoke were condensed for BD95W5 and BD90W10 compared to

BD100 and diesel (Rathinam, Justin Abraham Baby, Devarajan, & T, 2018). An experiment was carried out on a single cylinder variable CI engine with bio-fuels (Jojoba Methyl Ester (JME), Algae Methyl Ester (AME), Chocolate Waste Methyl Ester (CME)) and the performance were compared with diesel. The experiments were conducted by varying the load; injection timing and compression ratio and various emissions were analyzed. AME exhibited better performance in terms of emission compared to other fuels (Ospina, Selim, Al Omari, Hassan Ali, & Hussien, 2019).

The overview of production of green diesel from biomass and its performance was discussed (Douvartzides, Charisiou, Papageridis, & Goula, 2019) and suggested that green diesel is an admirable fuel for CI engines and extensively reduce the emission. The alkanes were produced from inedible biomass and found that blend of catalytic cracking as well as alkylation was able to convert the biomass into bio-fuel (Nakagawa, Tamura, & Tomishige, 2019). Fresh water Macroalgal biomass from fresh water was used to produce the bio-fuels. Macroalgal biomass produced the 18.6% of lipid. The engine performance can be significantly improved by adding the diesel and butanol to algal biodiesel (Kumar et al., 2018).

This chapter shines its light on the synthesis of advanced second generation bio-fuels from biomass through different routes, biomass fuel upgradation techniques, and emission trends while incorporating biomass fuels in IC engines, storage stability of biomass based fuels and modern combustion strategies available to reduce the emission of biomass based fuels. Since the research works in adapting advanced biomass-based fuels in engines are in booming stage, this chapter will give a greater insight to the readers regarding biomass fuel production, upgradation and its emission aspects. This in turn effectively helps the research scientists to develop novel methodologies to overcome the difficulties faced in commercializing the biomass-based fuel in the fuel market. This chapter discusses the production of bio-fuel, upgradation of bio-fuel, testing the performance and emission characteristics with diesel engine.

BIOMASS FUEL UPGRADATION

From the previous section, it can be clearly seen that bio-oil obtained from biomass is a complex mixture of various organic compounds. Owing to its complex nature, the storage of bio-oil for long term in the view of using it in engines for emission control becomes a major concern among research community. Thermal and oxidative degradation of bio-oil during storage causes severe problems not only in property of fuel but also in the storage container. The former one leads to decomposition of volatile compounds in the bio-oil and the later is responsible for polymerization reactions in the oil at long storage. These cumulatively leads to increase in viscosity and solid formation in the bio-oil which in turn makes in unfit for use in engines. The ability of bio-oil to retain its native characteristics during storage is termed as storage stability. Several methods were reported to enhance the bio-oil stability namely physical and chemical methods which are discussed below.

PHYSICAL METHODS

Drying of Biomass

Usually biomass resources contain fair amount of moisture naturally which will be transferred to the extracted bio-oil during pyrolysis due to condensation reaction at storage. Whenever the level exceeds

Figure 3 Schematic diagram of fixed bed pyrolysis reactor

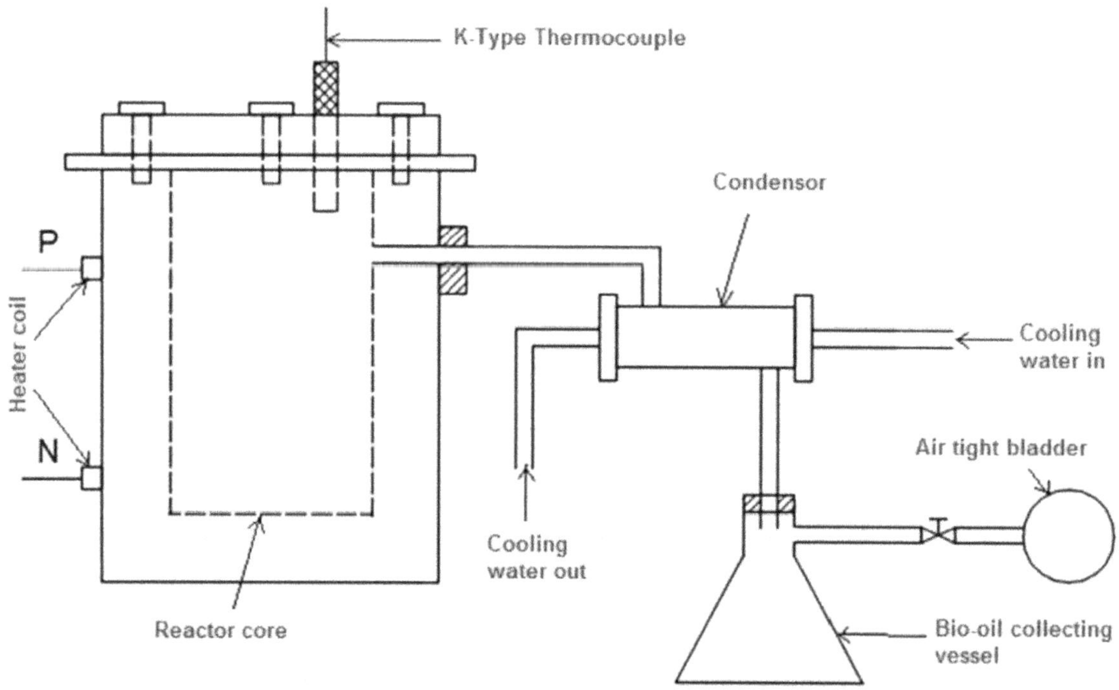

30% of its weight, phase separation occurs during storage. Therefore biomass drying is a typical pre-treatment process to maintain the moisture level below 10% to enhance the stability of bio-oil. Drying can be carried out under isothermal and non-isothermal conditions. The schematic representation of pyrolysis reactor was shown in Figure 3. Simple method like solar drying has also been carried out in certain reported works (Sakthivel, Ramesh, Shameer, & Purnachandran, 2018b).

Solvent Addition

Another most effective method reported so far is to add organic solvent with bio-oil to improve the storage stability of bio-oil at long run. Several solvents like acetone, methanol, ethanol and ethyl acetate can be used for the purpose based on the nature of the obtained bio-oil components and the nature of bonding present in the bio-oil. Diebold et al. revealed that addition of 10% methanol to pyrolysis oil improved its storage upto 96 h at 90 °C without any major changes in its composition (Diebold & Czernik, 1997). In addition to that Sakthivel et al. confirmed the effectiveness of methanol addition to Calophyllum inophyllum seed cake bio-oil improved its storage stability which is also validated through FTIR and GCMS. The authors reported that the induction period increased from 0.94 hours to 3.97 hours in Rancimat conditions after adding 10% methanol. Also the aging rate and insoluble solid formation are also minimized in the aged sample as compared to the original sample (Sakthivel, Ramesh, Mohamed Shameer, & Purnachandran, 2018a). Lu, Yang, and Zhu (2008) analyzed the viscosity changes of the bio-oil at 50 °C for 120 h and found decrement in viscosity of 22%, 16% and 12% after addition of 5%, 10% and 15% methanol by weight respectively.

Ash and Char Removing

Among the metal contents present in the ash like sodium, calcium, potassium and magnesium have the potential to induce the secondary reactions during high temperature pyrolysis process leading to production of undesirable compounds in the bio-oil. On the other hand, the catalysis also reduces the yield of bio-oil and tiny char particles will be included in the final bio-oil as residues. These residual particles form as solid core during condensation reaction at prolonged storage. Currently certain methods like acid pretreatment, gas filtration and cold liquid filtration are employed to remove the ash and char content preset in the biomass samples.

Emulsification

Biomass pyrolysis oil is generally not soluble in hydrocarbon fuels like diesel or petrol owing to its difference in physico chemical properties. To obtain homogeneous solution of bio-oil blends, third substance like surfactants can be employed which changes the surface properties of the bio-oil. In general, concentration of bio-oil, surfactant level and emulsification input energy are the major factor which affects the stability of the emulsion formed. Ikura et al, found that optimum surfactant ratio can be within 0.8-15% and bio-oil concentration be 10-20% above which causes instability to the emulsion (Ikura, Stanciulescu, & Hogan, 2003). The experiments performed by Chiaramonti et al, proved that stability of emulsion improves with concentration of surfactants where as it is limited to 0.5-2%. Also the authors suggested the selection of suitable surfactant is difficult and should be experience based for a particular bio-oil emulsion (Chiaramonti et al., 2003a, 2003b). Emulsification of bio-oil to diesel fuel still faces certain issues while used in engines like corrosion, pumping problems, injection clogging and atomization difficulties. In spite of these difficulties several works have been reported on successful usage of bio-oil emulsions in engines as renewable fuel which will be discussed in forthcoming topics.

Antioxidant Additives

Adding antioxidant additives is yet another effective method to enhance the storage stability of bio-oil in long run. Antioxidants may be either synthetic or natural which acts as chain terminators of active oligomers formed by polymerization or condensation process. Many phenolic based compounds are reported to capture the free radicals in bio-oil which catalyze the polymerization reaction of olefins. Owing to the lesser concentration of phenols in bio-oil, it is added externally to ensure the stability. Hydroquinone is one of the most widely used antioxidants in bio-oil to increase the stability. Baranitharan, Ramesh, and Sakthivel (2019b) found that addition of 1000 ppm tert butyl hydroquinone with Aegle marmelos bio-oil emulsion improved its oxidation stability and thermal stability by 22.33% and 27.05% respectively as compared to that of pure bio-oil emulsion (Paramasivam, K, & R, 2019).

CHEMICAL METHODS

Hydrogenation

Hydrogenation process mainly employed to improve the stability of bio-oil by reducing the concentration of acids and aldehydes as well as other reactive compounds which may cause instability of the oil. It involves series of reactions such as deoxygenation and hydrogenation in oxygenated atmosphere at elevated pressures. Oxygen content in the bio-oil is eliminated as water and carbon dioxide to obtain hydrocarbon fuel with high heating value and superior stability (Elliott, 2007). Since the reaction conditions are strict and high amount of hydrogen consumed, the cost of hydrogenation is significantly high. The catalyst for hydrogenation should be selected such that it possesses high catalytic activity as well as long life. Different catalysts employed for this purpose reported are Pt/Al_2O_3, HZSM-5, $Ru/\gamma\text{-}Al_2O_3$, Ni-Mo / $\gamma\text{-}Al_2O_3$, $RuCl_2$ $(PPh_3)_3$ etc. Since most of the reported catalysts are high temperature catalysts and thermal stability of bio-oil is usually poor, the development of low temperature reactive catalyst attracted the research community in recent days.

Cracking

The process of decomposing the bio-oil into smaller molecules with help of certain catalyst is termed as catalytic cracking. Oxygen is formally removed as water and oxides of carbon with help of catalysts like zeolites such as HY and HZSM-5. Zeolite catalysts have potential deoxygenation feature which can generate aromatic hydrocarbon fuels as main product. MnO was used as catalyst and obtained bio-oil with less water and higher calorific value which can be employed as potential fuel in IC engines for emission control. Even though catalytic pyrolysis proved to be effective way in reducing water content, the major setback lies in reduced bio-oil yield, loss of hydrocarbon elements and catalyst deactivation. The research can be focused on developing novel catalyst for pyrolysis with high selectivity, low coking rate and high bio-oil conversion rate.

Esterification

Esterification reaction is carried out in bio-oil by converting the reactive acids with presence of alcohols into stable esters. It directly reduces acidity of the bio-oil which addresses the corrosion problem in bio-oil storage. Solid catalyst, ionic catalyst and alkaline catalysts can be used for this purpose. While Zhang et al. utilized solid acid catalysts for esterification, it was observed that stability of bio-oil increased with the concentration of ester content in the bio-oil. Meanwhile the density of bio-oil decreased and calorific value increased by 22.58% and 50% respectively (Zhang, Chang, Wang, & Xu, 2006). Generally esterification consumes high dosage of alcohol additives and process carboxylic acids alone in the bio-oil. Moreover the water produced as the by-product affect the heating value as well as storage stability at long run. Therefore development of multi-functional catalyst to address these issues becomes crucial in the research arena in recent years.

BIOMASS FUELS IN IC ENGINES

With recent advancements in engine research, serious steps have been taken by the research community in finding alternative energy sources for engines to curb emissions in transportation sector. The emissions from CI engines play dominate effect since diesel traction is unavoidable in commercial sectors. In general, emissions can be decreased by engine modification and fuel modification in which the later method attracted researchers widely. Several renewable fuels were discussed in the literature in recent decades which can be widely branched into three categories namely first, second and third generation bio-fuels. The first generation bio-fuels are derived from edible resources such as vegetable oils which poses food versus fuel threat in the society.

The second generation fuel resources were obtained from non-edible sources which again indirectly affects the food chain by accommodating cultivation land for profit. Finally, the solution for fuel crisis is addressed by third generation bio-fuels which are derived from waste materials obtained from several sources like agriculture, industries and domestic settlements (Sakthivel, Ramesh, Purnachandran, & Shameer, 2017). The renewable biofuel utilization significantly reduces the accumulation of global greenhouse gases by constituting closed carbon cycle with some sacrifice in other emission levels. In the forthcoming sections the properties, performance and emission aspects of the biomass derived fuels are discussed which could enlighten the readers view towards biomass based fuels in emission reduction.

Physico-chemical Properties

As discussed in previous sections, biomass pyrolysis oil is complex mixture of various organic compounds. It is generally characterized by its dark colour and smoky odour in most of the cases. In order to apply a fuel in an engine, it has to possess certain physico-chemical property which is crucial in engine performance and emission. Most of the bio-oil by its nature are high viscous and corrosive in nature which can be upgraded for engine adaptation. Table 1gives an overview on properties of some typical biomass derived pyrolysis oils observed in literature.

Table 1. Physico-chemical properties of biomass pyrolysis oil (Alagu & Ganapathy Sundaram, 2018; Baranitharan, Ramesh, & Sakthivel, 2019a; Hossain et al., 2013; Pradhan, Bendu, Singh, & Murugan, 2017; Prakash, Singh, & Murugan, 2013; Sakthivel & Ramesh, 2017; R. Sakthivel et al., 2018b)

Feed	Kinematic Viscosity (cSt)	Density (g/cc)	Flash point (°C)	pH	Water content (%)	Cetane number	Calorific value (MJ/kg)
Aegle marmelos	7.83	0.845 g/cc	79.12	4.71	23.6	43	41.35
De-inking sludge	12.3	0.980	168	4.8	4	NR	37.04
Calophyllum bark	18.6	1.19	96	4.8	21.7	34	34.89
Calophyllum seed cake	12.2	1.17	88	4.3	23.3	40	36.84
Mahua seed	23.19	0.921	84	NR	NR	38	41.8
Neem seed	9.38	0.982	55	4.97	25.2	NR	20.8
Wood	25.3	NR	98	NR	15-30	25	20.58

*NR – Not Reported

From the above table, it is clear that all the bio-oil showed increased viscosity range which is not suitable for direct application in IC engines. Meanwhile the augmented water content and lower calorific value also affects the performance of engines if it is fuelled directly. It is worth to note that, the pH of all the samples lies in acidic owing to the presence of acids in the bio-oil which is harmful to the fuel storage and injector lines. The poor cetane rating of the bio-oil increases the ignition delay thereby affecting the combustion process to a greater extent. To avoid these problems, the bio-oil is upgraded before fuelling it into engines.

Performance Characteristics of CI Engine

Engine performance primarily depends on fuel employed. Among various performance attributes, brake thermal efficiency (BTE) and brake specific fuel/energy consumption (BSF/EC) are reported to be key factors to determine the effectiveness of the fuel used. BTE is the effective measure to determine the amount of fuel is converted to energy during combustion phase. Some of the authors reported that addition of bio-oil negatively affected the BTE of the engine. Pradhan et al. observed that highest BTE for 10% mahua seed bio-oil emulsion as 30.7% which is 31.4% for neat diesel. The authors also noted that the BTE deteriorated with augmentation of level of bio-oil in the fuel blend as the BTE was recorded to be 28.3% for the 40% mahua bio-oil emulsion (Pradhan et al., 2017). Similar results have been observed by R. Sakthivel et al. (2018a) for the Calophyllum inophyllum seed cake bio-oil emulsion fuels. It was reported that bio-oil level significantly reduced the quality of blends and negatively affects the performance of engine. Whenever the bio-oil level is increased from 10% to 15% in the blend, the BTE decreased from 30.87% to 30.7% at peak loading condition. The common reasons noted by most of the literatures in this scenario are high viscosity and calorific value of the bio-oil.

High viscosity leads to poor atomization which leads to inefficient burning and decreases the BTE. On the other hand, low calorific value leads to depreciated energy release during combustion which leads to lower BTE. Meanwhile the bio-oil of certain feeds augmented the BTE of engine due to its enriched oxygen content which leads to effective combustion as reported for Aegle marmelos seed cake and neem seed bio-oil emulsions (Alagu & Ganapathy Sundaram, 2018; Baranitharan et al., 2019b). BSFC and BSEC are another effective measure to evaluate the performance attribute of an engine based on fuel modification. It gives the amount of fuel or energy consumed to produce unit brake power in unit time. Generally BSEC is employed widely owing to different calorific values of the fuels in emulsion. The BSEC shows inverse trend as compared to that of BTE in most of the cases as reported in the literature. Volli, Singh, and Murugan (2014) reported that mustard seed bio-oil emulsions showed higher BSEC as compared to diesel fuel whereas Paramasivam et al. observed 10.06% lesser BSFC for vilvam seed cake bio-oil as compared to diesel fuel at peak loading conditions (Paramasivam, Kasimani, & Rajamohan, 2019; Volli et al., 2014). The authors reported similar reasons for the BSFC and BSEC trends as that of BTE in all the cases.

Emission Characteristics of CI Engine

As far as emission analysis of bio-oil fuelled engines are considered, bio-oil fuelled engines significantly reduces greenhouse gases by adapting the closed carbon cycle as compared to fossil fuels. Carbon dioxide is one of the major greenhouse pollutants observed at the tailpipe exhaust. While employing bio-oil derived from biomass as fuel, significant changes in CO_2 could be observed which is positive in some cases and negative in some other cases. The observance of more CO_2 at exhaust is the sign of complete combustion and its favourable condition especially when it is a biofuel powered engine. The presence of fuel bound oxygen in the bio-oil have a potential to oxidize the carbon present in the fuel molecules during the controlled combustion phase which inturn reduces harmful CO and augments CO_2 at the exhaust. It is a well known fact that CO and CO_2 have inverse trend since both depends on oxidation magnitude of carbon present in the fuel intake. Moreover other emissions also depend on the oxidation magnitude and the combustion temperature. In spite of the oxygenated nature of bio-oil, certain cases also reported where high water content in the bio-oil hinders the combustion efficiency thereby retarding the magnitude of CO_2 in the exhaust with augmented CO levels. The presence of various hydrocarbons in the fuel leads to emission. The by products produced from the reaction are given below.

- C (carbon solid) + CO_2 (gas) $\leftrightarrow$2CO (gas)
  (Boudourd gas reaction)
- CO (gas) +3H_2 (gas) $\leftrightarrow$ CH_4 (Methane gas) + H_2O (liquid)
  (Methanation gas reaction)
- CO (gas) + H_2O $\leftrightarrow$ CO_2 (gas) + H_2 (gas)
  (Water-gas shift gas reaction)
- N_2 (gas) + O (gas) $\leftrightarrow$ NO + N
  (Zeldovich Mechanism)

Figure 4. CI Engine emission characteristics (Baranitharan, Ramesh, & Sakthivel, 2019c)

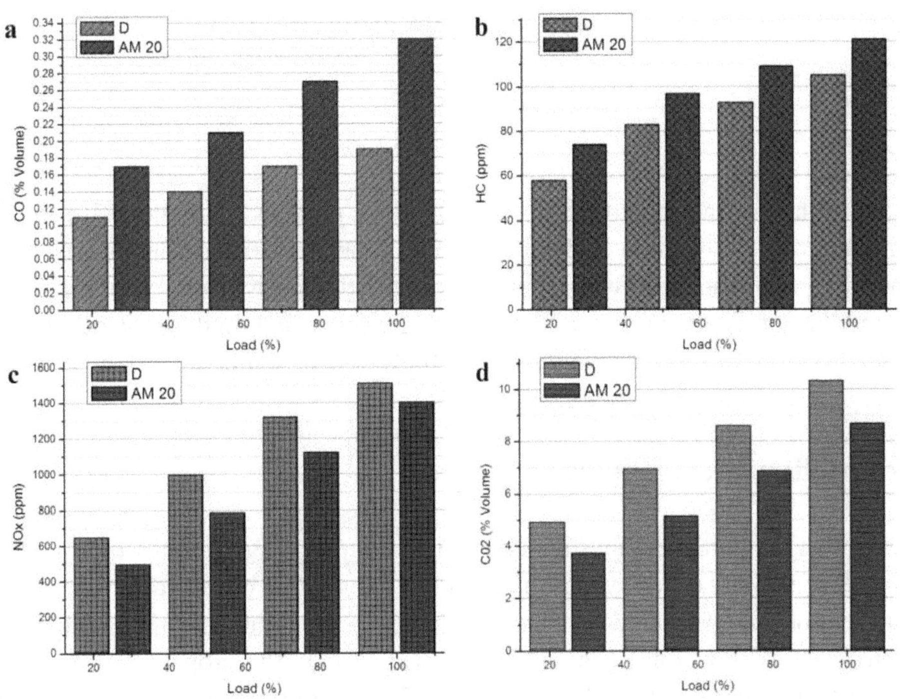

Next to that, unburnt hydrocarbon (HC) is major pollutant which causes serious health problems in human beings. It is generally formed by incomplete oxidation of fuel, fuel spray and combustion chamber geometry. As bio-oil emulsions are high viscous in nature, it becomes difficult to spray to achieve proper atomization. Owing to this difficulty, wall impingement effect occurs which in turn increases HC during combustion duration. This is one of the major reasons to limit the concentration of bio-oil in the fuel blend. Also some crucial cases can also been seen with reduced HC levels which was clearly attributed to the oxygenated nature of bio-oil. The emission characteristics of CI engine fuelled with AM de-oiled seed cake through pyrolysis process was shown in Figure 4.

Oxides of nitrogen (NOx) is one of the major and harmful pollutants of CI engines which is caused mainly due to the elevated combustion temperature i.e., thermal NOx. With wide analysis in the engine research, NOx emission is the predominate problem faced by the researchers while employing biodiesel fuels in engines as it causes high in-cylinder temperature which in turn augments NOx at the exhaust (Balamurugan, Tamilvanan, Anbarasu, Akil, & Srihari, 2013; Bhuiya et al., 2016; Tamilvanan, Balamurugan, & Vijayakumar, 2019). This problem of NOx can be effectively addressed by employing bio-oil fuels. The experimental results revealed that biomass fuels effectively reduced NOx levels due to the quenching effect during combustion phase as compared to other bio-fuels. The combustion quenching directly reduces the in-cylinder temperature which is the chief reason for thermal NOx whereas some cases show that bio-oil fuel records more NOx magnitude due to the fuel bound nitrogen present in the bio-oil.

Smoke emission from engines depends on variety of factors like engine operating conditions, fuel bound oxygen, combustion duration, temperature etc. The application of bio-oil showed assorted results in smoke emission throughout the engine studies reported. The fuel bound oxygen is main reason reported behind the reduction of smoke emission while utilizing bio-oil emulsions in engines. On the other hand, smoke emission is observed to be increased in some case due to the poor volatility of the bio-oil fuel at high ratios which hinders the combustion. Also it is worth to note that smoke emission of bio-oil fuels are reported to be higher than that of diesel fuel in all the cases (Baranitharan et al., 2019b; Pradhan et al., 2017; Sakthivel, Ramesh, Joseph John Marshal, & Sadasivuni, 2019).

SYNGAS

Syngas (mixture of hydrogen and carbon monoxide), which obtained through gasification process, is found to replace conventional fuels in IC engines. Syngas is obtained from different types of feed stocks ranging from coal to waste biomass. Syngas can be combusted along with diesel, petrol or natural gas in an IC engine and it compensates portion of the demand for fossil fuel. In this type of dual fuel engines, very small amounts of syngas mixing with common fuels or a large amount as the principal source of energy in the combustion chamber is used.

Syngas Production via Gasification

Gasification involves the partial combustion (limited amount of oxygen) of the biomass which yields a gaseous product commonly called syngas that contains mainly hydrogen and carbon monoxide along with concentrations of carbon dioxide, methane, and nitrogen. The schematic representation of biomass gasification process was shown in Figure 5 (Zeng et al., 2017). The properties of hydrogen, carbon monoxide and methane are given in Table 2. Typical LHV values of syngas produced in a gasifier using

Table 2. Properties of hydrogen, methane and carbon monoxide (Sridhar et al., 2005)

	H_2	CO	CH_4
LHV (MJ/kg) [MJ/Nm^3]	(121) [10.8]	(10.2) [12.7]	(50.2) [35.8]
Air-Fuel Ratio (mass) [mole]	(34.4) [2.38]	(2.46) [2.38]	(17.2) [9.52]
Peak flame Temp (K) @ 1 atm	2378	2384	2223
Flammability Limit φ(Lean/Rich)	0.01 / 7.17	0.34 / 6.80	0.54 / 1.69
Flame Speed (cm/sec)	270	45	35

air or air/steam as the oxidizing agent are 4-7 MJ/Nm^3. The complex and competing reactions involved in the gasification is given in Table 3.

Syngas Applications

Syngas with a variety of H_2 / CO composition can be used with many downstream processes (Balat, 2009) power generation, synthetic natural gas production, (Gassner & Marēchal, 2012) and H_2 production (Guoxin & Hao, 2009). In fuel cell units for electricity production, the gas having very high H_2/CO molar ratio (rich in hydrogen content) can be utilized. For Fischer - Tropsch Synthesis, the ratio of CO and H_2 having approximately 1: 2 can be used as feedstock. Gases rich in CH_4 can be employed as heating fuel (Chaudhari, Bej, Bakhshi, & Dalai, 2001). If the calorific value of syngas is above 4.7 MJ/Nm^3, it can be used for operating syngas engines (Kim et al., 2013).

DUAL FUEL ENGINES

Producing hydrogen and syngas from various renewable resources coupled with their clean combustion properties has attracted many researchers to show significant interest to use in IC engines. Up to

Table 3. Various reactions during the gasification process (Ahmad, Zawawi, Kasim, Inayat, & Khasri, 2016)

Reaction	Heat of Reaction	Reaction name	Reaction number
Heterogeneous reactions:			
$C+0.5O_2=CO$	-111 MJ/kmol	Char particle Combustion	R1
$C+CO_2 \leftrightarrow 2CO$	+172 MJ/kmol	Boudouard	R2
$C+H_2O \leftrightarrow CO+H_2$	+131 MJ/kmol	Water-gas	R3
$C+2H_2 \leftrightarrow CH_4$	-75 MJ/kmol	Methanation	R4
Homogenous reactions:			
$CO+0.5O_2=CO_2$	-283 MJ/kmol	CO particle Combustion	R5
$H_2+0.5O_2=H_2O$	-242 MJ/kmol	H_2 particle Combustion	R6
$CO+H_2O \leftrightarrow CO_2+H_2$	-41 MJ/kmol	CO shift	R7
$CH_4+H_2O \leftrightarrow CO+3H_2$	+206 MJ/kmol	Steam-methane Reforming	R8

Figure 5. Schematic of the biomass gasification process

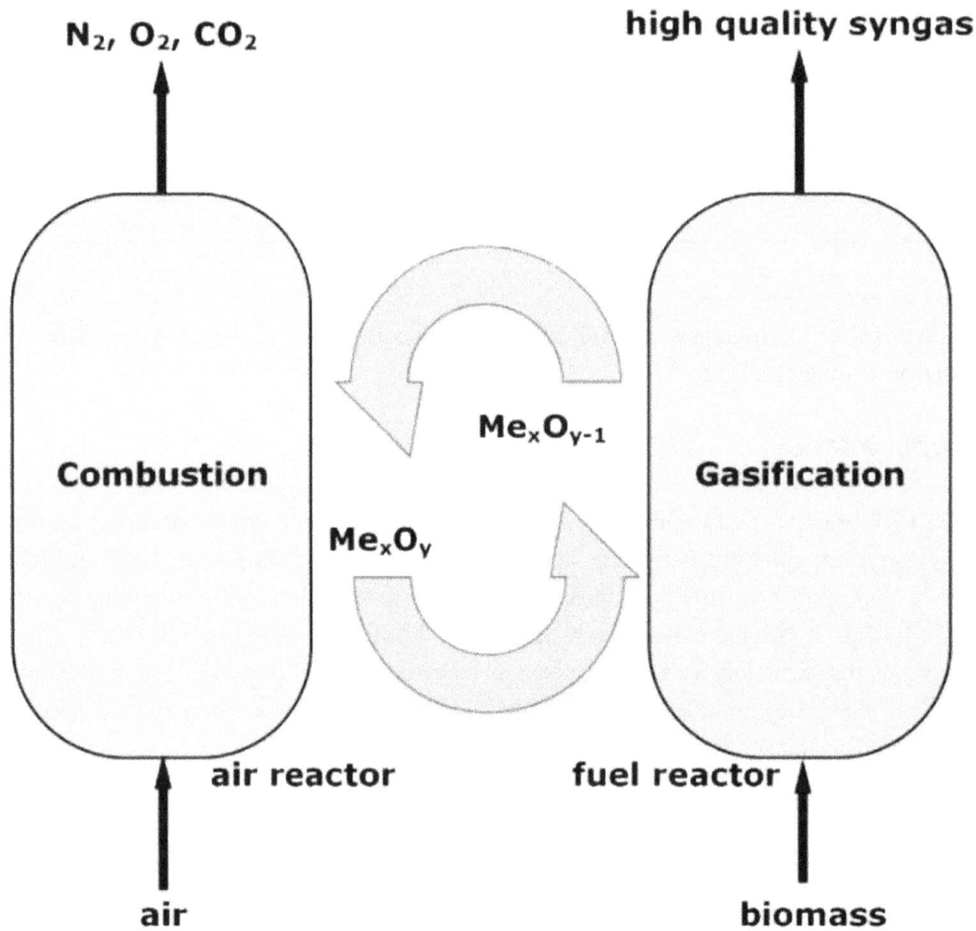

70% reduction of NOx emissions were reported by Lambe and Watson (1992) when hydrogen is used along with diesel in diesel engine. The dual fuel engines have the potential for controlling soot and NOx emissions but can cause increase of CO and HC emissions (Sahoo, Sahoo, & Saha, 2009). Researchers reported both increase and decrease of brake thermal efficiency with gaseous fuel injection in dual fuel engines. Xiao, Sohrabi, Galal, and Karim (2009) investigated the performance of diesel engine and found less engine thermal efficiency with hydrogen addition and 30-40% reduction in NOx emissions compared to straight diesel operation.

Syngas Use in Compression Ignition Engines

Dual fuel engine operation with syngas has been investigated by few researchers (Abu-Jrai, Tsolakis, & Megaritis, 2007; Garnier, Bilcan, Le Corre, & Rahmouni, 2005; Roy, Tomita, Kawahara, Harada, & Sakane, 2009). Syngas auto-ignition temperature (500°C) is higher than is achieved by the fuel charge in the diesel engine on the compression stroke. The combustion predictive model which describes the various stages of burning in a dual fuel syngas-diesel engine was established by Garnier et al. (2005).

Roy et al. (2009) varied the hydrogen concentration in the syngas from 13.7% to 20% and studied the performance and emissions of a dual fuel engine. The thermal efficiency and the NOx emissions were found higher for syngas with high hydrogen concentration.

Also results in the literature indicate that for the same thermal efficiency, dual-fuel engines generate low emissions of NOx and smoke compared to conventional diesel engines. High injection pressures reduce PM emissions at high loads. Injection pressure does not affect the PM emissions at low load conditions. Without EGR, higher injection pressures results in high NOx emissions, due to the faster and more intense combustion. Mahgoub, Sulaiman, Karim, and Hagos (2015) studied the emissions of CI engine at various engine speeds for varying syngas composition. It was found that CO_2 and NOx emissions reduced for all syngas compositions compared to diesel. The emissions of CO_2 and NOx were reduced at engine speed of 1200 rpm up to 1% and 108 ppm when Syngas with composition of 49% N_2, 12% CO_2, 25% CO, 10% H_2, and 4% CH_4 is used as fuel.

The concentration of H_2 in syngas composition is important as it accelerate the flame propagation, which boosts up the performance and reduces HC and CO_2 emissions and slight increase in NOx emissions. The CO content in the syngas helps in oxidation process inside the combustion chamber. Hence using syngas is more efficient than using H_2 alone for dual fuel combustion. The database for syngas composition required for combustion in CI engines need to be developed.

Syngas Use in Spark Ignition Engines

Studies were reported focusing on how the thermodynamic and chemical properties of syngas mixtures. Li, Karim, and Sohrabi (2003) studied the combustion and knock characteristics of hydrogen, carbon monoxide, and methane. They used syngas mixtures of 75% H_2 and 25% CO, by volume. They found that the base fuel auto-ignition properties determine whether the syngas acts through a thermodynamic effect or a chemical effect. Increasing syngas fraction, increases compression temperatures and advances combustion for the high octane fuel conditions. H_2 in the syngas blend reduces the radical pool build up in the primary ignition region, which reduces the temperature rise, causing a delay in the main ignition stage. High thermal efficiency is possible with syngas fueled SI engines. Without increasing the compression ratio a SI gasoline engine running on syngas is estimated to have a thermal efficiency of 10% to 15% as opposed to 15% - 20% running on gasoline due to the lower energy content of the syngas – air mixture.

Sobyanin et al. (2005) found from the experiments that SI engines operate safely in the idle mode using syngas air mixtures resulting in ultra low <10 ppm emissions of CHx and NOx along with 30% improvement in thermal efficiency. In the case of diesel engines, the level of NOx emissions was cut at least by several times even though hydrocarbons (CHx) and CO emission increases. Lee and Castaldi (2010) investigated the suitability of land fill gas (LFG) along with syngas (H_2/CO=2) in IC engine. In case of mixture with 5% addition of syngas to LFG at 0.8 kW load condition the concentrations of UHC, NOx and CO emissions reduced from 113.4ppm to 11.1ppm, from 99.8ppm to 64.5ppm and 802.1ppm to 203.1ppmrespectively. However, High percentage of syngas has insignificant effect on the reduction of emissions. Papagiannakis, Rakopoulos, Hountalas, and Giakoumis (2007) presented comparison of results between experimental and computed values for syngas fuelled SI engine. High NO and CO emissions were observed when compared to natural gas. Concerning engine efficiency, it is observed that the concentration of syngas fuel increases the engine efficiency. Samiran, Ng, Jaafar, Valera-Medina, and Chong (2016) studied the combustion performance and emissions for high (H_2/CO= 3) and moderate (H_2/CO = 1.2) H_2 rich syngases. Across all equivalence ratios tested, lower NOx emissions per kWh observed for higher H_2 fractions in the syngas and the CO emissions depend more on equivalence ratio and less influenced by the H_2 fraction in the syngas.

Indrawan, Thapa, Bhoi, Huhnke, and Kumar (2017) studied the engine power and emission when syngas generated from low density biomass is used as fuel. In this study the CO_2 emission decreased with increasing load for the engine running on syngas and opposite trend is observed when natural gas was used as fuel. The low value of LHV of syngas (6.47 MJ/Nm^3) when compared to natural gas (37.79 MJ/Nm^3) is attributed for this trend. Due to Complete combustion and lean fuel mixture, low concentration of CO emission using syngas was observed. The Lower flame temperature which results in lower temperature and pressure in the engine combustion chamber is the reason for low concentrations of NOx emission. The hydrocarbon (HC) emission using natural gas (1–1843 ppm) was higher compared to that using syngas (0–262 ppm), since syngas consists of low methane and high hydrogen concentrations. The SO_2 emission from an IC engine depends on fuel characteristics. The syngas contains low sulfur concentration and it generates a low of SO_2 emission as compared to natural gas. Gobbato, Masi, and

Table 4. Summary of the Emission studies by using syngas in IC engines

S.No	Author Name	Engine Type	Fuel Used	Emissions
1	(Papagiannakis et al., 2007)	S.I engine	Syngas	NO and CO emissions reduced
2	(Mahgoub et al., 2015)	S.I engine	Syngas	1. Reduced the emissions of CO_2 and NOx up to 1% and 108 ppm, respectively at engine speed of 1200 rpm. 2.The lowest emission of UHC and NOx was emitted when the engine was operating at speed of 2000 rpm
3	(Azimov, Tomita, Kawahara, & Harada, 2011)	Duel-fuel engine	Syngas	1. Increased H_2 content led to higher combustion temperatures and efficiency, lower CO and HC emissions, but higher NOx emissions. 2. NOx emissions decreased, when the CO_2 content of the syngas increased to 34%.
4	(Arroyo, Moreno, Muñoz, Monné, & Bernal, 2014)	S.I engine	Syngas	Nox emissions increased, CO_2 and CO emissions increased, HC emissions decreased
5	(Brusca, Chiodo, Galvagno, Lanzafame, & Garrano, 2014)	S.I engine	Syngas	CO&CO_2 emissions reduced
6	(Chacartegui et al., 2011)	gas turbine	Syngas	NOx emissions are higher in fuels with high CO %
7	(Indrawan et al., 2017)	I.C engine	Syngas	CO emission decreased, CO_2 emission decreased, lower NOx emission, lower HC emission, lower SO_2 emission
8	(Kirillov, Sobyanin, Kuzin, Brizitski, & Terentiev, 2012)	I.C & Diesel engine	syngas	1. In S.I engine CHx and NOx decreased . 2. In Diesel engine CO and CHx increased.
9	(Gobbato et al., 2015)	S.I engine	producer gas	NOx concentration reduced, CO emissions are very low
10	(Shaw, Jamal Akhtar, Priyam, & Kumar Singh, 2016)	4 stroke Diesel engine	producer gas	The CO_2 emission increased at high producer gas opening
11	(Samiran, Ng, Mohd Jaafar, Valera-Medina, & Chong, 2016)	swirl flame combustor	syngas	lower NOx emissions /KW.hr & CO emissions are equivalence ratio dependent
12	(Lee & Castaldi, 2010)	I.C engine	Land fill gas + syngas	reduces CO from 802.1 ppm to 203.1 ppm, UHC from 113.4 ppm to 11.1 ppm, NOx emission from 99.8 ppm to 64.5ppm

Benetti (2015) studied the emissions from a producer gas fuelled SI engine. NOx concentrations reduce because of the reduction in flame temperature. The summary of the literature is given in Table 4.

CONCLUSION

Nitrogen oxides (NOx), carbon monoxide (CO), hydrocarbons (HC) and particulate matters (PM) are important components among the exhaust gas emissions from engine vehicles. The addition of H_2 in gas engine can expand flammability limit and reduce NOx emission which is essential for lean burn engines and for meeting EURO-VI regulations. The advantage of low harmful emissions from engines using syngas derived from biomass gasification as fuel show potential to use biomass and waste resources for generating off–grid power in environmental friendly manner. Commercial natural gas engine can be coupled to feed syngas directly from gasifier with minimum engine modification. Reducing the residence time of the fuel in the combustion chamber by retarding the engine timing as well as lowering the engine compression are thought to lower the temperature of the fuel charge and hence its NOx emission. Meanwhile, utilization of bio-oil also reduces harmful NOx emission with some tradeoff towards the performance aspects. Meanwhile some authors pointed out the performance improve while using bio-oil due to its oxygenated nature as compared to traditional fossil diesel. The stability of bio-oil is poor as compared to that of biodiesel fuel owing to its complex nature which can be effectively improved by adding solvents and antioxidants. On the whole, biofuels obtained from biomass will play a vital role in forthcoming days in terms of emission reduction for developing health environment.

REFERENCES

Abu-Jrai, A., Tsolakis, A., & Megaritis, A. (2007). The influence of H2 and CO on diesel engine combustion characteristics, exhaust gas emissions, and after treatment selective catalytic NOx reduction. *International Journal of Hydrogen Energy*, *32*(15), 3565–3571. doi:10.1016/j.ijhydene.2007.02.014

Ahmad, A. A., Zawawi, N. A., Kasim, F. H., Inayat, A., & Khasri, A. (2016). Assessing the gasification performance of biomass: A review on biomass gasification process conditions, optimization and economic evaluation. *Renewable & Sustainable Energy Reviews*, *53*, 1333–1347. doi:10.1016/j.rser.2015.09.030

Alagu, R. M., & Ganapathy Sundaram, E. (2018). Preparation and characterization of pyrolytic oil through pyrolysis of neem seed and study of performance, combustion and emission characteristics in CI engine. *Journal of the Energy Institute*, *91*(1), 100–109. doi:10.1016/j.joei.2016.10.003

Arroyo, J., Moreno, F., Muñoz, M., Monné, C., & Bernal, N. (2014). Combustion behavior of a spark ignition engine fueled with synthetic gases derived from biogas. *Fuel*, *117*, 50–58. doi:10.1016/j.fuel.2013.09.055

Azimov, U., Tomita, E., Kawahara, N., & Harada, Y. (2011). Effect of syngas composition on combustion and exhaust emission characteristics in a pilot-ignited dual-fuel engine operated in PREMIER combustion mode. *International Journal of Hydrogen Energy*, *36*(18). doi:10.1016/j.ijhydene.2011.04.192

Balamurugan, K., Tamilvanan, A., Anbarasu, M., Akil, M., & Srihari, S. (2013). Nano-copper additive for reducing NOx emission in soya bean biodiesel-fuelled CI engine. *Journal of Biofuels*, *4*(1), 1–8. do i:10.5958/j.0976-4763.4.1.001

Balat, M. (2009). Gasification of biomass to produce gaseous products. *Energy Sources. Part A*, *31*(6), 516–526.

Baranitharan, P., Ramesh, K., & Sakthivel, R. (2019a). Analytical characterization of the Aegle marmelos pyrolysis products and investigation on the suitability of bio-oil as a third generation bio-fuel for C.I engine. *Environmental Progress & Sustainable Energy*, *38*(4). doi:10.1002/ep.13116

Baranitharan, P., Ramesh, K., & Sakthivel, R. (2019b). Measurement of performance and emission distinctiveness of Aegle marmelos seed cake pyrolysis oil/diesel/TBHQ opus powered in a DI diesel engine using ANN and RSM. *Measurement*, *144*, 366–380. doi:10.1016/j.measurement.2019.05.037

Baranitharan, P., Ramesh, K., & Sakthivel, R. (2019c). Multi-attribute decision-making approach for Aegle marmelos pyrolysis process using TOPSIS and Grey Relational Analysis: Assessment of engine emissions through novel Infrared thermography. *Journal of Cleaner Production*, *234*, 315–328. doi:10.1016/j.jclepro.2019.06.188

Bender, M. H. (2000). Potential conservation of biomass in the production of synthetic organics. *Resources, Conservation, and Recycling*, *30*(1), 49–58. doi:10.1016/S0921-3449(00)00045-8

Bhuiya, M. M. K., Rasul, M. G., Khan, M. M. K., Ashwath, N., Azad, A. K., & Hazrat, M. A. (2016). Prospects of 2nd generation biodiesel as a sustainable fuel – Part 2: Properties, performance and emission characteristics. *Renewable & Sustainable Energy Reviews*, *55*, 1129–1146. doi:10.1016/j.rser.2015.09.086

Boden, T., Marland, G., & Andres, R. (2017). *National CO2 emissions from fossil-fuel burning, cement manufacture, and gas flaring: 1751-2014*. Carbon Dioxide Information Analysis Center, Oak Ridge National Laboratory, US Department of Energy.

Brusca, S., Chiodo, V., Galvagno, A., Lanzafame, R., & Garrano, A. M. C. (2014). Analysis of reforming gas combustion in Internal Combustion Engine. *Energy Procedia*, *45*, 899–908. doi:10.1016/j.egypro.2014.01.095

Chacartegui, R., Torres, M., Sánchez, D., Jiménez, F., Munoz, A., & Sánchez, T. (2011). Analysis of main gaseous emissions of heavy duty gas turbines burning several syngas fuels. *Fuel Processing Technology*, *92*(2), 213–220. doi:10.1016/j.fuproc.2010.03.014

Chaudhari, S., Bej, S., Bakhshi, N., & Dalai, A. (2001). Steam gasification of biomass-derived char for the production of carbon monoxide-rich synthesis gas. *Energy & Fuels*, *15*(3), 736–742. doi:10.1021/ef000278c

Chiaramonti, D., Bonini, M., Fratini, E., Tondi, G., Gartner, K., Bridgwater, A. V., ... Baglioni, P. (2003a). Development of emulsions from biomass pyrolysis liquid and diesel and their use in engines—Part 1: Emulsion production. *Biomass and Bioenergy*, *25*(1), 85–99. doi:10.1016/S0961-9534(02)00183-6

Chiaramonti, D., Bonini, M., Fratini, E., Tondi, G., Gartner, K., Bridgwater, A. V., ... Baglioni, P. (2003b). Development of emulsions from biomass pyrolysis liquid and diesel and their use in engines—Part 2: Tests in diesel engines. *Biomass and Bioenergy*, *25*(1), 101–111. doi:10.1016/S0961-9534(02)00184-8

Chrysikou, L. P., Dagonikou, V., Dimitriadis, A., & Bezergianni, S. (2019). Waste cooking oils exploitation targeting EU 2020 diesel fuel production: Environmental and economic benefits. *Journal of Cleaner Production*, *219*, 566–575. doi:10.1016/j.jclepro.2019.01.211

Demirbas, A. (2011). Competitive liquid biofuels from biomass. *Applied Energy*, *88*(1), 17–28. doi:10.1016/j.apenergy.2010.07.016

Demirbas, M. F. (2006). Current technologies for biomass conversion into chemicals and fuels. *Energy Sources. Part A, Recovery, Utilization, and Environmental Effects*, *28*(13), 1181–1188. doi:10.1080/00908310500434556

Diebold, J. P., & Czernik, S. (1997). Additives To Lower and Stabilize the Viscosity of Pyrolysis Oils during Storage. *Energy & Fuels*, *11*(5), 1081–1091. doi:10.1021/ef9700339

Dinesh, K., Tamilvanan, A., Vaishnavi, S., Gopinath, M., & Mohan, K. R. (2019). Biodiesel production using Calophyllum inophyllum (Tamanu) seed oil and its compatibility test in a CI engine. *Biofuels*, *10*(3), 347–353. doi:10.1080/17597269.2016.1187543

Douvartzides, S. L., Charisiou, N. D., Papageridis, K. N., & Goula, M. A. (2019). Green Diesel: Biomass Feedstocks, Production Technologies, Catalytic Research, Fuel Properties and Performance in Compression Ignition Internal Combustion Engines. *Energies*, *12*(5), 809. doi:10.3390/en12050809

EIA, U. (2015). Total Petroleum and Other Liquids Production 2014.

Elliott, D. C. (2007). Historical Developments in Hydroprocessing Bio-oils. *Energy & Fuels*, *21*(3), 1792–1815. doi:10.1021/ef070044u

Garnier, C., Bilcan, A., Le Corre, O., & Rahmouni, C. (2005). Characterisation of a syngas-diesel fuelled CI engine: SAE Technical Paper.

Gassner, M., & Maréchal, F. (2012). Thermo-economic optimisation of the polygeneration of synthetic natural gas (SNG), power and heat from lignocellulosic biomass by gasification and methanation. *Energy & Environmental Science*, *5*(2), 5768–5789. doi:10.1039/c1ee02867g

Gobbato, P., Masi, M., & Benetti, M. (2015). Performance analysis of a producer gas-fuelled heavy-duty SI engine at full-load operation. *Energy Procedia*, *82*, 149–155. doi:10.1016/j.egypro.2015.12.007

Guoxin, H., & Hao, H. (2009). Hydrogen rich fuel gas production by gasification of wet biomass using a CO2 sorbent. *Biomass and Bioenergy*, *33*(5), 899–906. doi:10.1016/j.biombioe.2009.02.006

Hossain, A. K., Ouadi, M., Siddiqui, S. U., Yang, Y., Brammer, J., Hornung, A., ... Davies, P. A. (2013). Experimental investigation of performance, emission and combustion characteristics of an indirect injection multi-cylinder CI engine fuelled by blends of de-inking sludge pyrolysis oil with biodiesel. *Fuel*, *105*, 135–142. doi:10.1016/j.fuel.2012.05.007

IEA, C. (2017). Emissions from fossil fuel combustion. Highlights: OECD/IEA.

Ikura, M., Stanciulescu, M., & Hogan, E. (2003). Emulsification of pyrolysis derived bio-oil in diesel fuel. *Biomass and Bioenergy*, *24*(3), 221–232. doi:10.1016/S0961-9534(02)00131-9

Indrawan, N., Thapa, S., Bhoi, P. R., Huhnke, R. L., & Kumar, A. (2017). Engine power generation and emission performance of syngas generated from low-density biomass. *Energy Conversion and Management*, *148*, 593–603. doi:10.1016/j.enconman.2017.05.066

Kim, Y. D., Yang, C. W., Kim, B. J., Kim, K. S., Lee, J. W., Moon, J. H., ... Do Lee, U. (2013). Air-blown gasification of woody biomass in a bubbling fluidized bed gasifier. *Applied Energy*, *112*, 414–420. doi:10.1016/j.apenergy.2013.03.072

Kirillov, V. A., Sobyanin, V. A., Kuzin, N. A., Brizitski, O. F., & Terentiev, V. Y. (2012). Synthesis gas generation on-board a vehicle: Development and results of testing. *International Journal of Hydrogen Energy*, *37*(21), 16359–16366. doi:10.1016/j.ijhydene.2012.04.106

Kumar, V., Nanda, M., Joshi, H. C., Singh, A., Sharma, S., & Verma, M. (2018). Production of biodiesel and bioethanol using algal biomass harvested from freshwater river. *Renewable Energy*, *116*, 606–612. doi:10.1016/j.renene.2017.10.016

Lambe, S., & Watson, H. (1992). Low polluting, energy efficient CI hydrogen engine. *International Journal of Hydrogen Energy*, *17*(7), 513–525. doi:10.1016/0360-3199(92)90151-L

Lee, J., & Castaldi, M. J. (2010). A Study on Performance and Emissions of a 4-stroke IC Engine Operated on Landfill Gas with Syngas Addition. In *Proceedings 18th Annual North American Waste-to-Energy Conference*. 10.1115/NAWTEC18-3565

Li, H., Karim, G. A., & Sohrabi, A. (2003). Knock and combustion characteristics of CH4, CO, H2 and their binary mixtures: SAE Technical Paper.

Lu, Q., Yang, X., & Zhu, X. (2008). Analysis on chemical and physical properties of bio-oil pyrolyzed from rice husk. *Journal of Analytical and Applied Pyrolysis*, *82*(2), 191–198. doi:10.1016/j.jaap.2008.03.003

Mahgoub, B., Sulaiman, S., Karim, Z., & Hagos, F. (2015). Experimental study on the effect of varying syngas composition on the emissions of dual fuel CI engine operating at various engine speeds. In *Proceedings IOP Conference Series: Materials Science and Engineering*. 10.1088/1757-899X/100/1/012006

Nakagawa, Y., Tamura, M., & Tomishige, K. (2019). Recent development of production technology of diesel- and jet-fuel-range hydrocarbons from inedible biomass. *Fuel Processing Technology*, *193*, 404–422. doi:10.1016/j.fuproc.2019.05.028

Nanda, S., Azargohar, R., Dalai, A. K., & Kozinski, J. A. (2015). An assessment on the sustainability of lignocellulosic biomass for biorefining. *Renewable & Sustainable Energy Reviews*, *50*, 925–941. doi:10.1016/j.rser.2015.05.058

Nanda, S., Rana, R., Sarangi, P. K., Dalai, A. K., & Kozinski, J. A. (2018). *A Broad Introduction to First-, Second-, and Third-Generation Biofuels, Recent Advancements in Biofuels and Bioenergy Utilization* (pp. 1–25). Springer.

Ospina, G., Selim, M. Y. E., Al Omari, S. A. B., Hassan Ali, M. I., & Hussien, A. M. M. (2019). Engine roughness and exhaust emissions of a diesel engine fueled with three biofuels. *Renewable Energy*, *134*, 1465–1472. doi:10.1016/j.renene.2018.09.046

Papagiannakis, R., Rakopoulos, C., Hountalas, D., & Giakoumis, E. (2007). Study of the performance and exhaust emissions of a spark-ignited engine operating on syngas fuel. *International Journal of Alternative Propulsion*, *1*(2/3), 190–215. doi:10.1504/IJAP.2007.013022

Paramasivam, B., K, R., & R, S. (2019). Investigation and improvement on storage stability of pyrolysis oil obtained from Aegle marmelos de-oiled seed cake. *Energy Sources. Part A, Recovery, Utilization, and Environmental Effects*, 1–15. doi:10.1080/15567036.2019.1632989

Paramasivam, B., Kasimani, R., & Rajamohan, S. (2019). Experimental assessment and multi-response optimization of diesel engine performance and emission characteristics fuelled with Aegle marmelos seed cake pyrolysis oil-diesel blends using Grey relational analysis coupled principal component analysis. *Environmental Science and Pollution Research International*, *26*(7), 6980–7004. doi:10.100711356-019-04164-8

Pradhan, D., Bendu, H., Singh, R. K., & Murugan, S. (2017). Mahua seed pyrolysis oil blends as an alternative fuel for light-duty diesel engines. *Energy*, *118*, 600–612. doi:10.1016/j.energy.2016.10.091

Prakash, R., Singh, R. K., & Murugan, S. (2013). Experimental investigation on a diesel engine fueled with bio-oil derived from waste wood–biodiesel emulsions. *Energy*, *55*, 610–618. doi:10.1016/j.energy.2013.03.085

Rathinam, S., Justin Abraham Baby, S., Devarajan, Y., & T, A. (2018). Influence of water on exhaust emissions on unmodified diesel engine propelled with biodiesel. *Energy Sources. Part A, Recovery, Utilization, and Environmental Effects*, *40*(21), 2511–2517. doi:10.1080/15567036.2018.1503756

Roy, M. M., Tomita, E., Kawahara, N., Harada, Y., & Sakane, A. (2009). Performance and emission comparison of a supercharged dual-fuel engine fueled by producer gases with varying hydrogen content. *International Journal of Hydrogen Energy*, *34*(18), 7811–7822. doi:10.1016/j.ijhydene.2009.07.056

Sahoo, B., Sahoo, N., & Saha, U. (2009). Effect of engine parameters and type of gaseous fuel on the performance of dual-fuel gas diesel engines—A critical review. *Renewable & Sustainable Energy Reviews*, *13*(6-7), 1151–1184. doi:10.1016/j.rser.2008.08.003

Sakthivel, R., & Ramesh, K. (2017). Influence of temperature on yield, composition and properties of the sub-fractions derived from slow pyrolysis of Calophyllum inophyllum de-oiled cake. *Journal of Analytical and Applied Pyrolysis*, *127*, 159–169. doi:10.1016/j.jaap.2017.08.012

Sakthivel, R., Ramesh, K., Joseph John Marshal, S., & Sadasivuni, K. K. (2019). Prediction of performance and emission characteristics of diesel engine fuelled with waste biomass pyrolysis oil using response surface methodology. *Renewable Energy*, *136*, 91–103. doi:10.1016/j.renene.2018.12.109

Sakthivel, R., Ramesh, K., Mohamed Shameer, P., & Purnachandran, R. (2018a). Experimental investigation on improvement of storage stability of bio-oil derived from intermediate pyrolysis of Calophyllum inophyllum seed cake. *Journal of the Energy Institute*. doi:10.1016/j.joei.2018.02.006

Sakthivel, R., Ramesh, K., Purnachandran, R., & Shameer, P. M. (2017). A review on the properties, performance and emission aspects of the third-generation biodiesels. *Renewable & Sustainable Energy Reviews*.

Sakthivel, R., Ramesh, K., Purnachandran, R., & Shameer, P. M. (2018). A review on the properties, performance and emission aspects of the third-generation biodiesels. *Renewable & Sustainable Energy Reviews*, *82*, 2970–2992. doi:10.1016/j.rser.2017.10.037

Sakthivel, R., Ramesh, K., Shameer, P. M., & Purnachandran, R. (2018b). A Complete Analytical Characterization of Products Obtained from Pyrolysis of Wood Barks of Calophyllum inophyllum. *Waste and Biomass Valorization*. doi:10.100712649-018-0236-7

Samiran, N. A., Ng, J.-H., Jaafar, M. N. M., Valera-Medina, A., & Chong, C. T. (2016). H2-rich syngas strategy to reduce NOx and CO emissions and improve stability limits under premixed swirl combustion mode. *International Journal of Hydrogen Energy*, *41*(42), 19243–19255. doi:10.1016/j.ijhydene.2016.08.095

Sharuddin, S. D. A., Abnisa, F., Daud, W. M. A. W., & Aroua, M. K. (2018). Pyrolysis of plastic waste for liquid fuel production as prospective energy resource. *IOP Conference Series. Materials Science and Engineering*, *334*, 012001. doi:10.1088/1757-899X/334/1/012001

Shaw, D., Jamal Akhtar, S., Priyam, A., & Kumar Singh, R. (2016). Performance Study of Dual Fuel Engine Using Producer Gas as Secondary Fuel. *Carbon: Science and Technology*, *8*, 63–71.

Sobyanin, V., Sadykov, V., Kirillov, V., Kuzmin, V., Kuzin, N., Vostrikov, Z., ... Terentyev, V. (2005). Syngas as a fuel for IC and diesel engines: efficiency and harmful emissions cut-off. In *Proceedings of International Hydrogen Energy Congress and Exhibition IHEC*.

Sridhar, G., Sridhar, H., Dasappa, S., Paul, P., Rajan, N., & Mukunda, H. (2005). Development of producer gas engines. *Proceedings of the Institution of Mechanical Engineers. Part D, Journal of Automobile Engineering*, *219*(3), 423–438. doi:10.1243/095440705X6596

Tamilvanan, A., Balamurugan, K., & Vijayakumar, M. (2019). Effects of nano-copper additive on performance, combustion and emission characteristics of Calophyllum inophyllum biodiesel in CI engine. *Journal of Thermal Analysis and Calorimetry*, *136*(1), 317–330. doi:10.100710973-018-7743-4

Volli, V., Singh, R. K., & Murugan, S. (2014). The Use of Mustard Cake Pyrolytic Oil Blends as Fuel in a Diesel Engine. *Waste and Biomass Valorization*, *5*(4), 661–668. doi:10.100712649-013-9270-7

Xiao, F., Sohrabi, A., Galal, M., & Karim, G. (2009). The Performance of an IDI Diesel Engine Having Low Concentrations of Hydrogen in the Intake Air: SAE Technical Paper.

Zeng, J., Xiao, R., Zhang, H., Chen, X., Zeng, D., & Ma, Z. (2017). Syngas production via biomass self-moisture chemical looping gasification. *Biomass and Bioenergy*, *104*, 1–7. doi:10.1016/j.biombioe.2017.03.020

Zhang, Q., Chang, J., Wang, & Xu, Y. (2006). Upgrading Bio-oil over Different Solid Catalysts. *Energy & Fuels*, *20*(6), 2717–2720. doi:10.1021/ef060224o

Chapter 4
Effect of Additives on the Performance and Emission Attributes of Diesel Engines

Shanmuga Sundaram. N.
https://orcid.org/0000-0001-6757-8585
SVS College of Engineering, Amrita Vishwa Vidyapeetham University, India

Sivakumar Muthusamy
https://orcid.org/0000-0003-1761-4802
KSR College of Engineering, India

ABSTRACT

Automobile vehicles are the main sources of environmental pollution, especially those with diesel engines. They cause a number of health diseases and harm to the ecosystem. Biofuels are a suitable alternative fuel for IC engines which have potential to reduce engine emissions with more or less equal performance of the petroleum fuels. Though Biodiesel is suitable for Diesel engines, it suffers with high density, lesser calorific value, high fuel consumption and increased emissions of nitrogen. However, additives minimize the deteriorating factors of the Biodiesel and maintain the international pollution norms. Many different types of additives are used with the diesel and (or) biodiesel to enhance performance and to improve its quality. The researchers conclude that the use of additives along with diesel and biodiesel improves the performance and reduction in emission. This review discusses effects of additives with diesel and biodiesel on the performance and emission characteristics of Diesel engines.

INTRODUCTION

Much energy is consumed in transportation sector especially by diesel engines. The petroleum fuels play significant role in the development of transportation sector, industrial growth and agricultural sector as well as to fulfil other human needs. Generally, diesel engines are advantageous compared to gasoline engines due to their higher thermal efficiency, reliability and durability, less fuel consumption and less

DOI: 10.4018/978-1-7998-2539-5.ch004

Table 1. Nomenclature

ASTM	American Society for Testing and Materials	DMC	dimethyl carbonate
BP	brake power	DTBP	di-tertiary butyl peroxide
BSFC	brake specific fuel consumption	EHN	ethyl hexyl nitrate
BTE	brake thermal efficiency	EVA	ethylene vinyl acetane
BHA	butylatedhydroxyanisol	ETBE	etilterbutileter
BHT	butylatedhydroxytoluence	HC	hydro carbon
CO	carbon monoxide	IAN	iso-amyl nitrate
CFPP	cold filter plugging point	NPAA	nonylphenoxyl acetic acid
CNT	carbon nanotubes	NO_x	oxides of nitrogen
DEE	diethyl ether	TBHQ	tert-butylhydroquine

CO_2 emission (Kim H et al. 2010; Sahin et al. 2014; Lu et al. 2008). However, diesel engines also face major issues such as increased NO_x (Yang et al. 2013). The improvements in engine technologies, improving fuel performance and emission characteristics are regarded as a promising solution to reduce the pollution levels of automobile vehicle. In order to reduce the pollutants emitted by the automobile vehicles, the two strategies may be solution. First one is complete replacement of major petroleum-derived fuels used in the transportation sector with alternative fuels produced from renewable resources. Second one is partial replacement of these fossil fuels by using fuel additives. The term fuel additive used throughout this article refers to any substances added to IC engines fuels at any rates in order to alter the fuel properties.

Biodiesel is an alternative fuel that is capable to meet the present and future energy demand. It is formulated from vegetable oil and animal fat, which are non-toxic and bio-degradable (Zhang et al. 1998), eco-friendly and reliable (Demirbas, 2007). With respect to environmental pollution biodiesel is more adoptable compared to fossil fuel as it forms low carbon and smoke which are responsible for global warming (Rahman et al. 2013; Abbaszaadeh et al. 2012). Biodiesel is defined by ASTM as a fuel comprised of esters of long chain fatty acids derived from different animal fats and vegetables oils (Demirbas, 2007; Nigam et al. 2011). Biodiesel is an oxygenated fuel which contains 10-15% oxygen by weight (Shahabuddin et al. 2012) and it contains neither sulphur, nor aromatics. These factors lead biodiesel to enhance more complete combustion and less emission of particulate matter (PM), carbon monoxide (CO) and hydrocarbons (HC), while increase in nitrogen oxides (NO_x) emission (Puhan et al. 2007; Zheng et al. 2008). The most common biodiesel blends used worldwide are B20 (20% biodiesel, 80% diesel) and B05 (05% biodiesel, 95% diesel), but B05 is an ASTM- approved fuel for safe operation in any compression ignition engine designed to be operated on conventional diesel (EPA, 2002).

Many researchers investigated the engine performance and emission characteristics of diesel engine operated by using diesel-biodiesel blends. Ekrem (2010) studied experimentally with the effect of biodiesel extracted from rapeseed oil of on the performance and emission characteristics of diesel engine, and it is found that biodiesel produced 11% higher BSFC and higher NO_x, but with reduced other emission such as smoke, HC and CO. Kalligeroset al. (2003)investigated the biodiesel produced from sunflower and olive oils in a single cylinder diesel engine and observed lower particulate matter (PM), carbon monoxide (CO), unburned hydrocarbon (HC) and NO_x emissions with a slight increase in fuel consumption. Atul

Dhar et al. (2012) evaluated the performance and emission characteristics of biodiesel produced from neem oil in diesel engine; and observed higher brake thermal efficiency, higher BSFC, and higher NO_x emission as compared to diesel; and CO, HC emission were lower than diesel. Sivakumar et al. (2017) studied the effect of pongamia biodiesel on performance and emission characteristics of diesel engine, and observed low brake power, higher brake specific fuel consumption, higher brake thermal efficiency and higher NO_x emission compared to diesel. Chauhan et al. (2012) carried out the experiments on performance and emission characteristics of jatropha biodiesel blends and reported low CO, HC and smoke emission and higher BSFC, higher NO_x emission while compared to diesel. In general, it seems that the number of papers indicating that usage of biodiesel increased NO_x emission is more than those claiming otherwise (Hoekman et al., 2012).

OBJECTIVES

Overall, although many research works emphasize that the application of biodiesel has its unique advantages such as promoting sustainable rural development, cleaner environment by reducing emissions, still there is a gap to accept biodiesel as an ideal alternative fuel. This is due to the increased NO_x emission, low energy content which could in turn increase fuel consumption of the biodiesel (Demirbas, 2000). In order to overcome the issues with the biodiesel the various techniques are employed such as fuel properties modification, engine design alteration, exhaust gas treatment etc. The fuel properties modification improves the combustion that resulted in low fuel consumption and emissions without any modifications in the existing engine, fuel injection and exhaust systems. Numerous investigations (Gairing et al., 1995; Varatharajan et al., 2011) reported that the physical properties of biodiesel can be uplifted by using different additives including metal based additives, oxygenated additives and antioxidants, cold flow additives etc. These additives were used along with the biodiesel by various researchers to improve the performance and to reduce exhaust emissions from diesel engines.

This article is aimed to comprehend and to review the mechanism and impacts of different additives used in diesel and biodiesel fuels as well as their blends in order to enhance the fuel properties. Various additives that are used to improve brake power of the engine, reducing harmful emissions, improving fluid stability, decreasing wear, enhancing viscosity index, etc. are included in the article. The outcome of various research works are discussed and concluded.

Grouping of Additives

At present days, additives play a valuable and very cost effective role in reducing pollution levels and meeting fuel specification requirements. Additives are the chemicals that are used to promote the properties of the fuel. They are mixed with fuel such as diesel, biodiesel, gasoline, aviation oil etc. to improve the thermal efficiency and fuel economy (Rashedulet al.2014). The selection of additives for diesel/biodiesel fuel is based on the fuel blending properties, economic feasibility, additive solubility, viscosity of fuel blend and flash point of the fuel blend. The benefits of fuel additives are very significant that includes:

- Enhanced the nagging properties and immovability of the fuel
- Improving the combustion characteristics of the fuel
- Reducing the fuel consumption and improving the engine performance

- Reduction in the pollutants emitted
- Protecting the fuel tank, pipeline etc from corrosion

Additives can be grouped in terms of their application as listed below.

(i). Oxygenated additives
(ii). Metal based additives
(iii). Antioxidant additives
(iv). Cetane number improver additives
(v). Cold flow improver additives

Oxygenated Additives

The oxygenate additives generally used are alcohols, ether and ester. They contain oxygen that are used to enhance the octane rating and combustion quality of the fuel.Theyare capable of mixing with diesel and biodiesel fuels with any ratio without separation issues. Alcohols such as butanol, propanol, methanol and ethanol; ethers like diethyl ether, diiso-propyl ester, dimethyl ether, methyl tert-butyl ether, and ethyl tertiary butyl ether; esters such as dicarboxylic acid esters, acetoacetic esters and dimethyl carbonate esters are treated as efficient oxygenated additives (Rahmat et al., 2010). Oxygenated additives are added with biodiesel fuels for minimizing its ignition temperature. However, the emission of smoke depends on the oxygen content and molecular structure of the fuel (Ribeiro et al., 2007).

It is reported that the composition of biodiesel and the use of additives directly affect the properties such as viscosity, density, volatility, and the cetane number of the fuel (Lapuerta et al., 2010). Oxygenated additives improve the combustion flame formation and combustion process which resulted in completed combustion of fuel and least emission of harmful pollutants. Borris et al. (2014) used oxygenated additives of EtilTerbutilEter (ETBE) and Diglyme in proportions of 5%, 10% and 15% with diesel fuel in the diesel engine. The test results showed a significant increase in BTE and BSFC and also constant NO_x emissions depending on the amount of additives in the mixture, engine speed and load condition.Qi et al. (2011) performed a comparative study of addition of diethyl ether (DEE) and ethanol in biodiesel-diesel blends with that of without additives. The results indicated that DEE blended fuel gives lowered BSFC and NO_x compared to ethanol blended fuel.

Metal Based Additives

The addition of metalbased additives with biodiesel fuels leads to the complete combustion of fuel and reduced exhaust emissions. Metal based additives act as catalyst during the combustion of fuels. Commonly used additives include copper chloride ($CuCl_2$), iron chloride ($FeCl_3$), iron (Fe), barium, manganese (Mn) and calcium (kannan et al., 2011). The reduction of exhaust emission with the addition of metal-based additives may be due to the fact that the metals either react with water vapour to generate hydroxyl radicals or directly react with carbon atoms as a catalyst thereby releasing oxidation temperature (Yang et al., 1998). Usually, metal-based additives are added as a metal organic compound in the form of nanoparticle in the diesel engine (Jelles et al., 1999). Particulate traps are useful device for absorbing the soot emissions from diesel engines (Jung et al., 2005). Nano metal oxide as fuel additives with biodiesel playing an important role to enhance the engine performance and exhaust emissions of a diesel engine

(Sivakumar et al, 2018).Keskin et al. (2007) studied the effect of Mn and Ni based additives with the tall oil, and recorded a lower BSFC, lower viscosity and pour point of the biodiesel, lower smoke and NO_x emissions than diesel. MozhiSelvan et al. (2009) studied the effect of cerium oxide nanoparticles in diesel, diesel-castor oil, biodiesel-ethanol blends and found an increase in brake thermal efficiency and reduction in NO_x, CO, and smoke emissions.

Keskin et al. (2007) studied the effect of Mn and Ni based additives with the tall oil, and recorded a lower BSFC, lower viscosity and pour point of the biodiesel, lower smoke and NO_x emissions than diesel. Mozhi Selvan et al. (2009) studied the effect of cerium oxide nanoparticles in diesel, diesel-castor oil, biodiesel-ethanol blends and found an increase in brake thermal efficiency and reduction in NO_x, CO, and smoke emissions.

Antioxidant Additives

Antioxidants also known as free-radical quenching agents are characterized by their capabilities in inhibiting the oxidation of other molecules (Christensen et al., 2014). Antioxidants consist hindered phenols, aromatic diamines or mixture of alkyl phenols and aromatic diamines. Oxidation of fuel, also termed as instability, leads to deterioration, resulting in fuel darkening and the formation of gums and sediments. Antioxidants enhance biodiesel stability and inhibit its tendency to deteriorate in storage (Karavalakis et al., 2011). Antioxidant additives like butylated hydroxyl anisole (BHA), butylated hydroxyl toluene (BHT), pyrogallol (PL), diphenylamine (DPA), tert-butylhydroxyquinone (TBHQ) and propyl-gallate (PG) are commonly used. The presence of unsaturated fatty acid esters in the biodiesel initiate the oxidize process in the biodiesel if it is stored for long time.

After the oxidation of biodiesel and diesel blends the density, viscosity and acid value increased, as the iodine value diminished with increase in storage time (Jain et al., 2010). Unstable species in biodiesel produce free radicals during the chain reaction and react with olefinic compounds to form gums. This oxidative chain reaction increases at an exponential rate producing increased quantity of free radicals and peroxide species. The addition of antioxidant additives breaks the chain propagating steps, by decomposing the peroxides and acts as radical traps. Rizwanul Fattah et al. (2014) conducted experiments using biodiesel extracted from palm oil with two monophenolic antioxidants of 2,6-di-tert-butyl-4methylphenol (BHA) and 2(3)-tert-butyl 4-methoxy phenol (BHT) at 1000 ppm concentration. The addition of BHA and BHT resulted in a mean reduction in BSFC of 0.64% and 0.18% respectively. A considerable reduction in NO_x emission of 12.6% and 9.8% with BHA and BHT respectively was also observed.

Cetane Number Improver Additives

Delay period is the one of the most important role plays in diesel engine, in which the delay period varies with chemical and physical properties of the diesel or biodiesel fuels. Cetane quality of diesel engine fuel is defined in terms of ease with which the fuel ignites in a diesel engine with reduced delay period (Lu et al., 2005). The lower cetane number of the diesel and biodiesel fuels lead to higher noise of the engine, increased fuel consumption, poor starting problem in cold weather conditions and large harmful exhaust emissions (Ullman et al., 1994). The delay period and cetane number of the fuel are improved by adding cetane number improver additives. It minimizes the ignition delay period during the combustion of fuel in diesel engines and there by smooth engine operation. Cetane number improver additives predominantly used are alkyl nitrates, of which 2-ethyl hexyl nitrate (2-EHN) in order to increase the cetane number of biodiesel fuel. The use of cetane number improver additives is a cost-effective method to increase the cetane quality of biodiesel.

Cold Flow Improver Additives

Cold flow improver additives for biodiesel fuel typically utilize vinyl eater co-polymer such as ethylene vinyl acetate (EVA). Other additive chemistries employed olefin-ester co-polymers and dispersants which may be combined with EVA. Biodiesel has a higher pour point and cloud point than diesel fuel. The cold filter plugging point (CFPP) temperature is close to the actual cold weather operating temperature of biodiesel. When the temperature drops crystal grow in size and begin to adhere to each other and formed large lattices of crystal, that resulted in the blocking of fuel filters and feed lines, ultimately leading to power loss and possible engine shutdown (Chandler et al., 1992). Use of cold flow improvers additives during the distillation process prevent these problems and produces smaller quantity of residual fuel and thereby reduction in the overall cost of fuel production (Chastek, 2011).

EFFECT OF ADDITIVES ON THE ENGINE PERFORMANCE AND EMISSION

From the above discussion it is observed that the addition of different additives with biodiesel and diesel fuel resulted in the improved performance and emission characteristics of the diesel engine. But proper selection of the additives is more important to have the desired output which is based on the properties of the fuel. The effect additives in particular to the engine performance and emission characteristics are discussed further.

Performance Characteristics

Power and Torque

Literatures discussing about the effect of additives on the brake power and torque are detailed here. Fangsuwannarak et al (2013) studied the comparative analysis of the effect of different additives viz. polymer-based bio-solution, natural organic based bio-solution and nano-titanium metalloid (TiO_2) compound on performance and emission characteristics of a pickup diesel engine fueled with diesel and palm biodiesel. It is reported that TiO_2 additive is more effective for improving engine power than pure diesel and B5 fuels by 7.78% and 1.36% respectively. The engine torque is also increased by 1.01% and 1.53% respectively with the use of pure diesel and B5 with TiO_2 additive. Imtenan et al. (2014) conducted a comparative study on the performance and emission characteristics of palm biodiesel-diesel blended with ethanol, n-butanol and diethyl ester as additives. Among the additives diethyl ether showed highest improvement through its lower density and viscosity profile and with quite a high calorific value. The diethyl ester showed the highest increment of 6.2% in brake power with DP20. Taghizadeh et al. (2016) investigated the torque and power generated due to addition of ethanol with concentrations of 2, 4, 6, 8, 10 and 12% to diesel fuel in the six-cylinder CI engine. The results showed that the torque and power increased by 3.8% with 6% ethanol blend with diesel while compared to that of pure diesel.

Brake Specific Fuel Consumption

The brake specific fuel consumption is bound to increase due to the inclusion of an oxygenated additive owing to the lower heating value and energy content of the fuel blend requiring more fuel to be injected to obtain the same power output (Fang et al., 2013). The most commonly used oxygenated additives are alcohols. Yasin et al. (2014) investigated the influence of methanol additive by 5% volume in a B20 blend on performance and combustion characteristics of Mitsubishi 4D68 multi-cylinder DI diesel engine. An increase in BSFC of 6% was observed when engine was fuelled with B20 M5. Kannan et al. (2011) studied the effect of ferric chloride (metal based additive) added to waste cooking palm oil based biodiesel on performance, emission and combustion characteristics of a diesel engine and noted a decreased in BSFC of 8.6%. Sajith et al. (2010) investigated the addition of cerium oxide additive with biodiesel on the engine performance and emission and pointed outa reduction in BSFC. This is due to the oxidation of carbon deposit in the engine by the cerium oxide that leads to the smooth and efficient operation. Venu and Madhavan (2016) performed a comparative study of effect of diethyl ether with titanium oxide and zirconium oxide nano particle additives in a diesel engine fuelled with biodiesel-ethanol blend. It is reported that 50 ml diethyl ether addition with biodiesel-ethanol blend reduced the brake specific fuel consumption. Soudagar et al. (2019) studied the effects of graphene oxide nanoparticles on performance and emissions of a CI engine fueled with dairy scum oil biodiesel (DSOME). The results were notable enhancements in the performance characteristics, a reduction in BSFC by 8.34% for nanofuel blend (DSOME40).

Brake Thermal Efficiency

Venkatesan and Kadiresh (2016) conducted experiments in a CI engine using aqueous cerium oxide nanofluid as additive with diesel and diesel-biodiesel blends and compared their characteristics with that of pure diesel. It is noticed that the brake thermal efficiency was increased by 5.81% as compared to diesel. Xiaolu et al. (2006) studied the effect of dimethyl carbonate (DMC) as an oxygenated additive blended with diesel fuel, and noted that the DMC fuelled engine has 2-3% higher thermal efficiency than that of diesel fuelled engine. Huang et al. (2009) pointed out that brake thermal efficiency increased as a result of the addition of n-butanol to the diesel while ethanol to the n-butanol-diesel blend reduced the thermal efficiency. Sathiyamoorthi et al. (2016) found 1.29% increase in brake thermal efficiency of CI engine with the inclusion of antioxidant additives butylatedhydroxyanisole (BHA), butylatedhydroxytoluene (BHT)in biodiesel blends (LGO25).Nanthagopal et al. (2017) examined the influence of zinc oxide and titanium dioxidenanoparticles dispersed in the fuel CalophyllumInophyllum methyl ester (CIME) in a twin cylinder water cooled diesel engine. It is reported that the brake thermal efficiency increases by 5-17% compared to pure CIME fuel at maximum brake power due to more surface to volume ratio of nanoparticles which exhibit rapid evaporation and better atomization. Prabhu (2018) investigated the effect of nanoparticles (Alumina and cerium oxide) on performance, emission and combustion characteristics of single cylinder diesel engine with fuel blends such as B20, B20A30C30 (biodiesel/diesel-nanoparticles) and reported that the BTE for the nanoparticles dispersed fuel blend significantly improved by 12% as compared to that of B100.Soudagar et al. (2019) studied about the effects of graphene oxide nanoparticles on performance and emissions of a CI engine fueled with dairy scum oil biodiesel (DSOME). The results were notable enhancements in the performance characteristics, the BTE improved by 11.56% for nanofuel blend (DSOME40). Dineshashivakumar et al. (2019) investigated the effects of water emulsion and diethyl ether additive on the performance and emissions of a single cylinder diesel engine using biodiesel blends. The results showed that, compared to biodiesel operation, the BTE of the emulsified fuel increases slightly with the composition of 2% DEE.

Emission

NO_x Emission

One of the most dangerous emissions in diesel engine is NO_x. The formation of NO_x is highly depends on the temperature inside the cylinder, the concentration of oxygen, the residence time for reaction to takes place (Yilamz N et al., 2014; Atmanli et al., 2015). Sathiyamoorthi et al. (2016) studied the effect of antioxidant additives (butylatedhydroxyanisole (BHA)) on the exhaust emissions of CI engines, and reported that the addition of BHA at 2000 ppm with biodiesel blend (LGO25) exhibited a maximum reduction of NO_x by 11% than LGO25 without additives. Yang et al. (2014) have investigated using biodiesel from pine oil with the ignition promoters of IAN (iso-amyl nitrate) and DTBP (di-tertiary butyl peroxide). The test results showed a reduction of NO_x emission by 12.8% and 19.2% with IAN and DTBP promoters respectively. Rizwanul Fattah et al. (2014) have conducted tests using biodiesel from palm oil with two antioxidants of 2,6-di-tert-butyl-4methylphenol (BHA) and 2(3)-tert-butyl 4-methoxy phenol (BHT) at 1000 rpm concentration. The addition of BHA and BHT resulted in a considerable reduction in NO_x emission of 12.6% and 9.8% respectively. Phoon et al. (2017) reported the effects of 1-octanol blend with green diesel along with 2-ethylhexyl nitrate (2EHN) as cetane improver, and found that the addition of EHN results in reduction in NO_x emission. Prabu and Anand (2015) examined the influence of 10, 30 and 60 ppm of alumina and ceria nanoparticles in jatropha oil methyl ester on performance and emission behaviour of diesel engine and it resulted in 13% reduction in NO_x emission when nanoparticles was added to biodiesel. Basha and Anand (2014) investigated the performance and combustion behaviour of diesel engine using carbon nanotubes (CNT) blended jatropha methyl ester emulsion. It was observed that the addition of CNT to the emulsion fuel resulted in drastic reduction of NO_x emissions. Annamalai et al. (2016) investigated the effects of cerium oxide nanoparticle as additive in Lemongrass Oil (LGO) emulsion fuel in a single cylinder, constant speed diesel engine. The results showed that the reduction of NO_x emission for LGO nano emulsion fuel by 20.3% compared with neat diesel fuel due to its high latent heat of evaporation of water molecule present in the fuel. Prabhu (2018) investigated the effect of nanoparticles (Alumina and cerium oxide) and observed that 30% reduction in NO emission for the nanoparticles dispersed fuel blend. Dineshashivakumar et al. (2019) examined the effects of water emulsion and diethyl ether additive, and found a lower emission of oxides of nitrogen for the emulsified fuel with the composition of 2% DEE compared to biodiesel operation for all load conditions. Lele et al. (2019) studied the NO formation mechanism for biodiesel and biodiesel-methanol blend fuelled light duty diesel engine. The results showed that a significant decrease in NO concentration with addition of methanol to biodiesel, which is qualitatively captured by engine simulations.

CO Emission

Carbon monoxide (CO) a poisonous gas that is colourless and odourless. Itis produced as a result of incomplete combustion due to insufficient air and (or) improper air-fuel mixture. The level of CO emissions is controlled primarily by the air-fuel equivalent ratio (Heywood, 2011). Higher oxygen content reduced CO emission due to complete combustion (Agarwal et al., 2006). CO emission decreased with the increase of cetane number of biodiesel, engine load and engine speed. The decreased CO emission of biodiesel with different additives is reported by number of authors. Prabu and Anand (2015) examined the influence of 10, 30 and 60 ppm of alumina and ceria nanoparticles as additives in jatropha oil methyl ester on performance and emission behaviour of diesel engine and it resulted in 60% reduction in CO emission when nanoparticles was added to biodiesel. Annamalai et al. (2016) investigated the effects of cerium oxide nanoparticle as additive in Lemongrass Oil (LGO) emulsion fuel in a single cylinder, constant speed diesel engine. The results showed that the reduction of CO emission for LGO nano emulsion fuel by 26% compared with neat diesel fuel due to oxidation agent of cerium oxide nanoparticle. Mofijur et al. (2012) performed a comprehensive study on the effect of antioxidant additives added in biodiesel blends on performance and emission of a diesel engine. The authors found B35 fuel with 1% NPAA (NonlyPhenoxy Acetic Acid) additives gave 18% and 10% lower CO emission compared to diesel and biodiesel without additives respectively. Ma et al. (2013) determined the effect of a Fe-based homogenous combustion catalyst on the emission characteristics of single cylinder diesel engine. The results showed that the use of Fe based catalyst led to 21% reduction in CO emission owing to the improved brake specific fuel consumption. CO emission reduction was also reported by Venu and Madhavan (2016) who investigated the addition of Zr and Ti into biodiesel-ether fuel blend. Qibai et al. (2018) investigated the effects of carbon coated aluminium nanoparticles added to diesel-biodiesel blends as additives on engine performance and emissions. The results showed that the adding additive reduced the CO emission by 19%, compared with B10. Prabhu (2018) reported a 60% reduction in CO emission for the nanoparticles dispersed fuel blend as compared to that of B100. Soudagar et al. (2019) studied about the effects of graphene oxide nanoparticles on performance and emissions of a CI engine fueled with dairy scum oil biodiesel (DSOME). The results were notable enhancements in the emission characteristics, a reduction in CO emission by 38.6% for nanofuel blend (DSOME40). Dineshashivakumar et al. (2019) examined the effects of water emulsion and diethyl ether additive, and found a lower emission of oxides of nitrogen for the emulsified fuel with the composition of 2% DEE compared to biodiesel operation for all load conditions. Dineshashivakumar et al. (2019) noted a lower CO emission with emulsified fuel with the composition of 2% DEE.

HC Emission

The unburned fuel from the engine cylinder is emitted as HC emission along with the other pollutants. HC is mainly formed inside the engine cylinder at low-temperature regions, such as boundary layer, near the walls and crevice layer etc. (Heywood, 2011). Many investigators (Puhan et al., 2005; Sahoo et al., 2009) reported decreased HC emission when the engine was fueled with biodiesel instead of diesel. Few of the researchers (Buyukkaya, 2010; Ozsezen et al., 2009) reported that it was reduced by 45 - 65% while compared to diesel. Higher oxygen content and cetane number of the fuel also lead to the reduction of HC emission. Higher oxygen content results in complete combustion of fuel, and higher cetane number fuels burn more quickly which reduce unburned hydrocarbon. Few investigators investigated the effects of additives on HC emission. Annamalai et al. (2016) investigated the effects of cerium oxide nanoparticle as additive in Lemongrass Oil (LGO) emulsion fuel in a single cylinder, constant speed diesel engine. The results showed that the reduction of unburned HC emission for LGO nano emulsion fuel by 16% compared with diesel fuel due to oxidation agent of cerium oxide nanoparticle. Prabhu (2018) observed that 44% reduction in HC emission with Alumina and cerium oxide nanoparticles dispersed fuel blend as compared to that of B100.Soudagar et al. (2019) investigated and reported a reduction in unburned HC emission by 21.68% for nanofuel blend (DSOME40). Dineshashivakumar et al. (2019) determined the lower HC emission for the emulsified fuel with the composition of 2% DEE.

Particulate Matter

Particulate matter (PM) consists of soot and other liquid or solid phase components. Oxygenated compounds exert their impacts on the emitter PM by changing the properties of fuel. In number of studies, the higher level of oxygen has been highlighted for the significant reduction in PM emissions when alcohol-diesel blend was used (Imran et al., 2013; Shahir et al., 2015). The biodiesel-diesel and alcohol-diesel blends behave slightly inversely in terms of their reducing effect on PM emissions. This could be attributed to the fact that biodiesel-diesel blends advance the ignition timing while alcohol-diesel blends retard the ignition timing (Shahabuddin et al., 2013; Shahir et al., 2015).

The positive influence of metal-based additives on the emitted PM has been frequently reported (Song et al., 2006; Zhang et al., 2013). Addition of metallic additives into biodiesel-diesel blends have also been reported that results in lower PM emissions. Keskin et al.(2007) claimed the combustion of B60-Mn diminished PM emissions by up to 30% compared to diesel at full load condition. Zhang et al. (2013) investigated an iron-based catalyst blended with diesel with a focus on the impact of the additive on particle size distribution of soot. Chandler et al. (2007) revealed a remarkable reduction of between 7.3 and 39.5% in soot emissions. However, they reported that the metallic additive reduced larger soot particles more substantially than smaller ones.

Smoke

Among all the particulate matters emitted by the IC engine soot is the main reason to produce smoke opacity. Oxygenates has tremendous effect on the reduction of smoke when is added to diesel fuel. The formation of smoke strongly depends on the engine load, and as the load on the engine increases the air-fuel ratio decreases which results in higher smoke (Can et al., 2004). High density and higher viscosity of biodiesel fuel also led to increase the smoke opacity (Aydin H et al., 2010). Basha and Anand (2014) investigated the performance and combustion behaviour of diesel engine using carbon nanotubes (CNT) blended jatropha methyl ester emulsion. It was observed that the addition of CNT to the emulsion fuel resulted in drastic reduction of smoke emissions. Annamalai et al. (2016) investigated the effects of cerium oxide nanoparticle as additive in Lemongrass Oil (LGO) emulsion fuel in a single cylinder, constant speed diesel engine. The results showed that the reduction of smoke emission for LGO nano emulsion fuel by 19.8% compared with diesel fuel due to its micro explosion, secondary atomization and enhanced the combustion rate caused by the cerium oxide. Prabhu (2018) observed that 38% reduction in smoke emission for the nanoparticles dispersed fuel blend as compared to that of B100. Soudagar et al. (2019) studied and pointed out a reduction in smoke emission by 24.88% for nanofuel blend (DSOME40). Dineshashivakumar et al. (2019) claimed a lower smoke emission for the emulsified fuel with the composition of 2% DEE.

SUMMARY

- Adding oxygenated additives to the biodiesel blend reduces the density and viscosity as well as increase the oxygen content of the fuel blend. Oxygenated additives showed least improvement in the flash point. Among the additives, diethyl ester improved the low temperature properties of the biodiesel fuel blends. Ethanol and Diethyl ether is low viscous additive, has the ability to reduce the viscosity of the fuel when blended with other fuels such as diesel, biodiesel or fuel blends.
- The brake thermal efficiency of biodiesel blends is lower than with diesel due to high viscosity and poor volatility. Most of the cases oxygenated additive blended biodiesel showed higher brake thermal efficiency compared to diesel and biodiesel without additives.
- Adding oxygenated additives like ethanol, diethyl ether, iso-butanol in biodiesel blend decreases the brake specific fuel consumption compared to diesel and biodiesel without additives. If the ethanol content of the blend increases, the brake specific fuel consumption increases.
- Normally all exhaust emissions like CO, HC, PM and smoke are reduced greatly with the addition of oxygenated additives to biodiesel fuels. Especially, iso-butanol, ethanol, and diethyl ether are more effective to reduce emissions due to excess oxygen content. But oxygenated additives are not suitable for reducing NO_X emission due to more oxygen content present in the biodiesel.
- Specific fuel consumption of biodiesel fuels are normally higher but the addition of metallic additives showed significant decrease in specific fuel consumption due to their catalyst effect as compared to biodiesel without additives.
- Biodiesel has no significant influence on engine torque and power output, but TiO_2 additives is more effective for improving engine power. Brake thermal efficiency also increases with the addition of metal-based additives to biodiesel. Fuel born catalyst and cerium oxide additive with biodiesel are more effective for increasing brake thermal efficiency compared to biodiesel without additives.
- Exhaust gas emissions is improved with the addition of metallic additives (Co, Mg, Mn, Ni etc.) as compared with mineral diesel and biodiesel-diesel blend without additive. Slight increase in NO_X, CO and CO_2 emission was observed with fuel born catalyst added biodiesel compared to neat biodiesel at optimized operating condition, but other emissions like HC, smoke decreased with fuel born catalyst.
- Antioxidant additives are effective to increase flash point, cetane number as well as oxidation stability of biodiesel. A slight increase in brake thermal efficiency is observed with BHA and BHT compared to neat biodiesel.
- Antioxidants are quite effective for reducing brake specific fuel consumption compared to biodiesel without additives. NPAA, BHT and BHA with palm oil biodiesel shows higher brake power compared to biodiesel without additives and lower compared to diesel.
- The antioxidant additives are quite effective in controlling NO_X formation of biodiesel fuels. Among all antioxidants, NPAA showed best performance compared to others. The NO_X reduction efficiency of antioxidants is observed in the order NPAA>BHA>BHT>EHN>TBHQ. HC and CO emissions of all antioxidant added biodiesel was higher when compared to biodiesel without antioxidant.
- Cold flow additives are quite effective in reducing pour point and cloud point of biodiesel fuels. The average reduction of pour point and cloud point of B100 is 7°C and 0.6°C respectively. Cetane number improver additives increase the cetane number of biodiesel fuels.

CONCLUSION

From the research articles of the number of authors the following conclusion are made.

- In general almost all the authors reported an increase in the performance parameter and reduction in the emission of CI engines with the addition of different additives along with the fuel.
- Additives can be mixed almost all the fuels like diesel, biodiesel, diesel-biodiesel blends etc.
- Additives that can be added to the fuel of the CI engines are grouped in to five categories viz. (i) Oxygenated additives, (ii) Metal based additives, (iii) Antioxidant additives, (iv) Cetane number improver additives, (v) Cold flow improver additives.
- Blending of additives improves the performance parameters of the CI engine viz. brake power, torque, brake specific fuel consumption and brake thermal efficiency.
- The common pollutants emitted by the CI engines NOx, CO, HC and smoke are reduced with the use of additives.
- Oxygenated additives are not useful for increasing brake power but some antioxidant and metal based additives are more helpful for increasing brake power of the engine. Metal based additives and antioxidants are more efficient to reduce fuel consumption of biodiesel fuel. Antioxidants are quite effective in controlling NO_x emission. But metal-based additives and oxygenated additives are also useful to reduce NO_x emission from biodiesel fuel. CO and HC emission of biodiesel fuel reduced with the addition of metal-based additives as well as ethanol and methanol also further reduce CO and HC emission. Metal based additives are most useful to reduce smoke opacity of biodiesel due to their catalyst effect. Diethyl ether and ethanol are also useful to reduce smoke opacity.

RECOMMENDATION FOR FURTHER RESEARCH

Finally, it is well-documented that additives in biodiesel-diesel blend could remarkably improve the quality of combustion in diesel engine and improves the performance and emissions. However, future research works should be conducted to address the following issues pertaining to the real-life applications of additives in biodiesel-diesel blend.

- Further studies should be devoted to investigate the impacts of combustion conditions of diesel engines on engine performance and emission parameters when additives are used in fuel blends. This is important since thermochemical characteristics of fuel blends are changed by the inclusion of the additives into biodiesel-diesel blends.
- More investigation are required for the development of fuel born catalyst added biodiesel fuelled diesel engine on the emission control components like particulate filter and catalytic converter to reduce PM and NO_X emissions.
- The impact of diesel-biodiesel additives on engine corrosion should also be investigated in order to ensure their reliability.
- Metallic and non-metallic additives should further studied to understand their exact action mechanisms in combustion of fuel.

- Durability tests are missing in the published literature available on the application of diesel and biodiesel additives and should be taken into consideration by future studies.

REFERENCES

Abbaszaadeh, A., Ghobadian, B., Omidkhah, M. R., & Najafi, G. (2012). Current biodiesel production technologies: A comparative review. *Energy Conversion and Management*, *63*, 138–148. doi:10.1016/j.enconman.2012.02.027

Agarwal, D., Sinha, S., & Agarwal, A. K. (2006). Experimental investigation of controlNO_x emissions in biodiesel-fuelled compression ignition engine. *Renewable Energy*, *31*(14), 2356–2369. doi:10.1016/j.renene.2005.12.003

Annamalai, M., Dhinesh, B., Nanthagopal, K., SivaramaKrishnan, P., Lalvani, J. I. J., Parthasarathy, M., & Annamalai, K. (2016). An assessment on performance, combustion and emission behaviour of a diesel engine powered by ceria nanoparticle blended emulsified biofuel. *Energy Conversion and Management*, *123*, 372–380. doi:10.1016/j.enconman.2016.06.062

Armas, O., Yehliu, K., & Boehman, A. L. (2010). Effect of alternative fuels on exhaust emissions during diesel engine operation with matched combustion phasing. *Fuel*, *89*(2), 438–456. doi:10.1016/j.fuel.2009.09.022

Atmanli, A., Ileri, E., Yuksel, B., & Yilmaz, N. (2015). Extensive analyses of diesel-vegetable oil-n-butanol ternary blends in a diesel engine. *Applied Energy*, *145*, 155–162. doi:10.1016/j.apenergy.2015.01.071

Aydin, H., & Bayindir, H. (2010). Performance and emission analysis of cottonseed oil methyl ester in a diesel engine. *Renewable Energy*, *35*(3), 588–592. doi:10.1016/j.renene.2009.08.009

Banapurmath, N. R., Tewari, P. G., & Hosmath, R. S. (2008). Performance and emission characteristics of a DI compression ignition engine operated on Honge, Jatropha and sesame oil methyl esters. *Renewable Energy*, *33*(9), 1982–1988. doi:10.1016/j.renene.2007.11.012

Basha, J. S., & Anand, R. B. (2014). Performance, emission and combustion characteristics of a diesel engine using carbon nanotubes blended jatropha methyl ester emulsions. *Alexandria Engineering Journal*, *53*(2), 259–273. doi:10.1016/j.aej.2014.04.001

Bhaskar, K., Nagarajan, G., & Sampath, S. (2014). The performance and emission characteristics of fish oil methyl esters and diesel blends in a partially premixed charge compression ignition engine. *International Journal of Green Energy*, *11*(4), 389–403. doi:10.1080/15435075.2013.773514

Borris, C. C., Carlos, M., Aida, D.-S., & Paloma, A. (2014). Effects of the addition of oxygenated fuels as additives on combustion characteristics and particle number and size distribution emissions of a TDI diesel engine. *Fuel*, *132*, 93–100. doi:10.1016/j.fuel.2014.04.071

Buyukkaya, E. (2010). Effects of biodiesel on a diesel engine performance, emission and combustion characteristics. *Fuel*, *89*(10), 3099–3105. doi:10.1016/j.fuel.2010.05.034

Can, O., Celikten, I., & Usta, N. (2004). Effects of ethanol addition on performance and emissions of a turbocharged indirect injection diesel engine running at different injection pressures. *Energy Conversion and Management*, *45*(15-16), 2429–2440. doi:10.1016/j.enconman.2003.11.024

Chandler, H. F. G., & Brown, G. I. (1992). Effect of cold flow additives on low temperature operability of diesel fuel. SAE Paper 9291186.

Chandler, M. F., Teng, Y., & Koylu, U. O. (2007). Diesel engine particulate emissions: A comparison of mobility and microscopy size measurements. *Proceedings of the Combustion Institute*, *31*(2), 2971–2979. doi:10.1016/j.proci.2006.07.200

Chastek, T. Q. (2011). Improving cold flow properties of canola-based biodiesel. *Biomass and Bioenergy*, *35*(1), 600–607. doi:10.1016/j.biombioe.2010.10.024

Chauhan, B. S., Kumar, N., & Cho, H. M. (2012). A study on the performance and emission of a diesel engine fueled with jatropha biodiesel oil and its blends. *Energy*, *37*(1), 616–622. doi:10.1016/j.energy.2011.10.043

Choi, B., Jiang, X., Kim, Y. K., Jung, G., Lee, C., Choi, I., & Song, C. S. (2015). Effect of diesel fuel blend with n-butanol on the emission of a turbocharged common rail direct injection diesel engine. *Applied Energy*, *146*, 20–28. doi:10.1016/j.apenergy.2015.02.061

Christensen, E., & McCormick, R. L. (2014). Long-term storage stability of biodiesel and biodiesel blends. *Fuel Processing Technology*, *128*, 339–348. doi:10.1016/j.fuproc.2014.07.045

Demirbas, A. (2000). A direct route to the calculation of heating values of liquid fuels by using their density and viscosity measurements. *Energy Conversion and Management*, *41*(15), 1609–1614. doi:10.1016/S0196-8904(00)00012-1

Demirbas, A. (2007). Progress and recent trends in biofuels. *Progress in Energy and Combustion Science*, *33*(1), 1–18. doi:10.1016/j.pecs.2006.06.001

Demirbas, A., & Demirbas, I. (2007). Importance of rural bioenergy for developing countries. *Energy Conversion and Management*, *48*(8), 2386–2398. doi:10.1016/j.enconman.2007.03.005

Dhar, A., Kevin, R., & Agarwal, A. K. (2012). Production of biodiesel from high-FFA neem oil and its performance, emission and combustion characterization in a single cylinder DICI engine. *Fuel Processing Technology*, *97*, 118–129. doi:10.1016/j.fuproc.2012.01.012

Di, Y., Cheung, C., & Huang, Z. (2009). Experimental study on particulate emission of a diesel engine fuelled with blended ethanol-dodecanol-diesel. *Journal of Aerosol Science*, *40*(2), 101–112. doi:10.1016/j.jaerosci.2008.09.004

Dinesha, P., Kumar, S., & Rosen, M. A. (2019). Combined effects of water emulsion and diethyl ether additive on combustion performance and emission of a compression ignition engine using biodiesel blends. *Energy*, *179*, 928–937. doi:10.1016/j.energy.2019.05.071

Selvan, V. A. M., Anand, R. B., & Udayakumar, M. (2009). Effects of cerium oxide nanoparticle addition in diesel and diesel-biodiesel-ethanol blends on the performance and emission characteristics of a CI engine. *Journal of Engineering and Applied Sciences (Asian Research Publishing Network)*, *4*, 1819–1823.

EPA. (2002). A comprehensive analysis of biodiesel impacts on exhaust emissions. Draft Technical Report EPA420-P-02-001. Retrieved from http://www.epa.gov

Fang, Q., Fang, J., Zhuang, J., & Huang, Z. (2013). Effects of ethanol-diesel-biodiesel blends on combustion and emissions in premixed low temperature combustion. *Applied Thermal Engineering*, *54*(2), 541–548. doi:10.1016/j.applthermaleng.2013.01.042

Fangsuwannarak, K., & Triratanasirichai, K. (2013). Effect of metalloid compound and bio-solution additives on biodiesel engine performance and exhaust emissions. *American Journal of Applied Sciences*, *10*(10), 1201–1213. doi:10.3844/ajassp.2013.1201.1213

Gairing, M., Marriott, J., Reders, K., & Reglitzky, A. (1995). The effects of modern additive technology on diesel fuel performance. SAE Technical Paper.

Hazar, H. (2009). Effects of biodiesel on a low heat loss diesel engine. *Renewable Energy*, *34*(6), 1533–1537. doi:10.1016/j.renene.2008.11.008

Heywood, J. (2011). *Internal Combustion Engine Fundamentals*. New Delhi, India: Tata McGraw Hill.

Hokeman, S. K., & Robbins, C. (2012). Review of the effects of biodiesel on NO_x emissions. *Fuel Processing Technology*, *96*, 237–249. doi:10.1016/j.fuproc.2011.12.036

Huang, J., Wang, Y., Li, S., Roskilly, A. P., Yu, H., & Li, H. (2009). Experimental investigation on the performance and emissions of a diesel engine fuelled with ethanol-diesel blends. *Applied Thermal Engineering*, *29*(11-12), 484–490. doi:10.1016/j.applthermaleng.2008.12.016

Imran, A., Varman, M., Masjuki, H., & Kalam, M. (2013). Review on alcohol fumigation on diesel engine: A viable alternative dual fuel technology for satisfactory engine performance and reduction of environment concerning emission. *Renewable & Sustainable Energy Reviews*, *26*, 739–751. doi:10.1016/j.rser.2013.05.070

Imtenan, S., Masjuki, H. H., Varman, M., Arbab, M. I., Sajjad, H., Fattah, I. M. R., ... Hasib, A. S. M. (2014). Emission and performance improvements analysis of biodiesel-diesel blends with additives. *Procedia Engineering*, *90*, 472–477. doi:10.1016/j.proeng.2014.11.759

Jain, S., & Sharma, M. P. (2010). Stability of biodiesel and its blends: A review. *Renewable & Sustainable Energy Reviews*, *14*(2), 667–678. doi:10.1016/j.rser.2009.10.011

Jelles, S., Krul, R., Makkee, M., & Moulijn, J. (1999). The influence of NO_x on the oxidation of metal activated diesel soot. *Catalysis Today*, *53*(4), 623–630. doi:10.1016/S0920-5861(99)00150-9

Jung, H., Kittelson, D. B., & Zachariah, M. R. (2005). The influence of a cerium additive on ultrafine diesel particle emissions and kinetics of oxidation. *Combustion and Flame*, *142*(3), 276–288. doi:10.1016/j.combustflame.2004.11.015

Kalligeros, S., Zannikos, F., Stournas, S., & Teas, C. (2003). An investigation of using biodiesel/marine biodiesel blends on the performance of a stationary diesel engine. *Biomass and Bioenergy*, *24*(2), 141–149. doi:10.1016/S0961-9534(02)00092-2

Kannan, G., Karvembu, R., & Anand, R. (2011). Effect of metal-based additive on performance emission and combustion characteristics of diesel engine fuelled with biodiesel. *Applied Energy*, *88*(11), 3694–3703. doi:10.1016/j.apenergy.2011.04.043

Karabektas, M. (2009). The effects of turbocharger on the performance and exhaust emissions of a diesel engine fuelled with biodiesel. *Renewable Energy*, *34*(4), 989–993. doi:10.1016/j.renene.2008.08.010

Karavalakis, G., Hilari, D., Givalou, L., Karonis, D., & Stournas, S. (2011). Storage stability and ageing effect of biodiesel blends treated with different antioxidants. *Energy*, *36*(1), 369–374. doi:10.1016/j.energy.2010.10.029

Keskin, A., Guru, M., & Altiparmak, D. (2007). Biodiesel production from tall oil with synthesized Mn and Ni based additives: Effects of the additives on fuel consumption and emissions. *Fuel*, *86*(7-8), 1139–1143. doi:10.1016/j.fuel.2006.10.021

Kim, H., & Choi, B. (2010). The effect of biodiesel and bioethanol blended diesel fuel on nanoparticles and exhaust emissions from CRDI diesel engine. *Renewable Energy*, *35*(1), 157–163. doi:10.1016/j.renene.2009.04.008

Kumar, N., Varun, S. R., & Chauhan, S. R. (2013). Performance and emission characteristics of biodiesel from different origins: A review. *Renewable & Sustainable Energy Reviews*, *21*, 633–658. doi:10.1016/j.rser.2013.01.006

Lapuerta, M. N., Garcia-Contreras, R., Campos-Femandez, J., & Dorado, M. P. (2010). Stability lubricity viscosity and cold flow properties of alcohol-diesel blends. *Energy & Fuels*, *24*(8), 4497–4502. doi:10.1021/ef100498u

Lele, A., Soni, K., Narayanaswamy, K., & Krishnasamy, A. (2019). Experimental and Modeling Investigation of NO formation mechanism for biodiesel and its blends with methanol. SAE International, Technical Paper 2019-01-0217, 13, DOI: doi:10.4271/2019-01-0217

Lu, X., Ma, J., Ji, L., & Huang, Z. (2008). Simultaneous reduction of NO_x emission and smoke opacity of biodiesel-fueled engines by port injection of ethanol. *Fuel*, *87*(7), 1289–1296. doi:10.1016/j.fuel.2007.07.006

Lu, X. C., Yang, J. G., Zhang, W. G., & Huang, Z. (2005). Improving the combustion and emissions of direct injection compression ignition engines using oxygenated fuel additives combined with a cetane number improver. *Energy & Fuels*, *19*(5), 1879–1888. doi:10.1021/ef0500179

Ma, Y., Zhu, M., & Zhang, D. (2013). The effect of a homogeneous combustion catalyst on exhaust emissions from a single cylinder diesel engine. *Applied Energy*, *102*, 556–562. doi:10.1016/j.apenergy.2012.08.028

Mat Yasin, M. H., Yusaf, T., Mamat, R., & Fitri Yusop, A. (2014). Characterization of a diesel engine operating with a small proportion of methanol as a fuel additive in biodiesel blend. *Applied Energy*, *114*, 865–873. doi:10.1016/j.apenergy.2013.06.012

Mofijur, M., Masjuki, H., Kalam, M., & Shahabuddin, M. (2012). Experimental study of additive added palm biodiesel in a compression ignition engine. *Energy Education Science and Technology, Part A. Energy Science and Research*, *30*, 737–748.

Nanthagopal, K., Ashok, B., Tamilarasu, A., Johny, A., & Mohan, A. (2017). Influence on the effect of zinc oxide and titanium dioxide as an additive with calophylluminophyllum methyl ester in a CI engine. *Energy Conversion and Management*, *146*, 8–19. doi:10.1016/j.enconman.2017.05.021

Nigam, P. S., & Singh, A. (2011). Production of liquid biofuels from renewable resources. *Progress in Energy and Combustion Science*, *37*(1), 52–68. doi:10.1016/j.pecs.2010.01.003

Ozsezen, A. N., Canakci, M., Turkcan, A., & Sayin, C. (2009). Performance and combustion characteristics of a DI diesel engine fueled with waste palm oil and canola oil methyl esters. *Fuel*, *88*(4), 629–636. doi:10.1016/j.fuel.2008.09.023

Phoon, L., Mustaffa, A., Hashim, H., Mat, R., Manan, Z., & Yunus, N. (2017). Performance and emission characteristics of green diesel blends containing diethyl-succinate and 1-octanol. *Journal of Cleaner Production*, *161*, 1192–1202. doi:10.1016/j.jclepro.2017.06.219

Prabhu, A. (2018). Nanoparticles as additive in biodiesel on the working characteristics of a diesel engine. *Ain Shams Engineering Journal*, *9*(4), 2343–2349. doi:10.1016/j.asej.2017.04.004

Prabu, A., & Anand R. B. (2015). Emission control strategy by adding alumina and cerium oxide nanoparticle in biodiesel. *Journal of Energy Institute,* 1-7.

Puhan, S., Nagarajan, G., Vedaraman, N., & Ramabramhmam, B. V. (2007). Mahua oil derivatives as a renewable fuel for diesel engine systems in India, a performance and emissions comparative study. *International Journal of Green Energy*, *4*(1), 89–104. doi:10.1080/15325000601015627

Puhan, S., Vedaraman, N., Sankaranarayanan, G., & Ram, B. V. B. (2005). Performance and emission study of Mahuca oil (madhucaindica oil) ethyl ester in a four stroke natural aspirated direct injection diesel engine. *Renewable Energy*, *30*(8), 1269–1278. doi:10.1016/j.renene.2004.09.010

Qi, D. H., Chen, H., Geng, L. M., & Bian, Y. Z. (2010). Experimental studies on the combustion characteristics and performance of a direct injection engine fuelled with biodiesel/diesel blends. *Energy Conversion and Management*, *51*(12), 2985–2992. doi:10.1016/j.enconman.2010.06.042

Qi, D. H., Chen, H., Geng, L. M., & Bian, Y. Z. (2011). Effect of diethyl ether and ethanol additives on the combustion and emission characteristics of biodiesel-diesel blended fuel engine. *Renewable Energy*, *36*(4), 1252–1258. doi:10.1016/j.renene.2010.09.021

Qibai, W., Xialin, X., Yaodong, W., & Roskilly, T. (2018). Effect of carbon coated aluminium nanoparticles as additive to biodiesel-diesel blends on performance and emission characteristics of diesel engine. *Applied Energy*, *221*, 597–604. doi:10.1016/j.apenergy.2018.03.157

Rahman, S. M. A., Masjuki, H. H., Kalam, M. A., & Sajjad, H. (2013). Impact of idling on fuel consumption and exhaust emission and available idle reduction technologies for diesel vehicles – a review. *Energy Conversion and Management*, *74*, 171–182. doi:10.1016/j.enconman.2013.05.019

Rahmat, N., Abdullah, A. Z., & Mohamed, A. R. (2010). Recent progress on innovative and potential technologies for glycerol transformation into fuel additives: A critical review. *Renewable & Sustainable Energy Reviews*, *14*(3), 987–1000. doi:10.1016/j.rser.2009.11.010

Ramadhas, A. S., Muraleedharan, C., & Jayaraj, S. (2005). Performance and emission evaluation of a diesel engine fueled with methyl esters of rubber seed oil. *Renewable Energy*, *30*(12), 1789–1800. doi:10.1016/j.renene.2005.01.009

Rashdul, H. K., Masjuki, H. H., Kalam, M. A., & Ashraful, A. M. (2014). Theeffect of additives on properties, performance and emission of biodiesel fuelled compression ignition engine. *Energy Conversion and Management*, *88*, 348–364. doi:10.1016/j.enconman.2014.08.034

Ribeiro, N. M., Pinto, A. C., Quintella, C. M., da Rocha, G. O., & Teixeria, L. S. (2007). The role of additives for diesel and diesel blended (ethanol or biodiesel) fuels: A review. *Energy & Fuels*, *21*(4), 2433–2455. doi:10.1021/ef070060r

Rizwanul Fattah, I. M., Masjuki, H. H., Kalam, M. A., Mofijur, M., & Abedin, M. J. (2014). Effect of antioxidant on the performance and emission characteristics of a diesel engine fueled with palm biodiesel blends. *Energy Conversion and Management*, *79*, 265–272. doi:10.1016/j.enconman.2013.12.024

Sahin, Z., Tuti, M., & Durgun, O. (2014). Experimental investigation of the effects of water adding to the intake air on the engine performance and exhaust emissions in a DI automotive diesel engine. *Fuel*, *115*, 884–895. doi:10.1016/j.fuel.2012.10.080

Sahoo, P. K., Das, L. M., Babu, M. K. G., Arora, P., & Sing, V. P. (2009). Comparative evaluation of performance and emission characteristics of jatropha, karanja, and polanga-based biodiesel as fuel in a tractor engine. *Fuel*, *88*(9), 1698–1707. doi:10.1016/j.fuel.2009.02.015

Sajith, V., Sobhan, C. B., & Peterson, G. P. (2010). Experimental investigations on the effects of cerium oxide nanoparticle fuel additives on biodiesel. *Advances in Mechanical Engineering*, 1–6.

Sathiyamoorthi, R., & Sankaranarayanan, G. (2016). Effect of antioxidant additives on the performance and emission characteristics of a DICI engine using neat lemongrass oil-diesel blend. *Fuel*, *174*, 89–96. doi:10.1016/j.fuel.2016.01.076

Shahabuddin, M., Kalam, M., Masjuki, H., & Mofijur, M. (2012). An experimental investigation into biodiesel stability by means of oxidation and property determination. *Energy*, *44*(1), 616–622. doi:10.1016/j.energy.2012.05.032

Shahabuddin, M., Liaquat, A., Masjuki, H., Kalam, M., & Mofijur, M. (2013). Ignition delay, combustion and emission characteristics of diesel engine fuelled with biodiesel. *Renewable & Sustainable Energy Reviews*, *21*, 623–632. doi:10.1016/j.rser.2013.01.019

Shahir, S., Masjuki, H., Kalam, M., Imran, A., & Ashraful, A. (2015). Performance and emission assessment of diesel-biodiesel-ethanol/bioethanol blend as a fuel in diesel engines: A review. *Renewable & Sustainable Energy Reviews*, *48*, 62–78. doi:10.1016/j.rser.2015.03.049

Sivakumar, M., Shanmugasundaram, N., Rameshkumar, R., & Syed Thasthagir, M. (2018). Effects of aluminium oxide nanoparticles blended pongamia methyl ester on performance, combustion and emission characteristics of diesel engine. *Renewable Energy*, *116*, 518–526. doi:10.1016/j.renene.2017.10.002

Sivakumar, M., Shanmugasundram, N., & Rameshkumar, R. (2017). Effects of pongamia methyl esters and its blends on a diesel engine performance, combustion and emission characteristics. *Environmental Progress & Sustainable Energy*, *36*(1), 269–276. doi:10.1002/ep.12492

Song, J., Wang, J., & Boehman, A. L. (2006). The role of fuel-borne catalyst in diesel particulate oxidation behaviour. *Combustion and Flame*, *146*(1-2), 73–84. doi:10.1016/j.combustflame.2006.03.012

Soudagar M. E. M., Ghazali, N. N., Kalam, M. A., & Badruddin, I. A. (2019).The effects of grapheme oxide nanoparticle additive stably dispersed in dairy scum oil biodiesel-diesel fuel blend on CI engine: performance, emission and combustion characteristics. Fuel, 257, ID:116015, . doi:10.1016/j.fuel.2019.116015

Srinivasa Rao, M., & Anand, R. B. (2015). Production characterization and working characteristics in DICI engine of pongamia biodiesel. *Ecotoxicology and Environmental Safety*, *121*, 16–21. doi:10.1016/j.ecoenv.2015.07.031 PMID:26242967

Taghizadeh-Alisaraei, A., & Rezaei-Asl, A. (2016). The effect of added ethanol to diesel fuel on performance, vibration, combustion and knocking of a CI engine. *Fuel*, *185*, 718–733. doi:10.1016/j.fuel.2016.08.041

Ullman, T. L., Spreen, K. B., & Mason, R. L. (1994). Effects of cetane number, cetane improver, aromatics and oxygenates on 1994 heavy-duty diesel engine emissions. SAE Technical Paper.

Utlu, Z., & Kocak, M. S. (2008). The effect of biodiesel fuel obtained from waste frying oil on direct injection diesel engine performance and exhaust emissions. *Renewable Energy*, *33*(8), 1936–1941. doi:10.1016/j.renene.2007.10.006

Vallinayagam, R., Vedharaj, S., Yang, W. M., Saravanan, C. G., Lee, P. S., Chua, K. J. E., & Chou, S. K. (2014). Impact of ignition promoting additives on the characteristics of a diesel engine powered by pine oil-diesel blend. *Fuel*, *117*, 278–285. doi:10.1016/j.fuel.2013.09.076

Varatharajan, K., Cherlathan, M., & Velraj, R. (2011). Mitigration of NO_x emissions from a jatropha biodiesel fuelled DI diesel engine using antioxidant additives. *Fuel*, *90*(8), 2721–2725. doi:10.1016/j.fuel.2011.03.047

Venkatesan, S. P., & Kadiresh, P. N. (2016). Influence of an aqueous cerium oxide nanofluid fuel additive on performance and emission characteristics of a compression ignition engine. *International Journal of Ambient Energy*, *37*(1), 64–67. doi:10.1080/01430750.2014.882863

Venu, H., & Madhavan, V. (2016). Effect of nano additives (titanium and zirconium oxides) and diethyl ether on biodiesel-ethanol fuelled CI engine. *Journal of Mechanical Science Technology*, *30*(5), 2361–2368. doi:10.100712206-016-0446-5

Xiaolu, L., Hongyan, C., Zhiyong, Z., & Zhen, H. (2006). Study of combustion and emission characteristics of a diesel engine operated with dimethyl carbonate. *Energy Conversion and Management*, *47*(11-12), 1438–1448. doi:10.1016/j.enconman.2005.08.021

Xue, J., Grift, T. E., & Hansen, A. C. (2011). Effect of biodiesel on engine performances and emissions. *Renewable & Sustainable Energy Reviews*, *15*(2), 1098–1116. doi:10.1016/j.rser.2010.11.016

Yang, H. H, Lee, W. J., Mi, H. H, Wong, C. H., & Chen, C. B. (1998). PAH emission influenced by MN-based additive and turbocharging from a heavy-duty diesel engine. Environment inter, 24, 389-403.

Yang, W., An, H., Chou, S., Vedharaji, S., Vallinagam, R., Balaji, M., ... Chua, K. J. E. (2013). Emulsion fuel with novel nano-organic additives for diesel engine application. *Fuel*, *104*, 726–731. doi:10.1016/j.fuel.2012.04.051

Yilmaz, N., & Vigil, F. M. (2014). Potential use of a blend of diesel, biodiesel, alcohols and vegetable oil in compression ignition engines. *Fuel*, *124*, 168–172. doi:10.1016/j.fuel.2014.01.075

Zhang, D., Ma, Y., & Zhu, M. (2013). Nanostructure and oxidative properties of soot from a compression ignition engine: The effect of a homogeneous combustion catalyst. *Proceedings of the Combustion Institute*, *34*(1), 1869–1876. doi:10.1016/j.proci.2012.05.096

Zhang, X., Reece, D., Haws, R., & Moller, G. (1998). Biodegradability of biodiesel in the aquatic environment. ASAE, 41-0(5), 1423. doi:10.13031/2013.17277

Zheng, M., Mulenga, M. C., Wang, M., & Tjong, J. (2008). Biodiesel engine performance and emissions in low temperature combustion. *Fuel*, *87*(6), 714–722. doi:10.1016/j.fuel.2007.05.039

Chapter 5
Influence of Nano-Particle Additives on Bio-Diesel-Fuelled CI Engines:
A Review

Tamilvanan A.
https://orcid.org/0000-0001-9406-5005
Kongu Engineering College, India

K. Balamurugan
https://orcid.org/0000-0002-9463-4439
Institute of Road and Transport Technology, India

T. Mohanraj
https://orcid.org/0000-0002-4866-1428
Amrita Vishwa Vidyapeetham University, India

P. Selvakumar
https://orcid.org/0000-0002-4197-9986
Kongu Engineering College, India

B. Ashok
Vellore Institute of Technology, India

R. Sakthivel
https://orcid.org/0000-0002-0766-0285
Amrita Vishwa Vidyapeetham University, India

ABSTRACT

Biodiesel is proven to be the best substitute for petroleum-based conventional diesel fuel in existing engines with or without minor engine modifications. The performance characteristics of biodiesel as a fuel in CI engine are slightly lower than that of diesel fuel. The emission characteristics of biodiesel are better than diesel fuel except NO_X emission. The thermo-physical properties of biodiesel are improved by suspending the nano metal particles in the biodiesel, which make them an observable choice for the use of nanoparticles-added fuels in CI engine. High surface area of nanoparticles that promotes higher operating pressure and heat transfer rates that further quicken the combustion process by providing better oxidation. Thus, it has been inferred that addition of nanoparticles as an additive to biodiesel fuel blends in diesel engines and its effects on performance, combustion, and emission characteristics are discussed in this chapter.

DOI: 10.4018/978-1-7998-2539-5.ch005

INTRODUCTION

Every year there is a rise in energy demand due to the rapid development in industrialisation and automotive sector, demand and depletion of fossil fuels, fuel price instabilities, diminished energy security, ambiguity in oil supply to the consuming nations, fuel import costs, increased harmful environmental effects due to various pollutants are the important driving forces to search for new alternative fuels that are renewable, eco-friendly and harmless in now-a-days. This precarious situation has directed many researchers to go a better substitute for conventional fuels in engines.

In recent decades, the biodiesel has received significant consideration, because it is biodegradable, economically viable, non-toxic, environmentally acceptable, easily accessible and technically suitable and also it has received much attraction for conventional diesel engines and can noticeably reduce exhaust emissions from the engine (Balamurugan, Tamilvanan, Anbarasu, Akil, & Srihari, 2013). Biodiesel has numerous advantages compared to conventional diesel fuel. The major advantages are 10–11% oxygen content which results in better combustion characteristics and increase the engine efficiency and also high cetane number concerning 60–65% which reduces the ignition delay (Dinesh, Tamilvanan, Vaishnavi, Gopinath, & Mohan, 2019). Biodiesel reduces carbon dioxide emission by 78% on a life cycle basis while compared to fossil diesel fuel and reduces smoke by reason of free soot. Most of the studies has exposed that, make use of biodiesel in compression ignition (CI) engine can reduce emissions of hydrocarbon (HC), particulate matter (PM), carbon dioxide (CO_2) and carbon monoxide (CO) but oxides of nitrogen (NO_X) emissions are raised. Another issues like high consumption rate, lower performance, improper combustion, clogging and stability with long-term practice of biodiesel blends restricts their usage as a prospective replacement in diesel engine (Sakthivel, Ramesh, Purnachandran, & Shameer, 2018).

The research community claimed that enhancement in the performance, combustion and emission characteristics and improvement in fuel economy of the pure diesel, biodiesel and its blends in CI engines could be achieved either by modifying the engine design or fuel modification with the application of nano metal additives. Fuel additives are reviewed and classified into different categories such as nanoparticles-based additives, oxygen-rich additives, tocopherol additives, water, and polymer based additives (Khalife, Tabatabaei, Demirbas, & Aghbashlo, 2017). Compared to engine modification technique, fuel modification technique is one of the cheaper and easier which are normally favored in CI engine without any additional equipment and engine modifications (Ying, Longbao, & Hewu, 2006). On that basis, nano fuel techniques (application of nano scale metal particle additives in diesel and biodiesel fuel) have been employed for improving the biodiesel performance and reducing its emissions by using fuel additives. Such formulated fuels have a lot of advantages compared to neat biodiesel and its blends (Mehta, Chakraborty, & Parikh, 2014).

Nano fuels comprise a nano-sized metal particle having size ranging from 1 to 100 nm is mixed inside the base fluid (diesel/biodiesel) by means of ultrasonication process. The nano fuel additive has a higher surface to volume ratio and act as a catalyst that results in enhanced characteristics of fuels which leads to enhanced performance and combustion characteristics (Tamilvanan, Balamurugan, & Vijayakumar, 2019; Tomar & Kumar, 2019). Metallic particle additives like cerium (Ce), iron (Fe), platinum (Pt), manganese (Mn) barium (Ba), cerium oxide (CeO), aluminum (Al), aluminum oxide (Al_2O_3), copper (Cu) and titanium oxide (TiO_2) are identified as combustion improving catalyst which could results in reduced fuel consumption and emissions (Hosseinzadeh-Bandbafha, Tabatabaei, Aghbashlo, Khanali, & Demirbas, 2018). The results obtained by adding metal additives into biodiesel fuel blends were promising as a result of improvement in chemical and thermo-physical properties of modified fuel (Saxena, Kumar, & Saxena, 2017).

In spite of having all the benefits and dominance, somewhat ambiguous and conflicting results are established in the previous literatures, additionally the experimental outcome of diverse researchers are not general so far as to attain at widespread idea about this novel technique of fuel modification. Keeping all these essentials in mind, a severe effort has been made to review the major published work on various aspects of nanoparticle loaded biodiesel blends, and its special effects on fuel properties and overall engine characteristics. The main objective of this chapter is to emphasis a pathway to carry out further investigation in this area for effective utilization of possible nanoparticle fuel technology and to give a capable future fuel for CI engine.

Background

Generally, biodiesel has been assessed to be oxygenated fuel that consists of an oxygen molecule which diesel does not have. Based on the source of production, there are different biodiesels such as jatropha, palm oil, soybean, and canola. Compared to diesel, the elements of sulfur and aromatic compounds are not presented in the biodiesel which resulted in lower emissions. From the combustion point of view, it has a higher cetane number and its properties are comparable to that of diesel so that it can be directly put into the existing conventional diesel engine without any modifications. On the other hand, it has many drawbacks such as high viscosity, cold start problems at low temperature and shortened life of pump and filter due to the acidification during long-term storage. (Kim, Suh, Park, & Lee, 2008). The higher oxygen content in biodiesel leads to considerable reduction of PM, HC and CO emission. The increase in NO_X is primarily due to higher oxygen content available in biodiesel (Ayyasamy, Balamurugan, & Duraisamy, 2018).

During last few decades, there was a lot of advancement has been implemented in the field of nanotechnology. The nano metal particles suspending in base fluids results in enhancement of thermo- physical properties for using various commercial applications including engineering, biotechnology, agriculture, transportation and medical sciences (Saxena et al., 2017; Tamilvanan, Balamurugan, Ponappa, & Kumar, 2014). Distinctive physic-chemical characteristics (e.g., thermal, electrical, magnetic, and optical features) of nanoparticles have made their usage extensively as fuel catalyst to increase brake thermal efficiency (BTE) and also reduce specific fuel consumption (SFC), ignition delay, and harmful emission from engine (Kumar, Dinesha, & Bran, 2017).

MAIN FOCUS OF THE CHAPTER

In recent times, a lot of experimental works were carried out using nano-sized, organic, non-metallic and metallic particles in biodiesel and diesel. The results acquired were encouraging as a result of enrichment in thermo-physical properties and also superior heat and mass transport properties owing to high thermal conductivity, higher reactive medium for combustion, high surface to volume ratio, enhancement in flash point, fire point, pour point etc. The nanoparticles can be used as an additive in diesel and biodiesel as nano fuel blends. An extensive literature analysis on physical and chemical properties, performance, and combustion and emission characteristics of nano fuels have been carried out in this chapter. This chapter emphasizes the researchers to utilize the different metals and metal oxide nano-particles as a fuel additive in the application of biodiesel.

Synthesis of Nanoparticles

There are various techniques involved in synthesis of nanoparticles that are typically classified as bottom-up (chemical methods) and top-down (physical methods). Copper (Cu) nanoparticles were synthesized by electrochemical method and particle size obtained was 30-40 nm. X-ray powder diffraction (XRD) and scanning electron microscope (SEM) analysis were used to determine nature and size of the nanoparticles (Tamilvanan, Balamurugan, Ponappa, & Madhan Kumar, 2016). Sol–gel method was used for synthesis of manganese oxide (MnO_2) and copper oxide (CuO) nano metal additives (Lenin, Swaminathan, & Kumaresan, 2013). Sol-gel combustion method was adopted to prepare cerium oxide (CeO_2) nanoparticles (Annamalai et al., 2016). Hybrid aluminium oxide –copper (Al_2O_3–Cu) powder was synthesized by a thermo-chemical method (Suresh, Venkitaraj, & Selvakumar, 2011). Titanium dioxide (TiO_2) nanoparticles were prepared by sol–gel method in the presence of titanium tetra iso-propoxide $Ti\{OCH(CH_3)_2\}_4$ which was treated with deionized water (H_2O), hydrochloric acid (HCl) and ethanol (Venu, Subramani, & Raju, 2019). The other materials like Magnesium oxide (MgO), Silicon dioxide (SiO_2), Zinc oxide (ZnO), Ferric oxide (Fe_2O_3), Cobalt oxide (Co_3O_4) and Magnalium (Al-Mg) also used as fuel additive. The properties of different nano particles are shown in the Table 1.

Preparation of Nano fuel Blends

The nano metal additives were mixed with the fuel in lesser proportion ranging from 20 to 500 ppm. On the other hand, additives on micro-scale range have complexities like agglomeration, sedimentation and non-uniform size distribution. Because of the above mentioned problems particle size within 100 nm was used as additive for fuel and that can be directly used in diesel engine (Shaafi & Velraj, 2015). The preparation of nanofuel was achieved by mixing the nano-additives with fuels through ultrasonication process. In this process, high-frequency sound waves were used to scatter the nano particles within the biodiesel (Khalife et al., 2017). But, occasionally the agglomeration of nano particles inside the base fuel was a main problem in stability and poor performance of additives. Hence to avoid these issues,

Table 1. Properties of different nanoparticles

Name of additive	Properties
CeO_2	Good Thermal Stability and better oxidizing agent
Cu, Al_2O_3, CuO	Oxidizing agent, high thermal conductivity and heat transfer rate
TiO_2	To catalyze reactions with other molecules at lower temperatures
MgO	Odorless, non-toxic and high catalytic activity
SiO_2	Higher stability and low toxicity
ZnO	High heat capacity and heat conductivity
Fe_2O_3	High catalytic activity
Co_3O_4	Oxidizing agent and catalyst

surfactants and dispersants were used to stabilize the solution (Fuskele & Sarviya, 2017). The purpose of using surfactant is to decrease the surface tension between liquid and solid or between liquids. Tween 80, span 80, ethanol and iso-propanol are the common surfactants favored for blending of additives with liquid fuels (Soni, Rathod, & Goswami, 2015). Compared with conventional fluids, nanofluids showed enhanced stability as a result of Brownian motion and the size effect of nanoparticles. The ultrafine nanoparticles facilitate the fluids to flow easily without any restrictions in micro sized channels (Basu & Miglani, 2016).

An aluminium oxide (Al_2O_3) nanoparticle of 0.025 g was added to formulate the dosing level of 25 ppm. Therefore, for dosing level of 50 ppm, 0.05 g/l was added with biodiesel. After the addition of Al_2O_3 nanoparticles, it was shaken well and it was poured into mechanical homogenizer apparatus for agitation process about 30 min for making uniform suspension of nanoparticles (Aalam, Saravanan, & Kannan, 2015). The Lemaon Grass oil (LGO) was prepared by the following way. First a quantity of 30 ppm of CeO_2 nanoparticle was dispersed with water with the aid of stirrer and ultrasonicator. Then the prepared nanofluid was mixed with neat LGO and surfactant (Spam 80) with the aid of mechanical stirrer. The prepared fuel was LGO nano emulsion fuel (Annamalai et al., 2016)

Impact of Nano Metal Additives on Properties of Biodiesel Blends

Addition of various metal and metal oxide in nano-powder form into the liquid fuel may improve the properties (Nanthagopal, Ashok, Tamilarasu, Johny, & Mohan, 2017). The nano-sized solid metallic particles are able to improve the thermal conductivity of the fuels which act as significant part in heat transfer characteristics as well as flow properties of fuels (Zhu et al., 2009). Addition of CuO and Al_2O_3 nanoparticles (50 ppm each) with diesel results in improvement in density, flash point and cetane index while the flow property viscosity remains same (Gumus, Ozcan, Ozbey, & Topaloglu, 2016). Similarly, addition of CeO_2 nanoparticles with neat biodiesel (jatropha methyl ester) improves the properties such as viscosity and flash point (Sajith, Sobhan, & Peterson, 2010). Al_2O_3 nanoparticles in the mass fraction of 50 and 100 ppm with soyabean oil leads to increase in density and its value enhances exponentially with concentration (Ghogare & Kale, 2016). The density shows an increase of 2.3% for 100 ppm and 3.1% for 200 ppm of TiO_2 nano-additive with mustard biodiesel. A decreasing trend of calorific value was obtained and an overall improvement of cetane index about 3.5% was observed (Yuvarajan, Babu, BeemKumar, & Kishore, 2018).

The outcome of two different nano-additives, MgO and SiO_2 with rapseed biodiesel was examined (Özgür, Özcanli, & Aydin, 2015). Different proportion of additive say 25 and 50 ppm were added. A reduction in density of $1 kg/m^3$ was obtained for 50 ppm of MgO additive and no change was observed with 25 ppm. On the other hand, SiO_2 particles exhibited the reverse trend, for 25 ppm concentration, the density was decreased and for 50 ppm concentration, no change was observed. The density, flash point and calorific values of diesel-biodiesel-ethanol fuel blends were improved with addition of 250 ppm ZnO nanoparticles. But, cetane number and kinematic viscosity were decreased due to the addition of nanoparticles (Prabakaran & Udhoji, 2016).

Because of the superior energy density of nanoparticles, calorific value of diesel/biodiesel could be enhanced. During combustion reactions, the nanoparticles work as a catalyst and had a constructive effect on ignition properties (Gupta, Agrawal, & Mathur, 2012). An increase of calorific value from 42.5 MJ/kg to 43.1 MJ/kg was observed for mahua biodiesel with 50 ppm of CuO nanoparticles (Chandrasekaran, Arthanarisamy, Nachiappan, Dhanakotti, & Moorthy, 2016). It was concluded that compared to diesel, fuels blended with Al_2O_3 and CuO nanoparticles improve ignition possibility at low temperatures and nanoparticle additives have better uniqueness to improve fuel properties like ignition delay, flash point and cetane index (Gumus et al., 2016).

Impact of Nano Metal Additives on the Performance Characteristics of Engine

The addition of nanoparticles to biodiesel and conventional diesel is the most hopeful fuel additive technique for getting a significant enhancement in the performance (Saraee, Jafarmadar, Taghavifar, & Ashrafi, 2015). The variation of BTE and SFC for previous works using different nano particles were shown in Table 2.

Brake Thermal Efficiency (BTE)

A better enhancement in thermal efficiency was obtained for biodiesel operation with additive like cobalt oxide (CO_3O_4) and Magnalium (Al-Mg) than neat biodiesel. Around 1% enhancement in thermal efficiency was obtained for Al-Mg additive compared to neat biodiesel (Ganesh & Gowrishankar, 2011). Brake thermal efficiency was higher for 50 ppm nano fuel blends than mahua oil (20MEOM) (Chandrasekaran et al., 2016). Addition of 50 ppm Al_2O_3 nanoparticles in biodiesel–diesel blend results in improvement of 2.5% thermal efficiency (Aalam et al., 2015). The BTE of B10 (soya bean biodiesel) with 40 nm copper particle fuel blends was better than B10 with 50 nm copper particle, pure B10 and pure diesel. Compared with pure diesel, a maximum enhancement of 1.03% in BTE was obtained for B10 with 40 nm copper particle. In addition, it was noticed that, reduction in the size of nanoparticles provides more surface area for reaction and enhances the efficiency. A maximum enhancement of 1.01% in BTE was observed when the particle size was reduced from 50 to 40 nm (Balamurugan et al., 2013).

The BTE of *Calophyllum inophyllum* biodiesel and their blends with nano copper additive was investigated. The BTE was higher compared to biodiesel without additives but fairly lesser than diesel. An enhancement of 1.2% was obtained for B10 and B20 with copper additive and 3.8% for B100 with additive. The enhancement in BTE was owing to the presence of nano copper additive which promotes superior oxidation as well as higher heat release rate (Tamilvanan et al., 2019). BTE of B20 (jatropha oil) with each 30 ppm of Al_2O_3 and CeO_2 was increased by 3-4%. This enhancement was due to the improved atomization and quick evaporation of nanoparticles dispersed test fuel, improved air fuel mixing, which permits more surface area of fuel to react with oxygen (Prabu, 2017). BTE of ZnO nano additives with palm oil methyl ester (POME) and ethanol blends were higher than POME and ethanol blends without additive. This is due to the addition of ZnO nano particles in the blend, which improves the higher surface to volume ratio and improves the average combustion temperature (Prabakaran & Udhoji, 2016).

The addition of Al_2O_3 and CuO yielded enhanced results in BTE. The BTE of nano diesel was elevated due to added nanoparticles that promotes oxidation rate and reduces the ignition temperature. Torque values were enhanced with the addition of CuO and Al_2O_3 nanoparticles. The maximum torque was increased about 1% and 3.28% for 50 ppm concentration of CuO and Al_2O_3 nano diesel fuels respectively (Gumus et al., 2016). BTE of diesel engine was improved by the addition of CeO_2 in jatropha fuel with the concentration of 20-80 ppm. The nanoparticles exists in the fuel encourage longer and more complete combustion, and it also acted as an oxygen buffer, as a result increase in efficiency. It was observed that, 1.5% efficiency was improved with the nanoparticles concentration of 80 ppm (Sajith et al., 2010). The engine performance fuelled by polonga oil diesel with Ferric oxide (Fe_2O_3) nanoparticles blends were closer to neat diesel. This enhancement in efficiency was due to the improvement in thermal properties with the addition of nano additives (Santhanamuthu, Chittibabu, Tamizharasan, & Mani, 2014).

BTE of LGO nano emulsion fuel was increased by 17.02% when compared to neat LGO and dropped by 7.7% when compared to diesel at full load condition. It was due to micro explosion and secondary atomization of emulsified fuel (Annamalai et al., 2016). BTE of neem oil methyl ester (NOME) with 100, 200 and 300 ppm of carbon nanotube (CNT) additive were improved by 2.12%, 4.17% and 3.43% respectively when compared to neat biodiesel at full load condition. This may be caused by the positive effects of nanoparticles on fuel physical properties and also reduced ignition delay time which leads to improved combustion (Balaji & Cheralathan, 2015). The BTE of CNT added water–diesel emulsion fuels was enhanced predominantly at higher loads. This could possibly be endorsed to high surface area and reactive surfaces of CNT that gave higher chemical reactivity to perform as potential catalyst (Basha & Anand, 2011).

Brake Specific Fuel Consumption (BSFC)

The value of BSFC for Mahua Oil Methyl Ester with Ferrofluid (MOMEF) added at all loads was 5.11% lesser than Mahua Oil Methyl Ester (MOME). This is because of reduced ignition delay caused by nano-sized ferro-additives (Devarajan, Munuswamy, & Mahalingam, 2017). A 2% reduction in brake specific energy consumption (BSEC) at full load was obtained by the addition of cobalt oxide (Co_3O_4) additive with Jatropha biodiesel. This is owing to catalytic chemical oxidation of biodiesel which enhances the combustion of biodiesel. A parallel trend of 3% reduction in BSEC was obtained with Magnalium (Al-Mg) nano fuel additive compare with biodiesel at 50% load (Ganesh & Gowrishankar, 2011). The addition of nano copper metal additives showed a considerable reduction of specific fuel consumption (SFC) at all loading condition. Compared to normal biodiesel blends at maximum load, a reduction of 3% to 6% was obtained for biodiesel blends with copper additives. This may be owing to the fact that, raise in combustion temperature brought out increase in the conversion efficiency of heat energy into mechanical work and results in drop of BSFC with respect to engine loads (Tamilvanan et al., 2019). BSFC of B20 and B100 (jatropha oil) with each 30 ppm of Al_2O_3 and CeO_2 was decreased by 1-2% compared to B20 and B100 fuels, by the reason of superior atomization property of nanoparticles resulting in enhanced combustion (Prabu, 2017).

ZnO nano particles in POME and ethanol blend supports the combustion process through catalytic action which allows enhanced combustion at elevated temperatures. Also, surface to volume ratio of nano particles are superior, enhanced atomization of fuel blend happened at high temperatures (Prabakaran & Udhoji, 2016). The BSFC decreases to 0.5% and 1.2% for CuO and Al_2O_3 additive. The better fuel economy was obtained due to oxygen and positive effects of nanoparticles to enhance in combustion efficiency (Gumus et al., 2016). The SFC was reduced with increasing the concentration of nanoparticles. CeO_2 oxidizes the deposited carbon atoms from the engine results in efficient functioning of engine and decreases the fuel consumption (Sajith et al., 2010).

The BSFC of Al_2O_3 nanoparticles with Zizipus jujube methyl ester (ZJME) blended biodiesel was lower than ZJME for all loads. Al_2O_3 nanoparticles oxidize the carbon deposited on the engine cylinder results in reduction of fuel consumption (Aalam et al., 2015). The LGO Nano emulsion exposed lower BSEC compared with neat LGO emulsion. This is owing to the faster evaporation rate of emulsion fuel because of the presence of cerium oxide nanoparticle. CeO_2 oxidizes the unburned hydrocarbon which was deposited in the engine cylinder wall. This results in reduced SFEC (Annamalai et al., 2016). The addition of CNT nano additive to neem biodiesel decreases the BSFC for all biodiesel blends compared to neat biodiesel. This is due to the positive effect of nanoparticles leading towards complete combustion (Balaji & Cheralathan, 2015).

Impact of Nano Metal Additives on Combustion Characteristics of Engine

In-Cylinder Pressure Variation

The addition of copper nanoparticle additives to *Calophyllum inophyllum* biodiesel blends revealed that, there was a slight increase in peak cylinder pressure compared to neat biodiesel blends. An increase of 1.88, 1.70 and 1.42% was obtained for B10, B20 and B100 biodiesel blends with Cu nano additives. This justifies the addition of Cu nano additive in biodiesel blends enhanced the oxidation reaction and led to acceleration in peak cylinder pressure (Tamilvanan et al., 2019). It was observed that, the maximum cylinder pressure of ZnO nano additive with POME and ethanol blends were higher than that of diesel, biodiesel and ethanol blends. This is a result of higher ceteane number and lower ignition delay of biodiesel with additives which lead to rapid premixed combustion phase (Prabakaran & Udhoji, 2016). Addition of Al_2O_3 nano particles with ZJME tends to decrease the ignition delay and improves the combustion. Compared to diesel, the cylinder pressure increases with increasing dosage levels of nanoparticles were noticed. The peak pressure was improved by shortening of the diffusion combustion of nanoparticles blended with biodiesel (Aalam et al., 2015).

Cylinder pressure of LGO nano emulsion fuel was lower than LGO emulsion fuel. This is because of the advancement of premixed combustion zone. At maximum load condition, cylinder pressure of diesel, neat LGO, LGO emulsion and LGO nano emulsion fuel were 71.443 bar, 64.948 bar, 68.50 bar, and 67.084 bar respectively (Annamalai et al., 2016). It was observed that, cylinder peak pressure for D2S5W (93% diesel + 2% surfactant + 5% water) was 75.4 bar compared to 73.5, 71.1 and 69.8 bar for D2S5W25CNT (93% diesel + 2% surfactant + 5% water + 25ppm of CNT), D2S5W50CNT and neat diesel respectively at full load condition. The reduced peak pressure for CNT blended water–diesel emulsion fuels was possibly due to the reduced ignition delay of CNT exist in water–diesel emulsion fuel (Basha & Anand, 2011).

Net Heat Release Rate (NHRR)

The combustion behavior of nano-scale metal and metal oxide particles of aluminium (n-Al and n-Al_2O_3) in ethanol was studied. It was concluded that, the amount of heat released increases linearly with n-Al concentration (above 3%). It was inferred that, heat release rate was insignificant during the addition of n-Al_2O_3 particles in ethanol as they did not contribute to the combustion process. In addition, other parameters such as ignition delay and mass burning rates were influenced by the existences of metal oxide particles (Jones, Li, Afjeh, & Peterson, 2011). The addition of copper nanoparticles to *Calophyllum inophyllum* biodiesel blends exhibited substantial enhancement in heat release rate, mainly 5-6% at 100% load and 4-5% at 75% load when compared to biodiesel blends. The enhancement in net heat release rate (NHRR) of biodiesel blends with copper nano additive was due to the improved oxidation reaction which happens inside the combustion chamber. It was decided that existence of nano copper metal additive would boost the oxidation reaction as well as combustion reaction (Tamilvanan et al., 2019).

It was observed that, heat release of ZnO with POME and ethanol blends was higher than diesel and biodiesel and ethanol blends. This is due to better ignition characteristics owing to higher surface area to volume ratio of ZnO nano particles results in higher oxygen content of biodiesel. Furthermore, the reactivity of oxygen was enhanced at higher temperatures (Prabakaran & Udhoji, 2016). The heat release rate (HRR) was found to be improved with the addition of Al_2O_3 nano particles to Zizipus jujube biodiesel. Maximum heat was released during the combustion due to addition of nanoparticles which accelerates the combustion process further. The tendency of the performance is owing to a higher calorific value of nano-biodiesel (Aalam et al., 2015).

The LGO nano emulsion fuel had lowered HRR compared with LGO emulsion fuel and diesel. The existence of Ce_2O_3 nano particles improves the cetane number and reduces the ignition delay period of LGO nano emulsion fuels. The presence of water molecules in LGO emulsions fuel reduces the cetane number and increases the ignition delay period. At full load, HRR of diesel, neat LGO, LGO emulsion and LGO nano emulsion fuel were 78.5, 63.7, 74, and 67.084 kJ/m^3°C respectively (Annamalai et al., 2016). It was observed that, HRR for D2S5W (water–diesel emulsion fuel) was 40.8 J/°CA compared to 36.6 J/°CA and 37.7 J/°CA for D2S5W50CNT and neat diesel respectively at full load condition. The least HRR for CNT blended water–diesel emulsion fuels was possibly due to reduced ignition delay of CNT in water–diesel emulsion fuel (Basha & Anand, 2011).

Impact of Nano Metal Additives on Emission Characteristics of Engine

The different emission characteristics such as NO_X, HC, CO & CO_2 emissions and smoke opacity for previous works using different nano particles were shown in Table 2.

NO_X Emission

The NO_x emission was reduced for the modified fuels with nanoparticle addition of 25 and 50 ppm SiO_2 and MgO to rape-seed oil trans-esterified with methanol (RME) fuel. A NO_x reduction of 7.2% and 9.4% for SiO_2 and 10.7% and 16.7% for MgO was observed in the presence of 25 and 50 ppm respectively (Özgür et al., 2015). For appending ferrofluid to MOME results in average reduction of NO_X emissions about 9.02%. It is due to the presence of ferrofluid which improves the thermal conductivity and facilitate to reduce the ignition delay (Devarajan et al., 2017). The addition of nano-fuel additives like Al-Mg and Co_3O_4 results in reduction of NO_X emission. Co_3O_4 shows enhanced reduction at all loads compared to Al-Mg additive (Ganesh & Gowrishankar, 2011).

Addition of 50 nm copper nano particles with soybean biodiesel (B10) results in 7.46% reduction of NO_X emissions at all levels of loads compared to pure diesel. It was noticed that, reduce the size of nanoparticles provides more surface area for reaction and reduces the NO_X emissions. A reduction of 16.33% in NO_X was achieved when using B10 with 40 nm copper (Balamurugan et al., 2013). The NO_X emission of biodiesel blends with Cu nano additive was lesser than pure diesel and it was higher than biodiesel blends at maximum load. This was due to the existence of nano-copper additive in biodiesel promotes oxidation reaction. This results in higher combustion chamber temperature and pressure which converts nitrogen to nitric oxide and therefore initiates higher rate of NO_X formation (Tamilvanan et al., 2019).

The NO_X emissions of CuO nano additives with 20MEOM were higher than 20MEOM fuel blends without additive. Because of complete combustion of biodiesel fuel with nano additives increases the

Table 2. The variation of performance and emission characteristics of previous works using different nano particles as fuel additive

Nanoparticles, Dosage & Size	Base Fuel	Engine & Operating condition	Results							Reference
			BTE	BSFC	HC	NO_X	CO	CO_2	Smoke	
Magnetite, 1% vol., 14 nm	mahua	2-cylinder 4S, DI, WC, 1300 rpm @ different loading	Increased by 2.27%	Decreased by 5.11%	Decreased by 16.72%	Decreased by 9.02%	Decreased by 32.06%		Decreased by 14.28%	Devarajan et al., (2017)
SiO_2, 25 ppm, < 30 nm	rapeseed	4-cylinder 4S, DI, WC, 1200–3200 rpm	Increased by 4.2%	-	-	Decreased by7.2%	Decreased by 12%	Increased by 10.7%	-	Özgür et al., (2015)
SiO_2, 50 ppm, < 30 nm			Increased by 4.8%	-	-	Decreased by 9.4%	Decreased by 12.7%	Increased by 10.8%	-	
MgO, 25 ppm, 30 nm			Increased by 6.8%	-	-	Decreased by10.7%	Decreased by 17.4%	Increased by 19.1%	-	
MgO, 50 ppm, 30 nm			Increased by 4.4%	-	-	Decreased by 16.7%	Decreased by 16.9%	Increased by 10%	-	
$Al_2O_3 + CeO_2$, each 30 ppm, 54 & 32 nm	jatropha	1-cylinder 4S, DI, AC, 1500 rpm	Increased by 3-4%	Decreased by 1-2%	Decreased by 33%	Decreased by 13-29%	Decreased by 40-50%	-	Decreased by 30-40%	Prabu, (2017)
TiO2, 100 and 200 ppm, 50 nm	mustard	1-cylinder 4S, DI, WC, 11000 rpm, different loading	-	-	Decreased by 8–13%	Reduced	Decreased by 4.2%	-	Reduced	Yuvarajan et al., (2018)
CuO, 50 ppm, <50 nm	mahua (B20)	1-cylinder 4S, DI, WC, 1500 rpm, different loading	Increased		Decreased by 5-10%	Increased by 2-5%	Decreased by 15-20%	-	Decreased by 15-25%	Chandrasekaran et al., (2016)
ZnO, 250 ppm, <100 nm	Diesel biodiesel ethanol	1-cylinder 4S, DI, WC, 1500 rpm	Increased	Decreased	Decreased	Increased	Decreased	-	Decreased	Prabakaran & Udhoji, (2016)
Fe_2O_3, 100;200 and 300 ppm, <50nm	polanga oil and diesel blend	1-cylinder 4S, DI, WC, 1500 rpm, different loading	slightly Increased	Decreased	Decreased	Decreased	Decreased	Decreased	10–15% lower at 300 ppm	Santhanamuthu et al., (2014)
Al_2O_3, 25 and 50 ppm, <100 nm	zizipus jujube biodiesel	1-cylinder 4S, DI, WC, 1500 rpm, different loading	Increased by 2.5%	Decreased by 6%	Decreased	Increased	Decreased	-	Reduced	Aalam et al., (2015)
CeO_2, 30 ppm, <17 nm	Lemongrass Oil emulsion	1-cylinder 4S, DI, WC, 1500 rpm, different loading	Increased by 17.02%	Decreased slightly	Decreased by 16%	Decreased by 24.8%	Decreased by 15.6%		Decreased by 6%	Annamalai et al., (2016)
Mno and CuO, 200 mg, 1-100 nm	Diesel	1-cylinder 4S, DI, AC, 1500 rpm	Increased by 4%		Decreased by 1%	Decreased by 4%	Decreased by 37%	-	-	Balaji & Cheralathan, (2015)

continues on following page

Table 2. Continued

Nanoparticles, Dosage & Size	Base Fuel	Engine & Operating condition	Results							Reference
			BTE	BSFC	HC	NO_x	CO	CO_2	Smoke	
CeO_2, 20–80 ppm, 10-20 nm	jatropha	1-cylinder 4S, DI, WC, 1500 rpm, different loading	Increased by 1.5%.	Decreased	Decreased by 25–40%	Decreased by 30%	Decreased	-	-	Sajith et al., (2010)
CuO, 40, 80 and 120 ppm, 50 nm	linseed oil	1-cylinder 4S, DI, WC, 1500 rpm, different loading	Increased by 3–4% for 80 ppm	Decreased	Decreased	Decreased at full load	Decreased	-	-	Jayanthi & Rao, (2016)
CuO, Al_2O_3, each 50 ppm, 30 to 50 nm, 27 to 43 nm,	diesel	200–3600 rpm with 250 interval	Increased	Decreased	Decreased	Decreased	Decreased	-	-	Gumus et al., (2016)
Al_2O_3 + CNT, 25 and 50 ppm, 51 nm	jatropha biodiesel	1-cylinder 4S, DI, AC, 1500 rpm, different loading	Maximum for Al_2O_3 + CNT blend	Decreased	Decreased	Minimum for Al_2O_3 + CNT blend	Decreased		Decreased	Basha & Anand, (2013)
MWCNT's, 10–50 ppm	jojoba	1- cylinder, DI, AC 1500, 2000 and 2500 rpm	Increased by 16% at 50 ppm	Decreased by 16% at 50 ppm	Increased by 60% at 20 ppm	Increased by 35% at 20 ppm	Increased by 50% at 20 ppm	-	-	EL-Seesy, Abdel-Rahman, Hassan, Ookawara, & Hawi, (2017)
Al_2O_3, 50 and 100 ppm,	Soyabean (B20 & B30)	1-cylinder 4S, DI, WC, 1500 rpm, different loading	No variation	No variation	-	-	-	-	Decreased at full load	Ghogare & Kale, (2016)
CNT, 25 and 50 ppm, 16 nm	water-diesel emulsion	1-cylinder 4S, DI, AC, 1500 rpm, different loading	Increased	Decreased	Decreased at higher load	Decreased	Decreased with higher level ppm		Decreased	Basha & Anand, (2011)
Al_2O_3, 100mg < 50 nm	Diesel + soya bean oil + ethanol	1-cylinder 4S, DI, AC, 1500 rpm	Increased by 17.9%	Decreased at higher load	Decreased	Increased	Decreased at higher load	-	-	Shaafi & Velraj, (2015)
ZnO, 100, 200 and 300 ppm, 35 nm	ethanol + butanol	1-cylinder 4S, DI, WC, 1500 rpm, different loading and pressure	Increased by 7.9% & 10.8% at 200 ppm & & 300 ppm	-	Decreased 48.5% at full load	Increased	Reduced slightly at full load	-	15.6% & 26.8% lower at 200 ppm & 300 ppm	Prabakaran & Vijayabalan, (2018)
Cu, 30 ppm, 20-40 nm,	Tamanu oil	1-cylinder 4S, DI, WC, 1500 rpm, different loading	Increased	Decreased	Decreased	Increased	Decreased	Decreased	Decreased	Tamilvanan et al., (2019)
Cu, 0.1g 40 and 50 nm	Soyabean biodiesel	1-cylinder 4S, DI, WC, 1500 rpm, different loading	1.01% increased for 40 ppm	-	No variation	Increased by 16% at 20 ppm	No variation	Increased	Decreased	Balamurugan et al., (2013)

AC - Air Cooled; WC –Water Cooled; DI – Direct Injection; 4S – Four Stroke.

combustion chamber temperature and pressure (Chandrasekaran et al., 2016). The presence of ZnO nano additives in POME and ethanol blends increases the calorific value and enhances the average combustion chamber temperature and pressure results in higher NO_X emissions compared to POME and ethanol blends and it was lower than diesel (Prabakaran & Udhoji, 2016). The NO_X emission of nano diesels containing Al_2O_3 and CuO nanoparticles was higher than neat diesel. Oxygenated additives improve the combustion results in higher combustion temperature and consequently increases the NOx emission (Gumus, Ozcan, Ozbey, & Topaloglu, 2016).

The addition of cerium oxide nanoparticles to jatropha biodiesel was found to reduce the NO_X emission, An average reduction of 30% was found with a dosing level of 80 ppm nanoparticles (Sajith et al., 2010). By the addition of Al_2O_3 nanoparticles, increases the diffusion, controlled combustion duration, results in reduction of NO_X emission. The addition of nanoparticles to 25% Zizipus jujube methyl ester (ZJME25) and ZJME50 up to part load results in gradual increase of NO_X emission, and higher NO_X emission was observed at maximum load. Among the blends, 25% of Al_2O_3 shows lower NO_X emission than other concentrations (Aalam et al., 2015).

Lesser NO_X emission was obtained with LGO nano emulsion fuel compared with all other fuel blends, since CeO_2 nanoparticles act as reducing agent which leads to the conversion of NO to N_2. At maximum load condition, compared with LGO and diesel, NO_X emission of LGO nano emulsion was decreased by 24.8% and 23% respectively (Annamalai et al., 2016). NO_X emission of NOME with 100, 200 and 300 ppm of CNT additive was reduced by 2.88%, 7.25% and 4.67% respectively compared to neat biodiesel at full load condition (Balaji & Cheralathan, 2015). For CNT blended water–diesel emulsion fuels, NO_X emission was distinctly reduced compared to neat diesel. The NO_X emission obtained at full load condition for D2S5W50CNT was 970 ppm, while it was 1340 and 1046 ppm for neat diesel and D2S5W respectively (Basha & Anand, 2011).

HC Emission

An average reduction of 16.72% in HC emissions were observed by the addition of ferrofluid to MOME. The variation in HC emissions of MOME with and without ferrofluid seems to vary from 0.033 g/kWh at lower loads to 0.073 g/kWh at higher loads respectively. Ferrofluid additive acts as an oxidation catalyst to reduce the carbon combustion activation temperature and improve HC oxidation (Devarajan et al., 2017). At 75% load, HC emission was reduced to 83% for Jatropha biodiesel with Co_3O_4 nano fuel additive compared to neat to Jatropha biodiesel. Because additive acts as oxygen buffer and donates its lattice oxygen and avoids fuel rich zone. Similar reduction up to 70% was observed with Al-Mg nano fuel additive at 50% load (Ganesh & Gowrishankar, 2011). The addition of copper nanoparticles with biodiesel results in additional reduction of HC emission where nano additives serve as oxidizing catalyst which supports complete combustion (Tamilvanan et al., 2019).

The substantial reduction of HC emissions was obtained for jatropha oil with each 30 ppm of Al_2O_3 and CeO_2 (Prabu, 2017). The catalytic activity of ZnO nano additives with POME and ethanol blends results in higher cylinder temperature at higher loads. This leads to more complete combustion occurred and reduces the HC emissions (Prabakaran & Udhoji, 2016). The amount of HC emission for Al_2O_3 nanoparticles diesel nanofuels was lower compared to CuO nanodiesel and neat diesel. This is due to condense the ignition delay and better ignition characteristics of Al_2O_3 nanoparticles (Gumus et al., 2016). HC emission of jatropha with CeO_2 was significantly reduced. The oxygen presented in CeO_2

helped to reduce the HC as well as the soot. A reduction of 25% to 40% in HC emissions was obtained for 40 to 80 ppm dosing of CeO_2 additive (Sajith et al., 2010).

HC was reduced significantly with the addition of Al_2O_3 nanoparticles to Zizipus jujube biodiesel. It was noticed that, HC emission reduced with increasing dosing level of Al_2O_3 nanoparticles in biodiesel (Aalam et al., 2015). HC emission was reduced for LGO nano emulsion compared with all other fuel blends, since cerium oxide nanoparticle performed as oxygen buffer. At full load condition, HC emission of LGO nano emulsion decreased by 16.03% and 35.5% compared with LGO and diesel fuels respectively (Annamalai et al., 2016).

CO Emission

A reduction of CO emission was noticed by the addition of 25 and 50 ppm dosage of SiO_2 and MgO nanoparticles with rapeseed methyl ester. The maximum CO emission reduction of 12% and 12.7% for modified fuels with SiO_2 and reduction of 17.4% and 16.9% with MgO nanoparticles in dosage level of 25 and 50 ppm was obtained respectively. The average CO emission reduction of 10.4% and 7.8% for the modified fuels with SiO_2 and reduction of 4% and 1% with MgO nanoparticles in dosage level of 25 and 50 ppm was obtained respectively (Özgür et al., 2015). Addition of ferrofluid to MOME, results in reduction of 32.06% CO emission. The difference in CO emissions of MOME with and without ferrofluid seems to vary from 0.13 g/kWh at lower loads to 1.55 g/kWh at higher loads. This was caused by its high surface-to-volume ratio, which activates the catalytic activity (Devarajan et al., 2017)

A lesser CO emission about 66% was noticed by adding Al-Mg at 50% load. Also 50% reduction at 75% load was noticed by adding Co_3O_4 additive because of the oxygen buffer of additives (Ganesh & Gowrishankar, 2011). Biodiesel with nano copper additives does not have any significant reduction in CO emission (Balamurugan et al., 2013; Tamilvanan et al., 2019). The reduction in CO emission was observed for B20 and B100 (jatropha oil) with each 30 ppm of Al_2O_3 and CeO_2. This is because of better spray atomization and quick evaporation of nano fuels (Prabu, 2017). CO emission for 20MEOM with CuO nano additive fuel blends was lower than normal biodiesel blend. In common CI engines work with a lean mixture and therefore CO emissions would be lower. This could be possibly by high catalytic activity of nano particles due to their higher surface to volume ratio and improves the mixing rate of air and fuel (Chandrasekaran et al., 2016).

For blends, ZnO nano additives with POME and ethanol, CO emissions were lesser than diesel and blends without additives at all loads. Therefore, addition of ZnO nano particles promotes the combustion and reduces the CO emissions (Prabakaran & Udhoji, 2016). CO emission decreases with addition of Al_2O_3 and CuO nanoparticles with neat diesel. In combustion chamber, fuel propagation may have affected by nanoparticles. Compared to CuO nanoparticles fuel blends, Al_2O_3 nanoparticles fuel blends further decreases the CO emission. This phenomenon is caused by the addition of Al_2O_3 results in more reduction of ignition delay time. The metal oxide nano additive acts as an oxygen donating catalyst, and provides oxygen for the oxidation of HC and CO (Gumus et al., 2016).

The existence of ferric oxide nanoparticles with polanga biodiesel results in reduced CO emission. This is because of catalyzing the combustion process which results in complete combustion (Santhanamuthu et al., 2014). Al_2O_3 nanoparticles blended fuels shows accelerated combustion because of the shortened ignition delay. Because of decreased ignition delay, fuel–air mixing was improved and uniformity in burning was occurred. As a result, there was a substantial reduction in CO emissions for Al_2O_3 blended biodiesel (Aalam et al., 2015). At full load condition, compared with LGO and diesel fuel, CO emission of LGO Nano emulsion decreased by 15.7% and 26% respectively, since cerium oxide nanoparticles act as oxidation catalyst which leads towards the complete conversion of CO to CO_2 (Annamalai et al., 2016).

CO_2 Emission

The CO_2 emissions values were increased for RME fuel with SiO_2 and MgO nanoparticles at the dosage of 25 and 50 ppm. The maximum increase in CO_2 emission about 10.7% and 10.8% for the modified fuels with SiO_2 and reduction of 19.1% and 10% with MgO nanoparticles in dosage level of 25 and 50 ppm respectively. The average increase in CO_2 emission about 2% and 2.7% for the modified fuels with SiO_2 and reduction of 7% and 1.6% with MgO nanoparticles in dosage level of 25 and 50 ppm respectively (Özgür et al., 2015). Polanga oil with ferric oxide nanoparticles results in lesser CO_2 emission. This due to the presence of oxygen content in the polanga oil and ferric oxide nanoparticles perform as catalyst for combustion of hydrocarbons (Santhanamuthu et al., 2014). Compared to diesel, a small increase in CO_2 emission was observed whereas using soybean biodiesel (B10) with nano-copper additive. This is due to the formation of copper oxide during the combustion process, which acts as oxygen buffer for carbon soot (Balamurugan et al., 2013).

A substantial drop in CO_2 emissions level of biodiesel with the addition of Cu nanoparticles was noticed. This was due to the catalytic action of nanoparticles, which improves the combustion characteristics of the fuels. The variation was large at high loads as a result of rapid oxidation reaction (Tamilvanan et al., 2019). CuO nano particles appropriately blended with mahua biodiesel blends promotes the enhanced combustion leading to elevated CO_2 emission than biodiesel without nano additive blends (Chandrasekaran et al., 2016). CO2 emission of polanga oil with ferric oxide nano particles was found to be slightly higher than neat diesel. This is due to the presence of oxygen content in the polanga oil and catalytic action of ferric oxide nanoparticles in the combustion of hydrocarbons (Santhanamuthu et al., 2014).

Smoke

An average reduction of 14.28% in smoke was noticed by adding ferrofluid to MOME. The addition of ferrofluid to MOME was resulted in better evaporation and ignition characteristics and shortened ignition delay (Devarajan et al., 2017). The addition of copper nano particle results in considerable reduction of smoke levels (Balamurugan et al., 2013). Compared to conventional diesel, the smoke opacity of biodiesel blends and copper nano additives was appreciably reduced. This is due to the intrinsic oxygen content and presence of nano copper additives. This results in better oxidation as well as complete combustion of fuel (Tamilvanan et al., 2019). Jatropha oil was treated with each 30 ppm of Al_2O_3 and CeO_2 nanoparticles results in reduction of 30-40% of smoke opacity when compared to normal biodiesel blends. This is owing to the better combustion characteristics of nanoparticles and promoting improved air fuel mixing inside the combustion chamber (Prabu, 2017).

The smoke of CuO nano additive with mahua fuel blends was lower than mahua fuel blends without additive. Due to the higher surface to volume ratio and shortened ignition delay, perfect air-fuel mixture, improved ignition characteristics and quick evaporation rate of nano additives were happened (Chandrasekaran et al., 2016). The smoke of ZnO nano additives with POME and ethanol blend was lower than that of diesel. This is due to better combustion and atomization of fuel in the combustion chamber (Prabakaran & Udhoji, 2016).

Increase in ferric oxide nanoparticles dosage in polanga oil biodiesel blends considerably reduces 10 – 15% of smoke density when compare to neat diesel. This is due to the presence of oxygen in polanga oil and ferric oxide in fuel blends (Santhanamuthu et al., 2014). The addition of 50 ppm Al_2O_3 nanoparticles considerably reduces smoke density for biodiesel blends with respect to all loading condition. This could be endorsed by presence of oxygen in the nanoparticles, which manipulate the reduction of smoke density. The reduction in smoke intensity of Al_2O_3 nanoparticles blended fuel was about 15–20% especially at maximum load (Aalam et al., 2015).

The LGO nano emulsion exhibited reduced smoke opacity compared with all other fuel blends, since CeO_2 nanoparticles improves the premixed combustion phase and it promotes rapid evaporation rate and better air-fuel mixing. At full load condition, the smoke opacity of LGO nano emulsion was decreased by 6.4% and 19.8% compared with LGO and diesel fuels respectively (Annamalai et al., 2016). The smoke opacity of CNT blended water–diesel emulsion fuels was lower than water–diesel emulsion fuel and neat diesel. Because of shorter ignition delay, enhanced air–fuel mixture was obtained. Because of this effect smoke emission was decreased for CNT blended water–diesel emulsion fuels (Basha & Anand, 2011).

Particulate Matter (PM)

Generally, metal based additive decreases the oxidation temperature which results in reduction of PM. The encouraging impacts of metal-based additives are normally connected with improving soot oxidation. During the reaction of metal additives with water molecules hydroxyl radicals were generated which oxidizes the existing soot. Metal additives acts as catalyst in reducing the temperature necessary for the carbon oxidation exists in the soot (Hosseinzadeh-Bandbafha et al., 2018). There was no significant variation in PM for Tall oil methyl ester (TOME) with Mg based additive at a speed of 2400 rpm (Keskin, Gürü, & Altıparmak, 2008). At full load condition, 54.6% of PM was reduced for ZnO metal additive with POME (Prabakaran & Udhoji, 2016). A reduction of 14.28% in PM was obtained for MOME in the presence of ferrofluid nano additive (Devarajan et al., 2017).

Impact of Nano Metal Additives on Tailpipe Emission and Engine Components

The presence of metal-based additives in exhaust of tailpipe was normally associated with oxidation of soot by two different mechanisms. First, the metal additives react with water molecules results in formation of hydroxyl radicals which consecutively oxidize the existing soot in the tailpipe. The addition of nanoparticles significantly reduces the friction between contact surfaces of the engine parts (Khan, Dewang, Raghuwanshi, & Sharma, 2019). During the combustion of biodiesel blends, sulfur in diesel reacts with additives and form sulfur oxides, which subsequently joined with moisture present in air and fuel of the intake system to generate sulfuric and sulphurous acids. These acids have an effect on piston rings, cylinder walls, and damages the engine bearings and exhaust valve guides (Soudagar et al., 2018). Even though catalytic converter and diesel particulate filter could be the solution to the issues related to naoparticle suspension in engine tailpipe emissions (Pandiaraj, Subbaiyan, Ayyasamy, & Nagarajan, 2019).

CONCLUSION

The present chapter provides an extensive description of enhancing the performance, combustion and emission characteristics of various biodiesel fuels and diesel with the application of various nanoparticles. The low-cost and trouble-free fuel reformulation method called nanofuels which is an appropriate way of making reliable and sustainable biodiesel for engine without any modifications in engine. The nanoparticles fuel additive in the fuel blends have greater potential to improve the stability and combustion aspects of various biodiesel blends. Metal and metal oxide nanoparticles like copper, alumina, carbon nano tubes, aluminium oxide, copper oxide, silicon dioxide, magnesium oxide, cerium oxide, zinc oxide and titanium dioxide are extensively used as nanofuel additive which improves the inbuilt properties of biodiesel which improves the performance as well as reduce the exhaust emissions of CI engine. Metal-based additives possibly diminish NO_x and smoke emissions connected with the biodiesel combustion and its blends by improving the cetane number. These additives could also reduce HC and CO emissions by enhancing the rate of fuel evaporation and oxidation reactions. Generally, selecting a suitable nano-additive for various biodiesel is the best possible solution to enhance the operating characteristics of CI engine without any engine modification.

REFERENCES

Aalam, C. S., Saravanan, C., & Kannan, M. (2015). Experimental investigations on a CRDI system assisted diesel engine fuelled with aluminium oxide nanoparticles blended biodiesel. *Alexandria engineering journal, 54*(3), 351-358.

Annamalai, M., Dhinesh, B., Nanthagopal, K., SivaramaKrishnan, P., Lalvani, J. I. J., Parthasarathy, M., & Annamalai, K. (2016). An assessment on performance, combustion and emission behavior of a diesel engine powered by ceria nanoparticle blended emulsified biofuel. *Energy Conversion and Management*, *123*, 372–380. doi:10.1016/j.enconman.2016.06.062

Ayyasamy, T., Balamurugan, K., & Duraisamy, S. (2018). Production, performance and emission analysis of Tamanu oil-diesel blends along with biogas in a diesel engine in dual cycle mode. *International Journal of Energy Technology and Policy*, *14*(1), 4–19. doi:10.1504/IJETP.2018.088279

Balaji, G., & Cheralathan, M. (2015). Effect of CNT as additive with biodiesel on the performance and emission characteristics of a DI diesel engine. *International Journal of Chemtech Research*, *7*(3), 1230–1236.

Balamurugan, K., Tamilvanan, A., Anbarasu, M., Akil, M., & Srihari, S. (2013). Nano-copper additive for reducing NOx emission in soya bean biodiesel-fuelled CI engine. *Journal of Biofuels*, *4*(1), 1–8. do i:10.5958/j.0976-4763.4.1.001

Basha, J. S., & Anand, R. (2011). An experimental investigation in a diesel engine using carbon nanotubes blended water–diesel emulsion fuel. *Proceedings of the Institution of Mechanical Engineers. Part A, Journal of Power and Energy*, *225*(3), 279–288. doi:10.1177/2041296710394247

Basha, J. S., & Anand, R. (2013). The influence of nano additive blended biodiesel fuels on the working characteristics of a diesel engine. *Journal of the Brazilian Society of Mechanical Sciences and Engineering*, *35*(3), 257–264. doi:10.100740430-013-0023-0

Basu, S., & Miglani, A. (2016). Combustion and heat transfer characteristics of nanofluid fuel droplets: A short review. *International Journal of Heat and Mass Transfer*, *96*, 482–503. doi:10.1016/j.ijheatmasstransfer.2016.01.053

Chandrasekaran, V., Arthanarisamy, M., Nachiappan, P., Dhanakotti, S., & Moorthy, B. (2016). The role of nano additives for biodiesel and diesel blended transportation fuels. *Transportation Research Part D, Transport and Environment*, *46*, 145–156. doi:10.1016/j.trd.2016.03.015

Devarajan, Y., Munuswamy, D. B., & Mahalingam, A. (2017). Performance, combustion and emission analysis on the effect of ferrofluid on neat biodiesel. *Process Safety and Environmental Protection*, *111*, 283–291. doi:10.1016/j.psep.2017.07.021

Dinesh, K., Tamilvanan, A., Vaishnavi, S., Gopinath, M., & Mohan, K. R. (2019). Biodiesel production using Calophyllum inophyllum (Tamanu) seed oil and its compatibility test in a CI engine. *Biofuels*, *10*(3), 347–353. doi:10.1080/17597269.2016.1187543

EL-Seesy. A. I., Abdel-Rahman, A. K., Hassan, H., Ookawara, S., & Hawi, M. (2017). *Investigation of the performance of a diesel engine fueled by biodiesel-diesel fuel mixture with addition of nanoparticles.* Paper presented at the ASME 2017 Power Conference Joint With ICOPE-17 collocated with the ASME 2017 11th International Conference on Energy Sustainability, the ASME 2017 15th International Conference on Fuel Cell Science, Engineering and Technology, and the ASME 2017 Nuclear Forum.

Fuskele, V., & Sarviya, R. (2017). Recent developments in nanoparticles synthesis, preparation and stability of nanofluids. *Materials Today: Proceedings*, *4*(2), 4049–4060.

Ganesh, D., & Gowrishankar, G. (2011). *Effect of nano-fuel additive on emission reduction in a biodiesel fuelled CI engine.* Paper presented at the 2011 International conference on electrical and control engineering. 10.1109/ICECENG.2011.6058240

Ghogare, P., & Kale, N. (2016). Experimental investigation on single cylinder diesel engine fuelled with soya bean biodiesel blends with nano additives. *International Journal of Pure and Applied Research in Engineering and Technology*, *4*(9), 247–257.

Gumus, S., Ozcan, H., Ozbey, M., & Topaloglu, B. (2016). Aluminum oxide and copper oxide nanodiesel fuel properties and usage in a compression ignition engine. *Fuel*, *163*, 80–87. doi:10.1016/j.fuel.2015.09.048

Gumus, S., Ozcan, H., Ozbey, M., & Topaloglu, B. (2016). Aluminum oxide and copper oxide nanodiesel fuel properties and usage in a compression ignition engine. *Fuel*, *163*, 80–87. doi:10.1016/j.fuel.2015.09.048

Gupta, H., Agrawal, G., & Mathur, J. (2012). An overview of Nanofluids: A new media towards green environment. *International Journal of Environmental Sciences (China)*, *3*(1), 433–440.

Hosseinzadeh-Bandbafha, H., Tabatabaei, M., Aghbashlo, M., Khanali, M., & Demirbas, A. (2018). A comprehensive review on the environmental impacts of diesel/biodiesel additives. *Energy Conversion and Management*, *174*, 579–614. doi:10.1016/j.enconman.2018.08.050

Jayanthi, P., & Rao, M. S. (2016). Effects of nanoparticles additives on performance and emissions characteristics of a DI diesel engine fuelled with biodiesel. *International Journal of Advances in Engineering and Technology*, *9*(6), 689.

Jones, M., Li, C. H., Afjeh, A., & Peterson, G. (2011). Experimental study of combustion characteristics of nanoscale metal and metal oxide additives in biofuel (ethanol). *Nanoscale Research Letters*, *6*(1), 246. doi:10.1186/1556-276X-6-246 PMID:21711760

Keskin, A., Gürü, M., & Altıparmak, D. (2008). Influence of tall oil biodiesel with Mg and Mo based fuel additives on diesel engine performance and emission. *Bioresource Technology*, *99*(14), 6434–6438. doi:10.1016/j.biortech.2007.11.051 PMID:18164614

Khalife, E., Tabatabaei, M., Demirbas, A., & Aghbashlo, M. (2017). Impacts of additives on performance and emission characteristics of diesel engines during steady state operation. *Progress in Energy and Combustion Science*, *59*, 32–78. doi:10.1016/j.pecs.2016.10.001

Khan, S., Dewang, Y., Raghuwanshi, J., & Sharma, V. (2019). Nanoparticles exceptional properties: Applications in internal combustion engines. In *AIP Conference Proceedings*. 10.1063/1.5098700

Kim, H. J., Suh, H. K., Park, S. H., & Lee, C. S. (2008). An Experimental and Numerical Investigation of Atomization Characteristics of Biodiesel, Dimethyl Ether, and Biodiesel−Ethanol Blended Fuel. *Energy & Fuels*, *22*(3), 2091–2098. doi:10.1021/ef700692w

Kumar, S., Dinesha, P., & Bran, I. (2017). Experimental investigation of the effects of nanoparticles as an additive in diesel and biodiesel fuelled engines: A review. *Biofuels*, 1–8. doi:10.1080/17597269.2017.1358948

Lenin, M., Swaminathan, M., & Kumaresan, G. (2013). Performance and emission characteristics of a DI diesel engine with a nanofuel additive. *Fuel*, *109*, 362–365. doi:10.1016/j.fuel.2013.03.042

Mehta, R. N., Chakraborty, M., & Parikh, P. A. (2014). Nanofuels: Combustion, engine performance and emissions. *Fuel*, *120*, 91–97. doi:10.1016/j.fuel.2013.12.008

Nanthagopal, K., Ashok, B., Tamilarasu, A., Johny, A., & Mohan, A. (2017). Influence on the effect of zinc oxide and titanium dioxide nanoparticles as an additive with Calophyllum inophyllum methyl ester in a CI engine. *Energy Conversion and Management*, *146*, 8–19. doi:10.1016/j.enconman.2017.05.021

Özgür, T., Özcanli, M., & Aydin, K. (2015). Investigation of nanoparticle additives to biodiesel for improvement of the performance and exhaust emissions in a compression ignition engine. *International Journal of Green Energy*, *12*(1), 51–56. doi:10.1080/15435075.2014.889011

Pandiaraj, S., Subbaiyan, D., Ayyasamy, T., & Nagarajan, S. (2019). Design, Development, and Analysis of Mullite Catalytic Converter for CI Engines: SAE Technical Paper.

Prabakaran, B., & Udhoji, A. (2016). Experimental investigation into effects of addition of zinc oxide on performance, combustion and emission characteristics of diesel-biodiesel-ethanol blends in CI engine. *Alexandria Engineering Journal, 55*(4), 3355-3362.

Prabakaran, B., & Vijayabalan, P. (2018). Influence of zinc oxide nano particles on performance, combustion and emission characteristics of butanol-diesel-ethanol blends in DI CI engine. In *Proceedings IOP Conference Series: Materials Science and Engineering*. 10.1088/1757-899X/377/1/012069

Prabu, A. (2017). *Nanoparticles as additive in biodiesel on the working characteristics of a DI diesel engine*. Ain Shams Engineering Journal.

Sajith, V., Sobhan, C., & Peterson, G. (2010). Experimental investigations on the effects of cerium oxide nanoparticle fuel additives on biodiesel. *Advances in Mechanical Engineering, 2*, 581407. doi:10.1155/2010/581407

Sakthivel, R., Ramesh, K., Purnachandran, R., & Shameer, P. M. (2018). A review on the properties, performance and emission aspects of the third generation biodiesels. *Renewable & Sustainable Energy Reviews, 82*, 2970–2992. doi:10.1016/j.rser.2017.10.037

Santhanamuthu, M., Chittibabu, S., Tamizharasan, T., & Mani, T. (2014). Evaluation of CI engine performance fuelled by diesel-polanga oil blends doped with iron oxide nano particles. *International Journal of Chemtech Research, 6*(2), 1299–1308.

Saraee, H. S., Jafarmadar, S., Taghavifar, H., & Ashrafi, S. (2015). Reduction of emissions and fuel consumption in a compression ignition engine using nanoparticles. *International Journal of Environmental Science and Technology, 12*(7), 2245–2252. doi:10.100713762-015-0759-4

Saxena, V., Kumar, N., & Saxena, V. K. (2017). A comprehensive review on combustion and stability aspects of metal nanoparticles and its additive effect on diesel and biodiesel fuelled CI engine. *Renewable & Sustainable Energy Reviews, 70*, 563–588. doi:10.1016/j.rser.2016.11.067

Shaafi, T., & Velraj, R. (2015). Influence of alumina nanoparticles, ethanol and isopropanol blend as additive with diesel–soybean biodiesel blend fuel: Combustion, engine performance and emissions. *Renewable Energy, 80*, 655–663. doi:10.1016/j.renene.2015.02.042

Soni, G. S., Rathod, P. P., & Goswami, J. J. (2015). Performance and emission characteristics of CI engine using diesel and biodiesel blends with nanoparticles as additive-A review study. *International Journal of Engineering Development and Research, 3*(4), 879–884.

Soudagar, M. E. M., Nik-Ghazali, N.-N., Kalam, M. A., Badruddin, I., Banapurmath, N., & Akram, N. (2018). The effect of nano-additives in diesel-biodiesel fuel blends: A comprehensive review on stability, engine performance and emission characteristics. *Energy Conversion and Management, 178*, 146–177. doi:10.1016/j.enconman.2018.10.019

Suresh, S., Venkitaraj, K., & Selvakumar, P. (2011). *Synthesis, Characterisation of Al_2O_3-Cu Nano composite powder and water based nanofluids*. Paper presented at the Advanced Materials Research.

Tamilvanan, A., Balamurugan, K., Ponappa, K., & Kumar, B. M. (2014). Copper nanoparticles: Synthetic strategies, properties and multifunctional application. *International Journal of Nanoscience*, *13*(02). doi:10.1142/S0219581X14300016

Tamilvanan, A., Balamurugan, K., Ponappa, K., & Madhan Kumar, B. (2016). Using response surface methodology in synthesis of ultrafine copper nanoparticles by electrolysis. *International Journal of Nanoscience, 15*(01n02).

Tamilvanan, A., Balamurugan, K., & Vijayakumar, M. (2019). Effects of nano-copper additive on performance, combustion and emission characteristics of Calophyllum inophyllum biodiesel in CI engine. *Journal of Thermal Analysis and Calorimetry*, *136*(1), 317–330. doi:10.100710973-018-7743-4

Tomar, M., & Kumar, N. (2019). Influence of nanoadditives on the performance and emission characteristics of a CI engine fuelled with diesel, biodiesel, and blends–a review. *Energy Sources. Part A, Recovery, Utilization, and Environmental Effects*, 1–18. doi:10.1080/15567036.2019.1623347

Venu, H., Subramani, L., & Raju, V. D. (2019). Emission reduction in a DI diesel engine using exhaust gas recirculation (EGR) of palm biodiesel blended with TiO2 nano additives. *Renewable Energy*, *140*, 245–263. doi:10.1016/j.renene.2019.03.078

Ying, W., Longbao, Z., & Hewu, W. (2006). Diesel emission improvements by the use of oxygenated DME/diesel blend fuels. *Atmospheric Environment*, *40*(13), 2313–2320. doi:10.1016/j.atmosenv.2005.12.016

Yuvarajan, D., & Babu, M. D. (2018). Experimental investigation on the influence of titanium dioxide nanofluid on emission pattern of biodiesel in a diesel engine. *Atmospheric Pollution Research*, *9*(1), 47–52. doi:10.1016/j.apr.2017.06.003

Zhu, D., Li, X., Wang, N., Wang, X., Gao, J., & Li, H. (2009). Dispersion behavior and thermal conductivity characteristics of Al_2O_3–H_2O nanofluids. *Current Applied Physics*, *9*(1), 131–139. doi:10.1016/j.cap.2007.12.008

Chapter 6
Simultaneous Reduction of NO_x and Smoke Emissions in Dual Fuel and HCCI Engines Operated on Biogas

Feroskhan M.
https://orcid.org/0000-0003-0974-9192
Vellore Institute of Technology, Chennai, India

Saleel Ismail
https://orcid.org/0000-0002-5849-4571
National Institute of Technology Calicut, Kozhikode, India

ABSTRACT

Biogas has emerged as a promising alternative to fossil fuels in internal combustion engines in recent times. It could be used as the primary fuel in Compression Ignition (CI) engines in combination with a small quantity of a high cetane fuel in two modes – dual fuel or Homogeneous Charge Compression Ignition (HCCI). This chapter compares the performance, combustion, and emission parameters of a CI engine operated with biogas in dual fuel and HCCI modes vis-à-vis conventional diesel operation. The effects of biogas composition (quantified in terms of the methane content), location of secondary fuel injection and engine load are investigated. It is observed that the use of biogas has the potential to reduce both NO_x and smoke emissions simultaneously, with HCCI mode offering ultra-low emissions. Operating the engine in dual fuel mode can provide high thermal efficiency and significant diesel substitution.

INTRODUCTION

Major fuel resources like such as coal, fossil fuels, nuclear fuel and natural gas are not renewable. Their fast decline, consequent price rise, environmental concerns, increased worldwide energy demand and emission norms have motivated the search for renewable, alternative energy sources such as solar, biofuels

DOI: 10.4018/978-1-7998-2539-5.ch006

and wind. Furthermore, petroleum reserves are mainly located in certain geographical regions. Countries in other areas face a serious problem in procuring fuel supplies to meet the ever-increasing energy demand. The combustion of fossil fuels also leads to pollution, acid rain, carbon dioxide build-up etc., thereby placing people and the environment at risk. Diesel fuelled Compression Ignition (CI) engines generate substantial amounts of NO_x and smoke emissions. Among alternative fuels, biofuels such as vegetable oils, bio-alcohols, biogas, and bio-diesels have earned significant attention because of their renewability and intrinsic ability to reduce net carbon dioxide (CO_2) emissions. Using biogas addresses three major challenges – harnessing methane, a powerful greenhouse gas that emanates from the decomposition of biomass, efficient disposal of biological waste and tapping a renewable power source. Biogas provides a number of benefits over other biomass-derived fuels (Tippayawong & Thanompongchart, 2010). Biogas burns quicker than solid fuels such as coal, is more eco-friendly and does not leave residues behind. Biogas is formed from biomass, which in turn is derived from vegetation, which absorb CO_2 during their lifetime. Thus biogas can be regarded as a CO_2-neutral fuel despite CO_2 emissions during combustion (Demirbas, 2004). Compared to other biofuels such as bio-diesels and bio-alcohols, biogas production also needs less handling effort and cost (Demirbas, Balat, & Balat, 2011).

Biogas has great potential for developing economies with rural background. India, for instance, had a cattle headcount of approximately 30 crores in 1996, about one-fifth of the global total. The annual production of dung in 1996 amounted to 980 million tons, capable of generating nearly 63.8 trillion liters of biogas, which could provide 1.3 trillion MJ of electricity (Vijay, 2007). In 2012, the cattle population increased to 512 million (Dept.of Animal Husbandry Dairying & Fisheries, 2012), showing a corresponding increase in biogas energy potential to 2.2 trillion MJ. Because of the variation in the composition of biomass, some characteristics of biogas samples, e.g. fractions of constituent elements and ignition temperature, differ from sample to sample (Demirbas, 2004).

The present work is an attempt to identify the most effective modes of operating a CI engine with biogas as the primary fuel. The impact of methane fraction of biogas and location of secondary fuel injection on engine performance, combustion and emissions are explored for the full load spectrum.

BIOGAS PRODUCTION

Different waste-to-energy (WTE) techniques have been explored to convert biological waste, such as industrial and municipal solid waste (MSW), into biogas (Demirbas et al., 2011; Gunaseelan, 1997). These techniques can be divided into four broad groups: a) hydrogenation, b) pyrolysis, c) gasification, and d) bio-conversion. A digester is one of the most common means of generating biogas, where bio-mass is subjected to anaerobic digestion. Anaerobic digestion is a three-stage method of bioconversion involving hydrolysis, acid formation and methane generation. The biogas thus produced is mainly a combination of CH_4, CO_2 and H_2S (Khoiyangbam, Gupta, & Kumar, 2011). Some important properties of biogas are presented in Table 1. The cost of methane generation by anaerobic digestion is similar to those of other sources of biomass energy such as synthesis gases and ethanol (Chynoweth, Owens, & Legrand, 2000). There are different types of anaerobic digesters available, e.g. floating drum and fixed dome designs (Khoiyangbam et al., 2011).

Biogas typically has 25 - 40% carbon dioxide by volume (Khoiyangbam et al., 2011). It can react with water to form corrosive carbonic acid, which can potentially damage pipelines and machinery. As carbon dioxide is non-combustible, it lowers the calorific value and energy density of the fuel. Compared

Table 1. Properties of biogas (Feroskhan & Ismail, 2017)

Composition (% by volume) (Ibrahim, Narasimhan, & Ramesh, 2015)	Methane (CH_4): 50 – 70 Carbon dioxide (CO_2): 25 – 50 Hydrogen (H_2): 1 – 5 Nitrogen (N_2): 0.3 – 3 Water vapour (H_2O) ~ 0.3 Hydrogen sulphide (H_2S) in traces
Self-ignition temperature (K) * (Bora, Saha, Chatterjee, & Veer, 2014)	1087
Flame speed (m/s) (Barik & Murugan, 2014a)	21
Density (kg/m^3) at 1 atm, 288 K * (Bora et al., 2014)	0.91
Stoichiometric air-fuel ratio (Cacua, Amell, & Cadavid, 2012)	6.05
Research octane number (Ibrahim et al., 2015)	130
Flammability limit (vol. % in air) (Barik & Murugan, 2014a)	7.5 – 11.7
Calorific value (MJ/kg) * (Bora et al., 2014)	20.67

*: for biogas with CH_4 and CO_2 content of 60% and 40% respectively.

to pure methane, the presence of CO_2 also results in decreased flame speed and flammability range. Due to its high auto-ignition temperature, biogas offers resistance to knocking in Spark Ignition (SI) engines. But this adversely affects ignition in CI engines. One feasible way to overcome these issues is to remove CO_2 from biogas, thus improving the methane fraction (Ryckebosch, Drouillon, & Vervaeren, 2011). This is called methane enrichment.

PURIFICATION OF BIOGAS

Various techniques exist to purify biogas by separating out the unwanted species, viz. selectively permeable membranes, gas-liquid adsorption membranes, water scrubbing, molecular sieves, chemical reagents and cryogenic separation. The conventionally used water scrubbing technique is a physical separation process in which compressed biogas is fed to the bottom of a packed column and pressurized water is sprayed at the top to form a counter-flow arrangement (Hullu, 2008). CO_2 dissolves at high pressure in water and can subsequently be released by decreasing the pressure, thus allowing water recycling. Bhattacharya, Mishra, & Singh (1988), Khapre (1989) and Dubey (2000) created various scrubbing processes with high CO_2 absorption. Vijay (2007) established a high-pressure bed scrubber system that removes up to 99% of CO_2 at 1 MPa working pressure.

In place of water, organic liquid reagents such as Triethanolamine (TEA), Monoethanolamine (MEA) and Diethanolamine (DEA) may be used as solvents for H_2S and CO_2 absorption; but this raises the price of the device. Likewise, calcium or sodium hydroxide may also be used to absorb CO_2, forming the corresponding carbonate salts in the process (Nijaguna, 2006). Biswas, Kartha, & Pundarikakhadu (1977) proved the efficacy of chemical absorption using the reagent mono-ethanolamine (MEA) which can be recovered by boiling. Savery & Cruzan (1972) used a combination of NaOH, $Ca(OH)_2$ and KOH as the solvent and discovered that agitation-generated turbulence increases absorption capacity.

The method of membrane separation has been in use for several decades (Guha, Majumdar, & Sirkar, 1991; Li & Teo, 1993). The fundamental concept of this technique is that molecules that are small and more soluble permeate faster through the membrane of approximately 1 mm thickness, while other molecules may not be allowed to pass. This process is significantly influenced by the permeability of individual species and difference in partial pressures. For this reason, CO_2 molecules permeate when compressed biogas is passed through a polymer membrane, while methane is retained owing to its large molecular size. This method requires pressure in the range 25–40 bar. By comparing the efficiency of various materials, Basu, Khan, Cano-Odena, Liu, & Vankelecom, (2010) have recommended the use of polyimides, cellulose acetate, silicon polycarbonate, ethyl cellulose, polysulfones, polymethylpentene and polydimethylsiloxane.

Molecular sieve technique, also known as Pressure Swing Adsorption (PSA), operates on the principle of selective attraction of molecules for adsorbent materials. This method comprises high pressure CO_2 adsorption, post-decompression recovery, and adsorption pressure build-up. High-surface porous materials such as silica gel, alumina, zeolites and activated carbon are used in PSA systems as adsorbents (Hullu, 2008). A naturally occurring zeolite called Neapolitan yellow tuff was used in one of the earliest recorded studies, in which the breakthrough curves showed high effectiveness of methane extraction (Pande & Fabiani, 1989).

Low temperatures (~ -90°C) and high pressures (~ 40 bar) are required for cryogenic separation (Hullu, 2008). Here, the CO_2 component is liquefied first and removed from the methane-rich gaseous part. This technique is still in the early phases of research.

One of the latest processes of purifying biogas utilises nano-structures such as Carbon Nanotubes (CNTs). Liu, Cooper, Dai, & Jiang (2012) used molecular dynamics to assess the permeability of CO_2 via CNTs. They observed windowed CNTs to be more efficient compared to normal CNTs. CNTs can also be used in blended matrix membranes as inorganic filler, improving the permeability and yield (Kusworo, Johari, & Widiasa, 2012).

A comparison of the different purification techniques described above is presented in Table 2. Scrubbing and membrane separation do not require the use of unique equipment or chemicals and are therefore the easiest to operate. While the operation of PSA is also simple, in order to replace the catalyst, the plant must be shut down several times a year as the adsorbent material gets degraded gradually by H_2S. The ultra-low temperatures and elevated pressures needed for cryogenic separation, on the other hand, require sophisticated devices, efficient insulation and sealing checks. Membrane systems have important benefits such as modularity and compactness. They can operate under moderate conditions

Table 2. Comparison of biogas purification methods (Feroskhan & Ismail, 2017)

Technique	Benefits	Disadvantages
Cryogenic cooling	High purity	Expensive operation
Membranes	Simple construction and operation Low cost	More CH_4 losses
Molecular sieves	Less energy usage Compactness	More CH_4 losses Expensive operation
Chemical reagents	Low CH_4 losses Efficiency	Expensive operation Prone to corrosion
Water scrubbing	Low CH_4 losses Simple and efficient	Expensive operation and investment High likelihood of clogging

and are energy efficient due to very low consumption of electricity and fuel (Hullu, 2008). Membranes with nano-particles are currently at the research level.

BIOGAS IN CI ENGINES

CI engines offer better performance and lower HC and CO emissions vis-a-vis SI engines. Biogas is generally used in CI engines in two modes – dual fuel and HCCI. These modes are discussed below.

Biogas in Dual Fuel Mode

Here, biogas is mixed with air in the intake manifold and induced into the cylinder(s) of the engine where the mixture is compressed. A small amount of diesel or bio-diesel (called pilot fuel) is injected when the piston is near the top dead center (TDC). The auto-ignition of the pilot spray initiates a flame that passes through the combustion chamber, consuming the intake charge (biogas-air). A comparison of the performance, emission and combustion indices of biogas-operated CI engines in dual fuel mode with standard diesel-only operation is presented below.

Combustion Characteristics

Dual fuel mode demonstrates patterns of performance comparable to those of SI engines (Lounici, Loubar, Tazerout, Balistrou, & Tarabet, 2014). The pilot diesel spray energy discharge is much greater than that of a spark, making it easier to ignite the biogas-air mixture. Because of the CO_2 content, biogas in dual fuel mode of has a longer Ignition Delay (ID) compared to diesel-only mode, resulting in higher Heat Release Rate (HRR) in initial stages. As a result, the maximum cylinder pressure becomes greater and occurs nearer to the TDC with rise in biogas addition (Barik & Murugan, 2014a; Königsson, Stalhammar, & Angstrom, 2011; Yoon & Lee, 2011b). Studies on an indirect injection engine operated with biogas and diesel demonstrated diesel replacement of up to 48%, shorter duration of combustion and consequently higher exhaust temperatures (Duc & Wattanavichien, 2007).

In a research on dual fuel operation with biogas and diesel, the peak net HRR was found to be about 30% higher than in diesel-only mode under comparable load and speed (Mustafi, Raine, & Verhelst, 2013). A higher quantity of pilot fuel can be used to offset the increase in ID. However, it is observed that the ID is almost independent of the amount of diesel injected for pure methane combustion (Polk, Gibson, Shoemaker, Srinivasan, & Krishnan, 2013). Bora et al. (2014) noted that, due to the shorter ID of biogas at elevated temperatures, the amount of pilot fuel provided could be lowered by using high compression ratios (16-18). Oxygen enrichment of the air supplied can also cause reduction in ID. This increases the speed of reaction and the propagation of flames. It has been noted that increasing the oxygen percentage in air from 21 to 27% reduces the ID by almost 3°CA (Cacua et al., 2012). The delay in ignition is smaller for dual fuel engines with thermal barriers compared to conventional dual fuel engines (Yilmaz & Gumus, 2017). Methane enrichment improves the peak pressure and duration of combustion and decreases ID (Verma, Das, & Kaushik, 2017). Bora et al. (2014) noted that the dual fuel mode with biogas and diesel reduces peak cylinder pressure and maximum HRR and reduces the ID relative to normal diesel operation.

Ray (2013) noted that the ID of the pilot fuel increased linearly with the biogas:diesel ratio. Pilot diesel injection amounting to about 10- 20% of the quantity used in conventional diesel mode is enough to operate the engine in dual fuel mode with biogas. Depending on the load, the flow of biogas must be controlled through a gas control valve. It was observed that the control of pilot injection quantity by the engine governor is sufficient to obtain the required performance in governed engines. In such instances, diesel replacement is comparatively small. Comparing diesel-biogas mode with diesel-petrol mode, Park & Yoon (2016) have shown that increasing the proportion of biogas input to the total energy input in the former case results in higher ID relative to diesel-petrol mode.

The study by Königsson ct al. (2011) on a dual fuel biogas-diesel engine has demonstrated that the combustion temperature and efficiency could be increased by advancing CA50, i.e. the crank angle at which the cumulative heat relase is half of the overall value. This expanded the lean limit of the engine as well. By raising the intake temperature, a comparable effect can be obtained. In dual fuel mode, as high as 40% EGR can be used while still enabling up to 95% diesel replacement. EGR decreases the operating limit and the effectiveness of burning. Also, using EGR favors near-stoichiometric operation. This enables a three-way catalyst to be used, making after-treatment more economical. The investigators recommended stoichiometric combustion with EGR and low intake temperature for dual fuel mode. Park, Yoon, Cha, & Lee (2014) studied dual-fuel engine with biogas as primary fuel and dimethyl ether (DME) as the pilot fuel. The contribution of biogas based on energy release ranged from 0 – 80%. Increasing the proportion of biogas induced a drop in HRR and peak cylinder pressure, in addition to causing unstable combustion, manifested as greater COV (Coefficient of Variance). For injection advance greater than 20°bTDC, ID was longer and SOI (Start of Ignition, described as the instance of 10% cumulative energy release) delayed for greater proportions of biogas. However, for retarded injection, both ID and SOI were almost independent of the biogas content.

Performance Characteristics

Bora et al. (2014) and Yoon & Lee (2011a) observed that dual fuel operation causes a fall in thermal efficiency and a rise in BSFC (Brake Specific Fuel Consumption) relative to diesel-only mode. This was due to low combustion temperature, advance pressure peak, low flame velocity and increased pumping required for the CO_2 component. Duc & Wattanavichien (2007) tested a dual-fuel biogas IDI engine with up to 48% diesel replacement. In full load operation, dual fuel and diesel engines showed nearly equivalent energy conversion efficiencies while part load efficiency was inferior for dual fuel mode. Bora et al. (2014) proposed to raise the compression ratio for partly negating the drop in brake thermal efficiency in dual fuel mode. Several researchers (Bedoya, Saxena, Cadavid, Dibble, & Wissink, 2012; Feroskhan & Ismail, 2016; Henham & Makkar, 1998; Kalsi & Subramanian, 2017) used simulated biogas (a formulated blend of methane and carbon dioxide) in dual fuel mode. Sorathia (2012) and Mustafi et al. (2013) reported that the brake thermal efficiency and specific energy consumption of diesel-only mode were closely matched in dual fuel mode. In the latter, the percentage of fuel energy transferred to the coolant was greater, whereas the loss to exhaust was smaller. For the dual fuel mode, exergy efficiency was observed to be greater and percentage exergy destruction smaller. Increasing the oxygen content from 21 to 27% enhanced the thermal brake efficiency from 15 to 18% in a biogas-diesel engine (Cacua et al., 2012). In a biogas-DME engine, Indicated Mean Effective Pressure (IMEP) was noted to decline with increased biogas content for delayed pilot injection, while an inverse trend was observed when injection advance was more than 20° (Park et al., 2014).

Experiments with biogas of various compositions have shown that a 7:3 proportion of methane:CO_2 offers the greatest thermal efficiency (Lounici et al., 2014). This was ascribed to the formation of O_2 and CO by dissociation of CO_2 at high flame temperatures. CO is a quick-burning gas that accelerates the rate of combustion. Additional concentration of oxygen also improves combustion. However, the dilution factor dominates for higher CO_2 concentration, reducing the thermal efficiency. Presenting a review of published literature, Sahoo, Sahoo, & Saha (2009) have reaffirmed that biogas comprising up to 20-30% CO_2 provides smaller BSFC vis-à-vis diesel-natural gas dual fuel mode, while BSFC rises with CO_2 content due to the inert gas effect. Rough engine operation was reported due to uneven combustion for more than 40% CO_2. As the CO_2 content increases, the flow rate of either pilot diesel or biogas could be raised to maintain constant engine speed and power.

Due to air displacement, dual fuel mode demonstrates slightly inferior volumetric efficiency compared to diesel-only mode. This effect is exacerbated by raising the CO_2 fraction. Luijten & Kerkhof (2011) noted that for biogas with 70% CH_4, volumetric efficiency fell by about 4 percentage points when the energy release from the two fuels were equal. They also observed that while biogas with 70% CH_4 can replace up to 55% diesel based on energy release, the replacement was limited to 35% for biogas with 40% CH_4.

Barik & Sivalingam (2013) performed studies on a biogas-diesel engine where the biogas flow rate was manually adjusted from 0 to 0.6 kg/h and the diesel supply rate was regulated by the governor. The maximum diesel replacement was 30% (on energy basis) at full load. Brake specific energy consumption was found to be greater and volumetric efficiency lower in dual fuel mode due to air displacement in the intake and consequent reduction in the rate of combustion. Barik & Murugan (2014a) also observed a decrease in thermal and volumetric efficiencies with increase in the biogas flow rate.

Emission Characteristics

Low temperatures induced by CO_2 in biogas increase HC and CO emissions, whereas particulate matter (PM) and oxides of nitrogen (NO_x) emissions are lower than diesel emissions (Barik & Murugan, 2014a; Nathan, Mallikrajuna, & Ramesh, 2009; Yoon & Lee, 2011b). Barik & Sivalingam (2013) observed that exhaust gas temperatures were less by 2.8%, HC and CO greater by 21% and 16% respectively, soot and NO_x smaller by 41.3% and 35% respectively, relative to diesel-only mode for maximum full-load diesel replacement. CO_2 stayed undissociated at greater CO_2 fractions, functioning as an inert gas and decreasing thermal efficiency. Similar to the effect generated by EGR, NO_x emissions reduce with increased CO_2 content, though HC and CO emissions tend to increase (Verma et al., 2017). CO and HC emissions can be reduced by raising the compression ratio, which increases the in-cylinder temperatures. This, however, creates a significant rise in emissions of CO_2 and NO_x (Bora et al., 2014). By changing the pilot fuel injection amount, the extent of variety of these parameters can be regulated (Mustafi et al., 2013; Polk et al., 2013). Barik & Murugan (2014b) noted that the optimal combination of performance and emissions was achieved using biogas consumption of 0.9 kg/h. This resulted in a substitution of 0.215 kg/h of diesel.

While oxygen enrichment reduces methane emission, it does not have a well-defined effect on CO. By reducing the instabilities during combustion, increasing the oxygen intake enables higher replacement of diesel by biogas (Cacua et al., 2012). EGR reduces NO_x emissions, but also deteriorates the lean operating limit and efficiency of combustion. In lean mixtures, the formation of NO_x can be ascribed to the diesel spray, while hot methane-air combustion is the major source under rich conditions (Königsson et al., 2011).

Biogas-DME dual fuel operation lowered Indicated Specific (IS) NO_x emissions, while ISCO and ISHC emissions increased with biogas supply. There were near zero emissions of soot (Park et al., 2014). In a biogas - Karanja Methyl Ester (KME) dual fuel engine, it was noted that at full load conditions, a biogas flow rate of 0.9 kg/h resulted in nearly 22% substitution of the pilot fuel (Barik & Sivalingam, 2014). The authors also observed that emissions of NO_x and PM for dual fuel operation could be lowered simultaneously, although emissions of CO and HC rise. Optimum timing of pilot injection was revealed as 24.5°CA bTDC with a decrease of 17.1% in CO emissions, 18.2% in HC emissions and 2.1% in smoke emissions relative to the timing of 23°CA bTDC (Barik & Murugan, 2016). Smoke emissions were found to be lower in dual fuel engines with thermal barrier coating (Yilmaz & Gumus, 2017). Increasing the compression ratio and advancing the injection timing in a dual fuel engine with rice-bran bio-diesel as pilot fuel lowered HC and CO emissions, while increasing CO_2 and NO_x (Bora et al., 2014; Bora & Saha, 2015).

Biogas in HCCI Mode

The Homogeneous Charge Compression Ignition (HCCI) mode is an alternative strategy for using biogas in a CI engine. This arose as a successful idea of mixing the advantages of SI and CI engines (Yao, Zheng, & Liu, 2009). By bringing biogas into the intake manifold, biogas and air are homogeneously mixed in the HCCI mode. The fully blended fuel-air mixture auto-ignites as a consequence of increase in temperature during compression. Characterised by lean mixtures and fast heat release rate, HCCI engines deliver high thermal efficiencies. These conditions also guarantee reduced emissions of NO_x and particulate matter (Yao et al., 2009). Despite these advantages, a significant drawback of HCCI engines is that unlike conventional SI or CI engines, the onset of combustion cannot be directly controlled. Various methods of combustion control such as varying compression ratio, pilot fuel injection, charge preheating, spark, exhaust gas recirculation (EGR), turbocharging, fuel reactivity control, fast thermal management etc. have been studied (Christensen & Johansson, 2000; Hyvönen, Haraldsson, & Johansson, 2003; Sankaralingam & Venugopal, 2016).

Combustion Characteristics

Nathan, Mallikarjuna, & Ramesh (2007, 2009 & 2010) have studied the HCCI operation with diesel alone, and with a biogas-diesel combination. They also compared these with biogas-diesel dual fuel operation. In the case of diesel HCCI, the injection timing during suction ranged from 5°bTDC to 20°aTDC. Large HRR values were reported at higher BMEP conditions. Manifold injection was used for both fuels in the biogas-diesel HCCI mode. The high ID is attributed to the high auto-ignition temperature of CH_4 and the inertness of the CO_2 fraction. Compared to diesel HCCI operation, HRR in later stages was noticed to be within safe limits in biogas-diesel HCCI case. Using lower quantities of biogas resulted in knocking, while higher quantities caused misfire.

Biogas-diesel Predominantly Premixed Charge Compression Ignition (PPCCI) mode was explored by Ibrahim et al. (2015). Advanced diesel injection timings were used to attain HCCI-like conditions. The best biogas energy ratio, intake charge temperature and injection timing were found to be 80%, 50-90°C and 55-70° bTDC respectively. Homogeneous lean mixtures resulted in lower energy release rates and delayed combustion. The effects of boost pressure, equivalence ratio and charge temperature in HCCI mode were investigated using a multi-cylinder biogas fuelled engine by Bedoya et al. (2012). Reducing the auto-ignition temperature employing high boost pressure and intake temperature was proposed to improve combustion (Bedoya et al., 2012). Jun, Ishii & Iida (2003) have noted that combustion efficiency can be improved by raising the in-cylinder temperature. Reduction in CH_4 fraction of biogas increases the BMEP without being subject to knock (Rahman & Ramesh, 2017).

Performance Characteristics

Compared to conventional diesel-only mode, due to improper combustion phasing and wall wetting, thermal efficiency is reduced from 30 to 20% at high BMEP in diesel HCCI mode (Nathan, Mallikarjuna, & Ramesh, 2007). While the use of biogas in both SI and CI engines usually provides lower thermal efficiency, high thermal efficiency has been noted in HCCI mode (Nathan, Mallikarjuna & Ramesh, 2010). Nathan, Mallikrajuna & Ramesh (2009) found that biogas-diesel HCCI mode provides lower thermal efficiency relative to dual fuel mode in particular, but the performance can be enhanced by pre-heating the intake mixtures to approximately 135°C and by using greater biogas energy ratios. A biogas energy ratio of 80% was observed to provide the highest thermal efficiency in PPCCI mode (Ibrahim et al., 2015). Indicated thermal efficiency and power output can be improved at high inlet pressure and temperature for lean mixtures (Bedoya, Saxena, Cadavid, & Dibble, 2012). Investigating the use of an ignition improver (diethyl ether or DEE) in an HCCI engine fuelled with biogas, Sudheesh & Mallikarjuna (2010) showed that the addition of DEE broadens the operating range and improves efficiency at all loads relative to dual fuel and SI modes. Bedoya et al. (2012) have experimentally assessed various approaches for extending the operating range of a biogas-based HCCI engine and to ensure stable combustion.

Emission Characteristics

Nathan et al. (2007) pointed out that NO_x emissions are lower in diesel HCCI mode compared to conventional CI operation due to low combustion temperatures. However, the engine emitted more smoke, HC and CO emissions in diesel HCCI mode. Non-homogeneous mixtures resulting from wall wetting and the loss of fuel via the exhaust during valve overlap were cited as the main causes of increased emissions. In subsequent studies, they reported that biogas-diesel HCCI mode has lower smoke and NO_x compared to CI and dual fuel modes (Nathan et al., 2009 & 2010). HC, CO, NO_x and smoke emissions in PPCCI mode were observed to be similar in magnitude to those of dual fuel mode on a brake-specific basis (Ibrahim et al., 2015). CO and HC emissions can be lowered using higher inlet temperature and charge pressure for lean mixtures (Bedoya et al., 2012). Jun et al. (2003) noted that by raising in-cylinder temperature, CO emissions can be lowered in CNG-based HCCI engines. Biogas in HCCI mode generates lower HC emissions than in SI mode (Sudheesh and Mallikarjuna, 2010). Kozlov, Chechet, Matveev, Titova, & Starik (2016) recorded low NO_x emissions from a biogas HCCI engine with n-heptane as ignition promoter. Lower methane content of biogas mitigates HC emissions (Rahman & Ramesh, 2017).

DEE AS A CI ENGINE FUEL

DEE is produced by dehydration of ethanol. Its superior cetane number and energy density make DEE an ideal ignition improver for CI engines (Bailey, Eberhardt, Goguen, & Erwin, 1997). In this study, DEE is used as secondary fuel in HCCI mode. The important properties of DEE are given in Table 3.

Table 3 Properties of DEE (Bailey et al., 1997)

Properties	Value
Formula	$C_2H_5OC_2H_5$
Boiling point (°C)	35
Cetane number	>125
Auto-ignition temperature(°C)	160
Stoichiometric air-fuel ratio	11.1
Density (kg/m^3)	712.5
Viscosity (N.s/m^2)	2.3x10^{-4}
LHV (MJ/kg)	33.9

EXPERIMENTAL SETUP

A conventional water-cooled four-stroke single-cylinder direct injection CI engine was used for this study. The important engine specifications are provided in Table 4. A test rig with biogas induction system was developed to carry out the tests in dual fuel and HCCI mode. The photograph and schematic of the setup are shown in Figure 1. Simulated biogas was prepared by mixing compressed CH_4 (99.5% pure) and CO_2 (99.9% pure), supplied from individual storage cylinders. Each cylinder was equipped with a flow regulator to independently vary the flow rates of the component gases. This allows the user control over

Table 4. Engine specifications

Parameter	Value
Engine	Kirloskar AV1XL model
Bore	0.0875 m
Stroke	0.080 m
Cubic capacity	481 cc
Peak pressure	73.6 bar
Number & arrangement of cylinders	1- Vertical
Working cycle	4-stroke diesel
Compression ratio	17:1
Maximum power	5.97 kW
Maximum torque	25 Nm
Rated speed	2200 rpm
Operating speed range	1500 – 2200 rpm
Fuel injection timing	28° bTDC
Inlet valve opening (IVO)	4.5° bTDC
Inlet valve closing (IVC)	35.5° aBDC
Exhaust valve opening (EVO)	35.5° bBDC
Exhaust valve closing (EVC)	4.5° aTDC

the composition as well as flow rate of simulated biogas. The extent of purity of biogas was represented by its methane fraction (by volume). Methane enrichment is achieved by reducing the CO_2 flow rate. Gas flow rates were measured using thermal mass flow meters. The gases were introduced into the manifold via a Y-shaped nozzle and mixed in a cylindrical chamber. Provision was made to inject DEE either at the intake port or 7 cm upstream in the manifold in order to facilitate HCCI operation. The biogas nozzle was located sufficiently far from the engine so that the flow oscillations induced by piston and valve motion do not affect the flow meter readings. Airflow rate was measured by an orifice meter connected to a differential water manometer, while a burette-stopwatch system was used to measure diesel flow. Temperatures at three locations (intake charge, engine coolant water entry and exit) were measured with the help of k-type thermocouples. In-cylinder pressure was sensed using a piezo-electric pressure transducer installed in the cylinder head. The electric charge developed in the sensor was converted into a voltage signal using a charge amplifier. An electromagnetic crank angle encoder was mounted on the engine output shaft. Signals from the charge amplifier and angle encoder were recorded using a data acquisition system and processed using a LabVIEW-based functional block program to obtain the pressure-crank angle history. Engine load was applied using an eddy current dynamometer coupled to an electronic control unit. Gaseous exhaust emissions, viz. CO, NO_x and HC, were measured using a portable gas analyser, while an opacity-based smoke meter was used to record particulate emissions. The DEE injection timing and quantity were regulated using an Arduino-based circuit which receives input from a camshaft position sensor. DEE consumption was measured using an electronic weighing balance.

METHODOLOGY

In the study presented here, simulated biogas is prepared as a mixture of CH_4 and CO_2 in varying proportions. It eliminates the requirement of biogas generation, purification and volumetric analysis, in addition to offering a direct means of controlling the composition. Various aspects of operating a CI engine fuelled with simulated biogas in the present study. Experimental investigations were carried out by adapting a conventional engine to operate on biogas as the primary fuel in dual fuel and HCCI modes. Diesel and DEE were used as the secondary fuels in these two modes respectively. Further, in HCCI mode, provision was made to inject DEE at either of two locations – at the intake port or 7 cm upstream in the intake manifold. The effects of applied load and biogas composition on various engine performance, combustion and emission indices were studied for dual fuel and HCCI modes at 12 lpm biogas flow rate and 35°C intake charge temperature at a constant speed of 1900 rpm. The lower limit of methane fraction, viz. 50% represents naturally occurring (raw) biogas while the upper limit of 100% represents pure methane. The torque values span 20 to 90% of the engine load range. Engine speed was maintained constant at 1900 rpm. Values of the input parameters used, constituting a total of 52 trials, are listed in Table 5. The output parameters studied were brake thermal and volumetric efficiencies, diesel consumption, secondary fuel energy ratio (SFER, defined as the ratio of energy release from diesel/DEE to the total energy release), overall equivalence ratio, HC, CO, NO_x and smoke emissions, maximum cylinder pressure, ignition delay and maximum HRR.

Figure 1. Photograph and schematic of the experimental setup

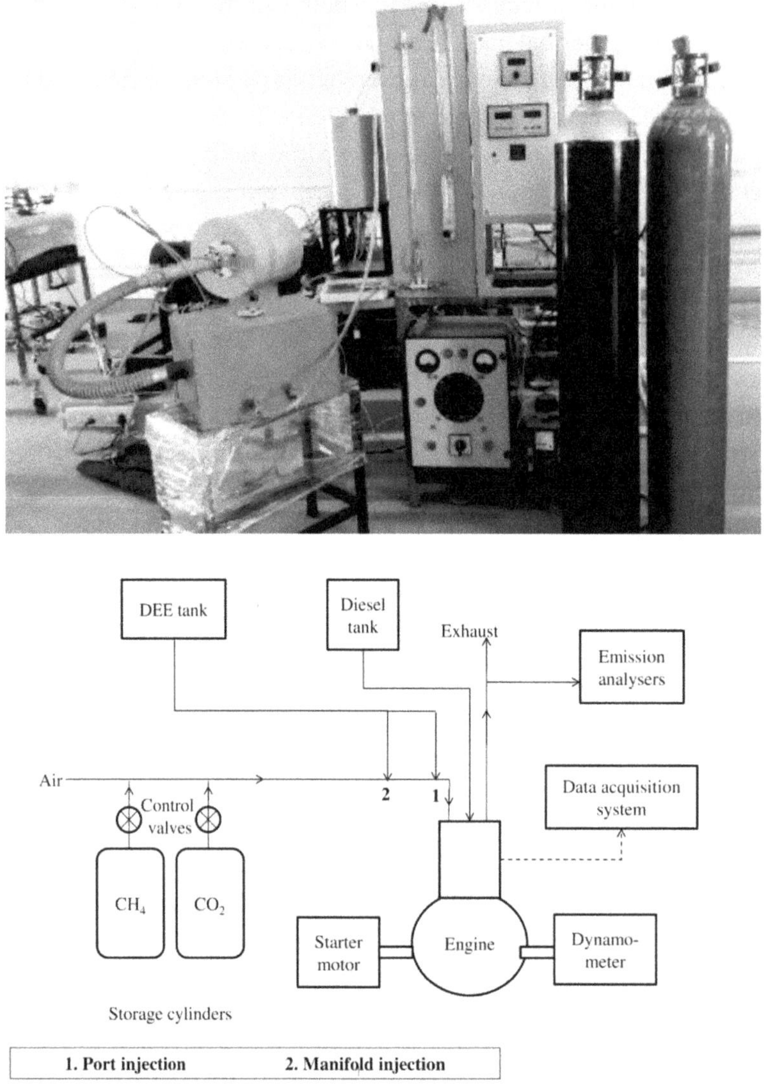

Table 5. Operating parameters for full factorial study

Torque, T (Nm)	Methane fraction (%)	Modes
5	50	Diesel-only
10	65	Dual fuel
15	80	HCCI (with port injection and manifold injection)
20	100	
Biogas flow rate, Q_{bg} = 12 lpm Intake temperature, T_{in} = 35°C Engine speed, N = 1900 rpm		

CALCULATIONS

The following formulae were used to calculate the performance indices:

Brake thermal efficiency,

$$\eta_{bt} = \frac{P_b}{\dot{E}_{total}} \tag{1}$$

where P_b is brake power, and $\dot{E}_{total}$ is total rate of fuel energy supplied, expressed as:

$$\dot{E}_{total} = \dot{E}_{diesel/DEE} + \dot{E}_{biogas} \tag{2}$$

The rate of energy release from each fuel is:

$$\dot{E} = \dot{m} \times LCV \tag{3}$$

where $\dot{m}$ is mass flow rate of the fuel and LCV is its lower calorific value.

Secondary fuel energy ratio (SFER),

$$SFER = \frac{\dot{E}_{diesel/DEE}}{\dot{E}_{total}} \tag{4}$$

Air-fuel ratio,

$$AFR = \frac{\dot{m}_{air}}{\dot{m}_{fuel}} \tag{5}$$

where $\dot{m}_{fuel}$ represents the combined mass flow rate of diesel/DEE and biogas.

Equivalence ratio,

$$ER = \frac{AFR_{stoich}}{AFR} \tag{6}$$

where AFR_{stoich} is the stoichiometric air-fuel ratio:

$$AFR_{stoich} = \frac{\dot{m}_{biogas}}{\dot{m}_{fuel}} * AFR_{stoich-biogas} + \frac{\dot{m}_{diesel/DEE}}{\dot{m}_{fuel}} * AFR_{stoich-diesel/DEE} \tag{7}$$

$$AFR_{stoich-diesel} = 14.5$$

$$AFR_{stoich-DEE} = 11.1$$

$$AFR_{stoich-biogas} = \frac{M_{O2}}{\left(M_{CH4} + \frac{\dot{V}_{CO2}}{\dot{V}_{CH4}} * M_{CO2}\right) * 0.232} \tag{8}$$

where M is molecular weight and $\dot{V}$ is volume flow rate.

Volumetric efficiency,

$$\eta_{vol} = \frac{\dot{V}_a}{\dot{V}_s} \tag{9}$$

where $\dot{V}_a$ is actual air flow rate and $\dot{V}_s$ is ideal air flow rate, expressed as:

$$\dot{V}_a = \dot{m}_{air} / \rho_{air} \tag{10}$$

$$\dot{V}_s = \frac{V_d * N}{120} \tag{11}$$

where V_d is the displacement volume and N is the engine speed in rpm.

ERROR ANALYSIS

Details of the measuring instruments used are provided in Table 6. Error estimates of the output parameters are obtained using the approach given by Moffat (1988) and are listed in Table 7.

COMPARISON OF ALL OPERATING MODES

The performance, emission and combustion characteristics of various modes viz. diesel-only, dual fuel, HCCI (with port injection and manifold injection), are discussed in the subsequent sections.

Brake Thermal Efficiency

The variations of brake thermal efficiency for the three modes and various methane fractions are shown in Figure 2. In all modes, thermal efficiency is low at light loads because of low power output and high

Table 6. Details of the measuring instruments

Quantity measured	Measuring device	Least count
Flow rate of CH_4	Thermal mass flow meter	0.1 litre/min
Flow rate of CO_2	Thermal mass flow meter	0.1 litre/min
Torque	Eddy current dynamometer	0.1 Nm
Air flow rate	Orifice meter and differential water manometer	0.0006 m^3/s
Diesel flow rate	Burette	0.2 ml
Crank angle	Angle encoder	1°CA
In-cylinder pressure	Pressure sensor	33 pC/bar
HC emissions	AVL DiGas 444N gas analyser	1 ppm
CO emission		0.01%
NO_x emissions		10 ppm
Smoke emissions	AVL 437C smoke meter	0.1%

Table 7. Error estimates of the output parameters

Output parameter	Error (±)
Secondary fuel energy ratio	1.33%
Brake thermal efficiency	2.46%
Volumetric efficiency	1.91%
Diesel consumption	1%
Equivalence ratio	1.51%
HC emissions	3%
CO emission	3%
NO_x emissions	1%
Smoke emissions	1%
Cylinder pressure	2%

coolant loss. Qualitatively similar results were reported by Bora et al. (2014), Yoon & Lee (2011a) and Duc & Wattanavichien (2007). Efficiency increases at high load due to improved combustion and energy conversion, corroborating the findings of . In biogas-diesel dual fuel mode, when a small quantity of diesel is injected at low loads, there is a large amount of excess air available and it forms very lean mixtures with biogas, which burns very late or stay unburned. Hence the efficiency is lower than that of the diesel-only mode. However, the surplus air is reduced at high loads and more combustible biogas-air mixtures are formed. This provides better brake thermal efficiency for dual fuel mode at high loads compared to diesel-only mode. The HCCI modes have inferior efficiency compared to dual fuel mode especially at low loads, confirming the findings of Nathan, Mallikrajuna & Ramesh (2009). This is possibly due to the over-lean mixtures which are difficult to auto-ignite at low temperatures and result in fuel loss. Manifold injection shows better efficiency than port injection. This could be attributed to reduced wall impingement of the injected DEE and higher homogeneity factor. Preheating DEE, modifying the

Figure 2. Comparison: Brake thermal efficiency

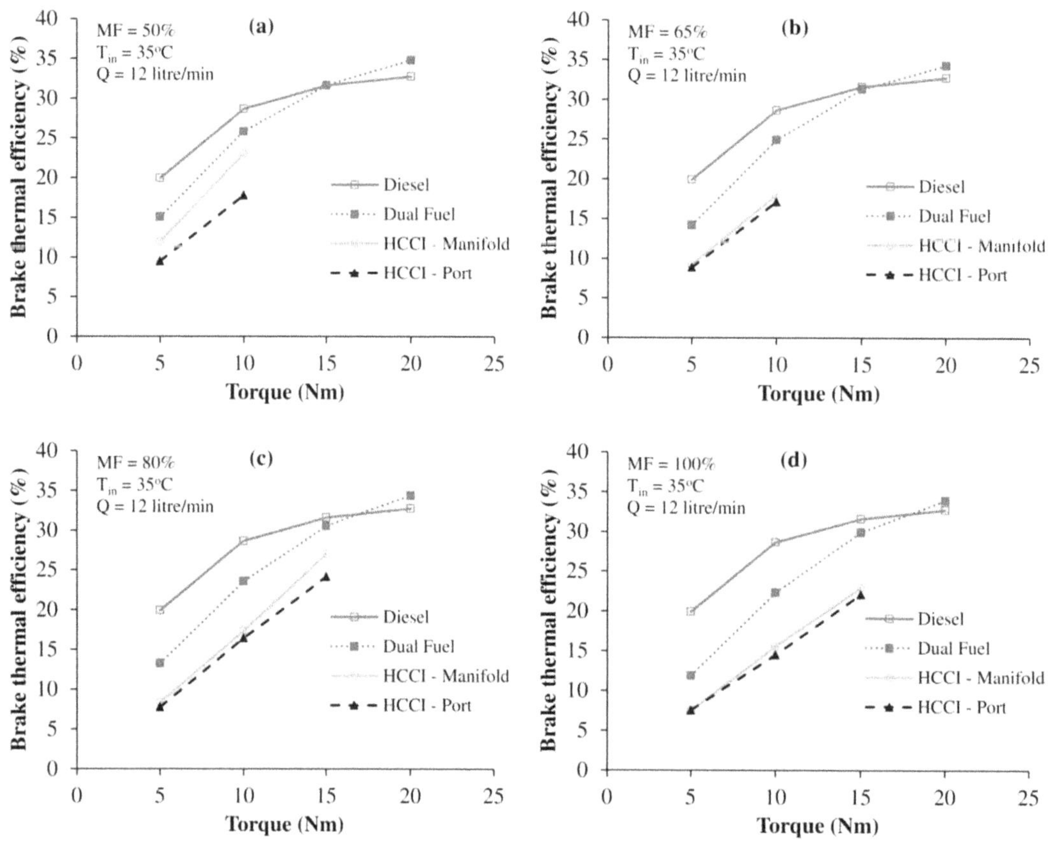

injection timings and manifold heating could improve the brake thermal efficiency of HCCI operation (Rahman & Ramesh, 2017; Ibrahim et al., 2015; Nathan et al., 2010). The operating range in HCCI mode is limited by the likelihood of knocking at high loads. Increasing the methane fraction extends the knock limit as shown in Figure 2. Higher methane fractions are associated with reduced brake thermal efficiency in general at low loads, as the fuel lost contains a greater combustible fraction.

Volumetric Efficiency

Figure 3 shows the correlation between volumetric efficiency and torque for different methane fractions and operating modes. Dual fuel mode and HCCI modes provide low volumetric efficiency due to displacement of air in the intake by the incoming biogas, confirming the observations of Luijten & Kerkhof (2011). Additionally in HCCI mode, especially with manifold injection, DEE injection poses resistance to the free inflow of air and results in the least volumetric efficiency among all modes. As air displacement is primarily dictated by the flow rate of biogas and not by its composition, methane fraction did not show a clear influence on volumetric efficiency.

Figure 3. Comparison: Volumetric efficiency

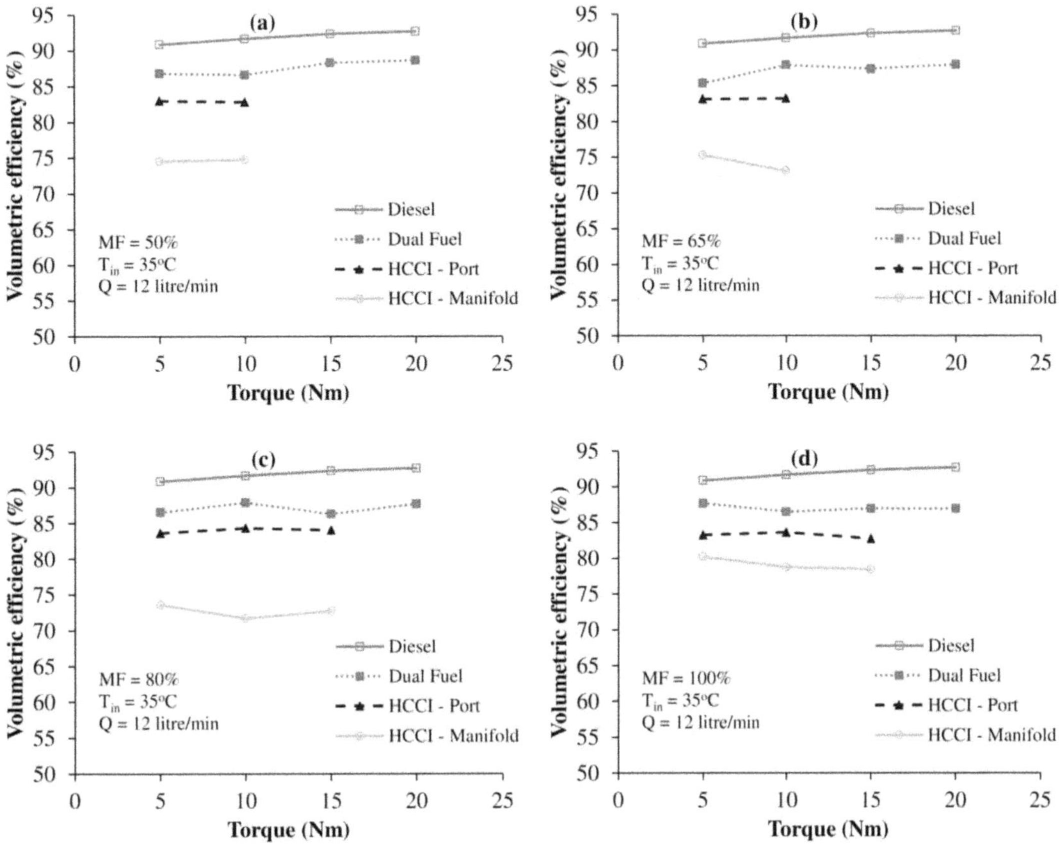

Diesel Consumption

Figure 4 depicts the relationship between diesel consumption and torque at various methane fractions for diesel-only and dual fuel modes. With the addition of methane, the proportion of energy supplied by diesel diminishes and the governor mechanism reduces the diesel supply to maintain constant engine speed. Thus, dual fuel mode shows lower diesel consumption. High torque requires more pilot fuel to maintain constant speed and hence diesel consumption increases with load as expected. Increase in methane fraction increases the energy content of biogas and lowers the requirement of diesel. A similar trend was reported by Luijten & Kerkhof (2011). As diesel was not used in HCCI operation, the corresponding characteristic is not plotted here.

Secondary Fuel Energy Ratio (SFER)

SFER (Secondary fuel energy ratio) is the ratio of energy supplied by the secondary fuel (diesel/DEE) to the combined energy release. Figure 5 shows variation of SFER with load and methane fraction for the three modes. In the diesel-only mode, the entire energy input is from diesel and hence SFER is unity. Biogas is the primary fuel in dual fuel and HCCI operation. SFER increases with load as the second-

Figure 4. Comparison: Diesel consumption

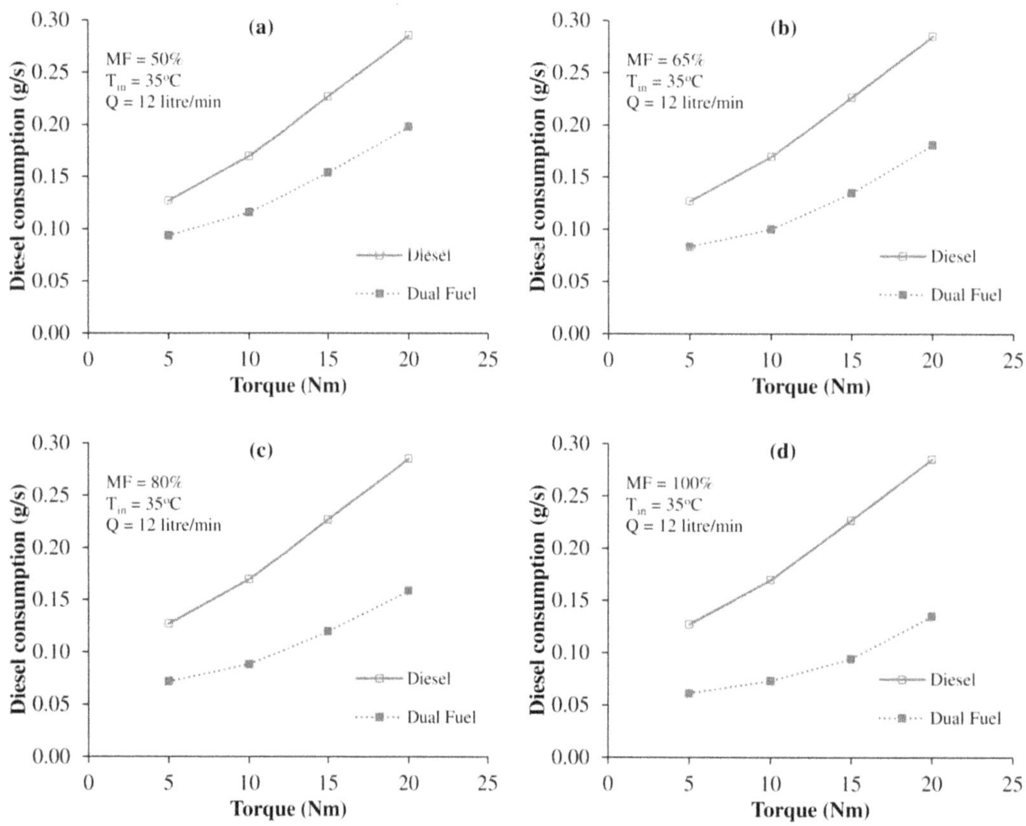

ary fuel consumption increases (refer Figure 4). Dual fuel mode has lower secondary fuel consumption compared to HCCI on account of higher thermal efficiency (refer Figure 2) and consequently lower SFER. Increase in methane fraction reduces secondary fuel consumption and SFER.

Equivalence Ratio

Figure 6 shows the variation of overall equivalence ratio with load and methane fraction for the three modes. Due to increased diesel consumption and reduced availability of excess air, the overall equivalence ratio increases with torque in diesel-only mode. Increase in secondary fuel with load is observed in dual fuel mode and HCCI modes as well. Here, the displacement of air in the intake by biogas causes a further increase in equivalence ratio vis-à-vis diesel-only mode. The difference between the three modes diminishes at high loads as a significant proportion of the total energy input occurs from secondary fuel. Due to the high secondary fuel consumption, HCCI mode has the highest equivalence ratio among all. With increase in methane fraction, the mass of biogas decreases, as methane is almost three times lighter than CO_2. Secondary fuel consumption also decreases, as shown in Figure 5. The reduction in primary and secondary fuel consumption causes the air-fuel ratio to increase. However, the stoichiometric air-fuel ratio increases at a faster rate, as methane requires more air for combustion on a per-kg basis compared to diesel. The net effect is a rise in equivalence ratio with methane fraction.

Figure 5. Comparison: SFER

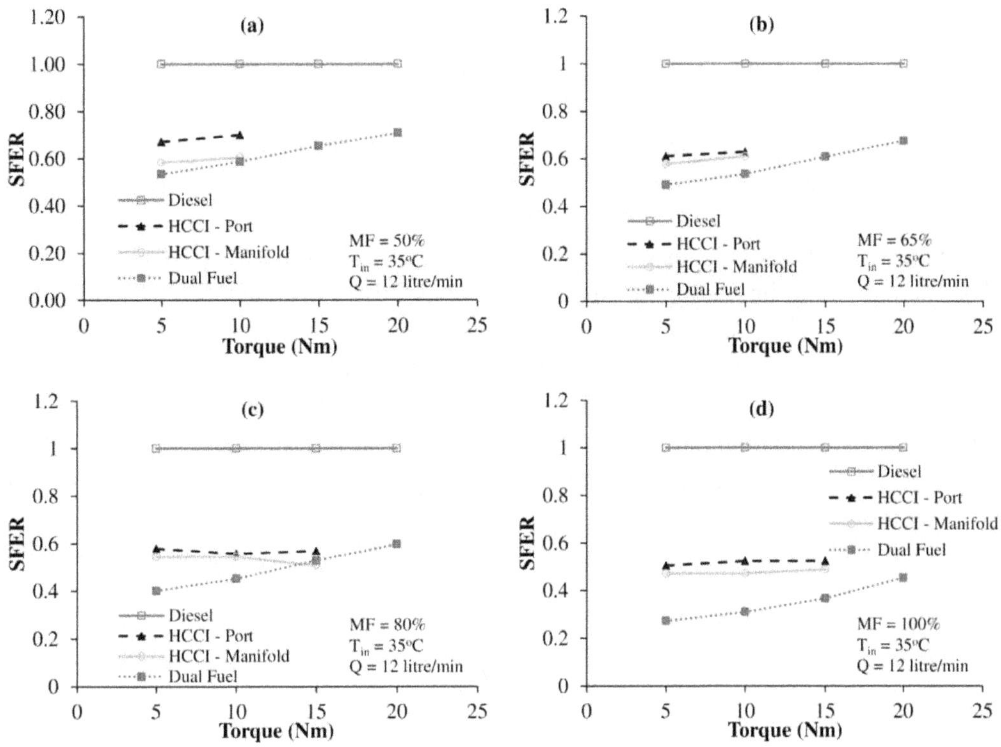

Figure 6. Comparison: Overall equivalence ratio

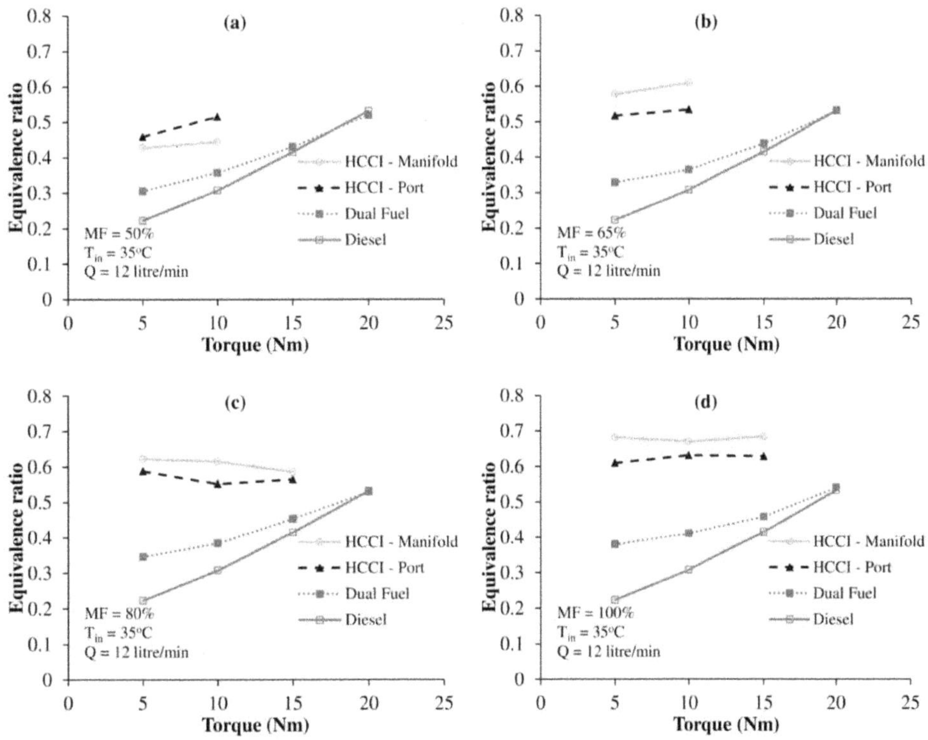

HC Emissions

Figure 7 depicts the variations in HC emissions for different modes and methane fractions. Brake specific HC emissions are high at light loads because of over-leaning, resulting in flame quenching and poor combustion. HC emissions decrease with load owing to better energy conversion efficiency and reduction of quenching at elevated operating temperatures. Induction of biogas in dual and HCCI modes results in higher overall equivalence ratios as discussed in previous section and a fraction of the supplied biogas escapes unburned via the exhaust, contributing to HC emissions. Diesel-only mode provides lowest HC emissions among all modes while the highest HC emissions occur in HCCI mode relative to other modes, corroborating the finding of Barik & Murugan (2014a), Nathan, Mallikrajuna, & Ramesh (2009) and Nathan et al. (2007). Manifold injection provides slightly lower HC emissions compared to port injection due to improved mixture distribution and combustion. Increase in methane fraction is reflected in a corresponding increase in HC emissions attributed to unburned biogas. With increase in methane content, the reduction in the quantity of injected DEE reduces the reactivity of the fuel-air mixture and affects the completeness of combustion, causing high HC emissions. A similar finding has been reported by Rahman & Ramesh (2017).

Figure 7. Comparison: HC emissions

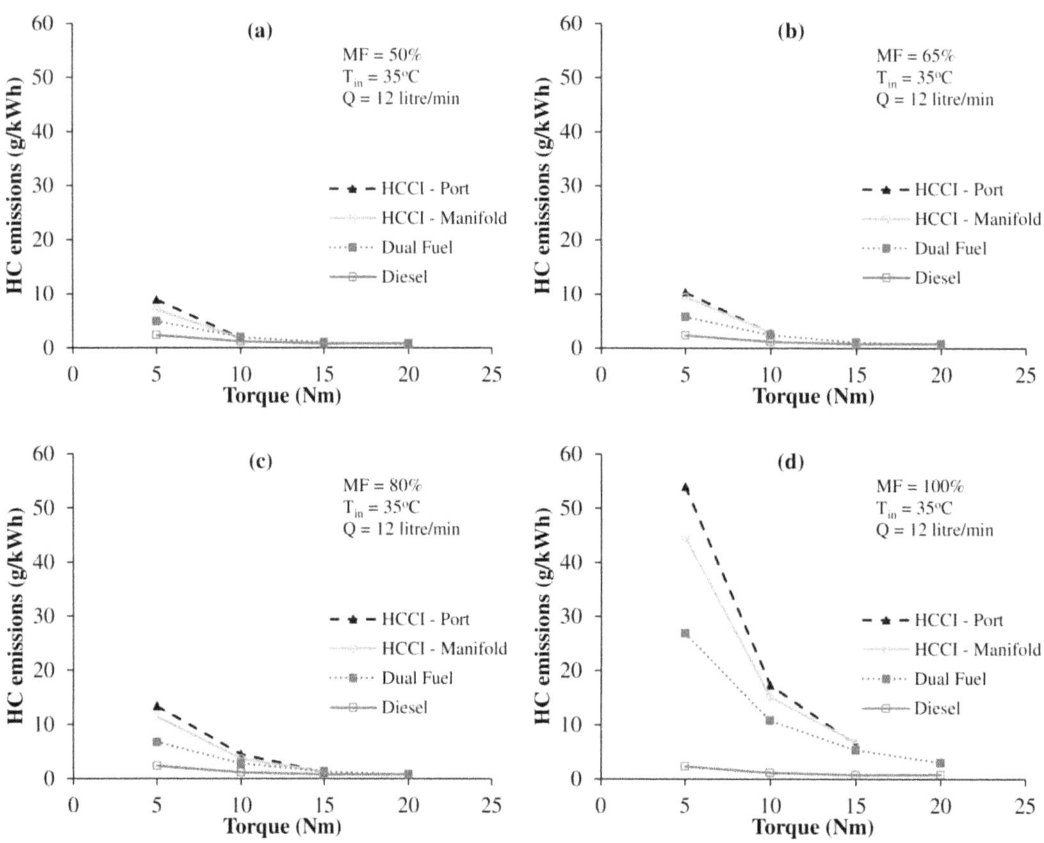

CO Emission

The variation of CO emission for various operating modes and methane fraction are shown in Figure 8. Although more CO is formed on a per-volume basis at higher loads, it has a diminishing trend when expressed per unit output work, i.e. in g/kWh. Compared to diesel-only mode, addition of biogas in dual fuel and HCCI modes causes a rise in CO emission owing to lack of air and the resultant high equivalence ratio. A similar trend has been reported by Yoon & Lee (2011b) and Nathan et al. (2007). Upon increasing the methane content of biogas, the fraction of fuel escaping as HC increases and reduces the fraction burned, such that less CO is formed.

NO_x Emissions

Figure 9 shows the comparison of NO_x emission for diesel-only, dual fuel and HCCI modes at various torques and methane fractions. NO_x emissions increase with load due to increased HRR and higher combustion temperatures. The presence of CO_2 in dual fuel mode suppresses the in-cylinder temperature and reduces NO_x emission, similar to the effect produced by EGR. The reduced availability of air due to displacement by biogas in the intake also helps in reducing NO_x formation. A similar trend has been observed by Barik & Sivalingam (2013). HRR and in-cylinder temperatures are least in HCCI

Figure 8. Comparison: CO emission

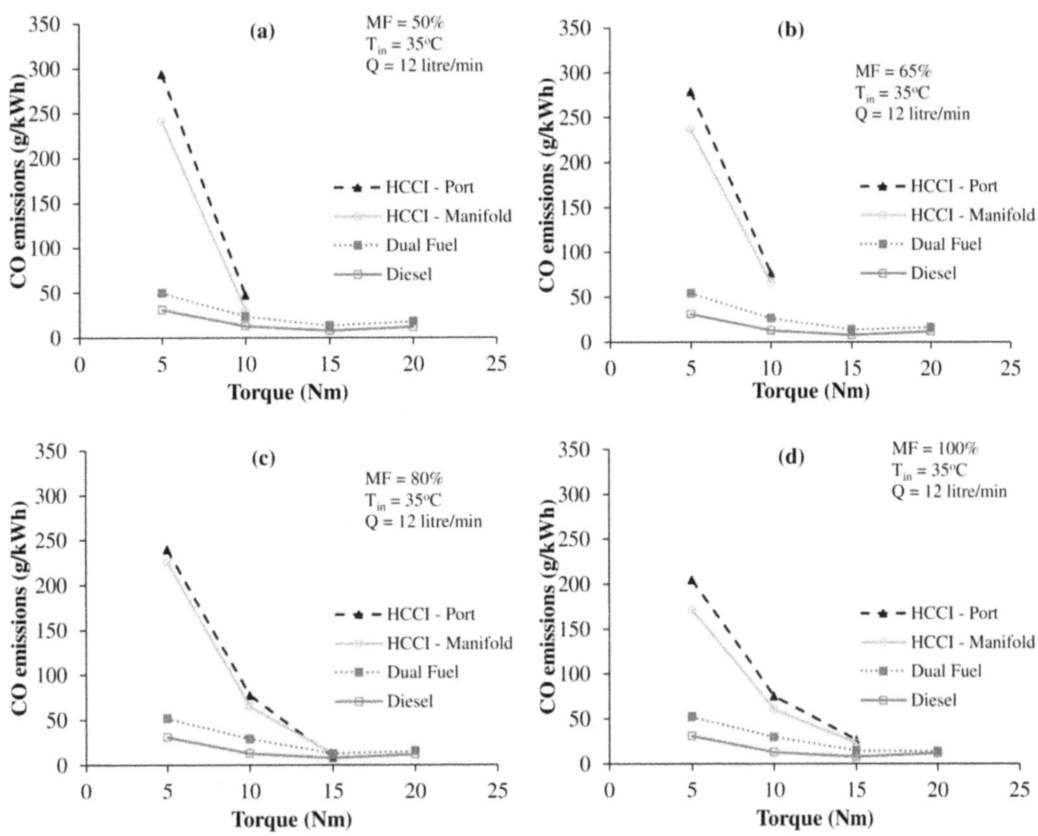

mode. Consequently, HCCI mode is characterized by ultra-low NO_x emissions which are 1 to 2 orders of magnitude lower than those of dual fuel mode. This translates into nearly 100% reduction compared to diesel-only mode. Several researchers have noted this as one of the highlights of HCCI operation. Increasing the methane fraction has two contrasting effects – the DEE supply is reduced, limiting the HRR. On the other hand, it reduces the diluent effect of CO_2 and tends to increase the temperature. Consequently, increasing the methane fraction does not significantly affect NO_x formation.

Smoke Emission

Figure 10 shows the comparison of smoke emission for various modes at different methane fractions and torques. Rich pockets of liquid fuels undergoing heterogeneous combustion are the primary sources for smoke emission. This scenario is more dominant in diesel-only mode especially at high load, entailing high smoke emission. Induction of biogas reduces liquid fuel consumption as discussed earlier. Consequently, smoke emissions are low in dual fuel and HCCI modes. In the latter, the homogenous mixture almost completely precludes soot formation, resulting in ultra-low smoke emission. Enhancement of methane percentage in biogas reduces smoke emission due to the reduction in liquid fuel injection. In

Figure 9. Comparison: NO_x emissions

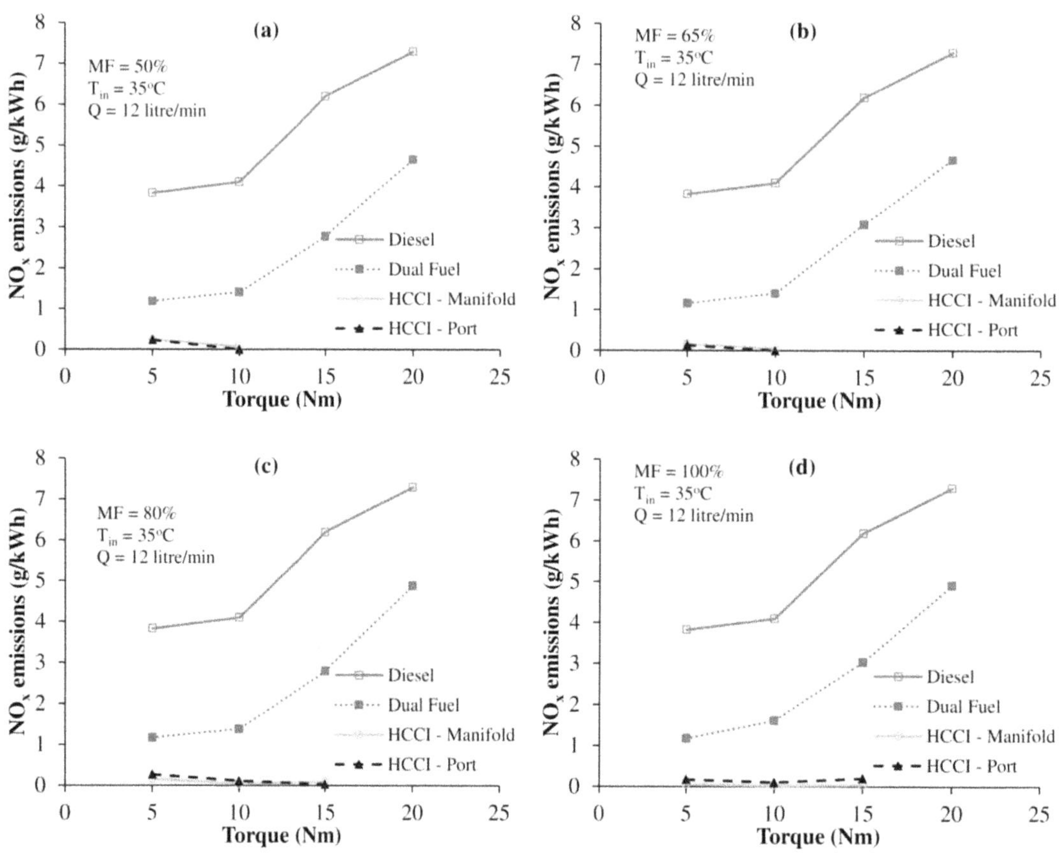

contrast to the conventional CI engine concept, dual fuel and HCCI modes are thus found to provide simultaneous reduction of NO_x and smoke emissions. This has been highlighted by several researchers in the past (Barik & Murugan, 2014a; Nathan et al., 2009 & 2010).

Maximum Heat Release Rate (MHRR)

Figure 11 shows representative net heat release rate histories of dual fuel mode for different methane fractions. MHRR values correspond to the peaks in these plots. Figure 12 depicts the relationship between MHRR and torque for various operating modes and methane fractions. The high reactivity of diesel results in the diesel-only mode possessing the highest values of MHRR. Induction of biogas or enhancement of methane fraction reduces the consumption of more reactive pilot fuels (diesel/DEE), lowering the HRR, confirming the findings of Barik & Murugan (2014a), Mustafi, Raine, & Verhelst (2013) etc. HCCI mode shows high MHRR at low methane fractions due to increased consumption of DEE, which is highly reactive and auto-ignites readily.

Figure 10. Comparison: Smoke emission

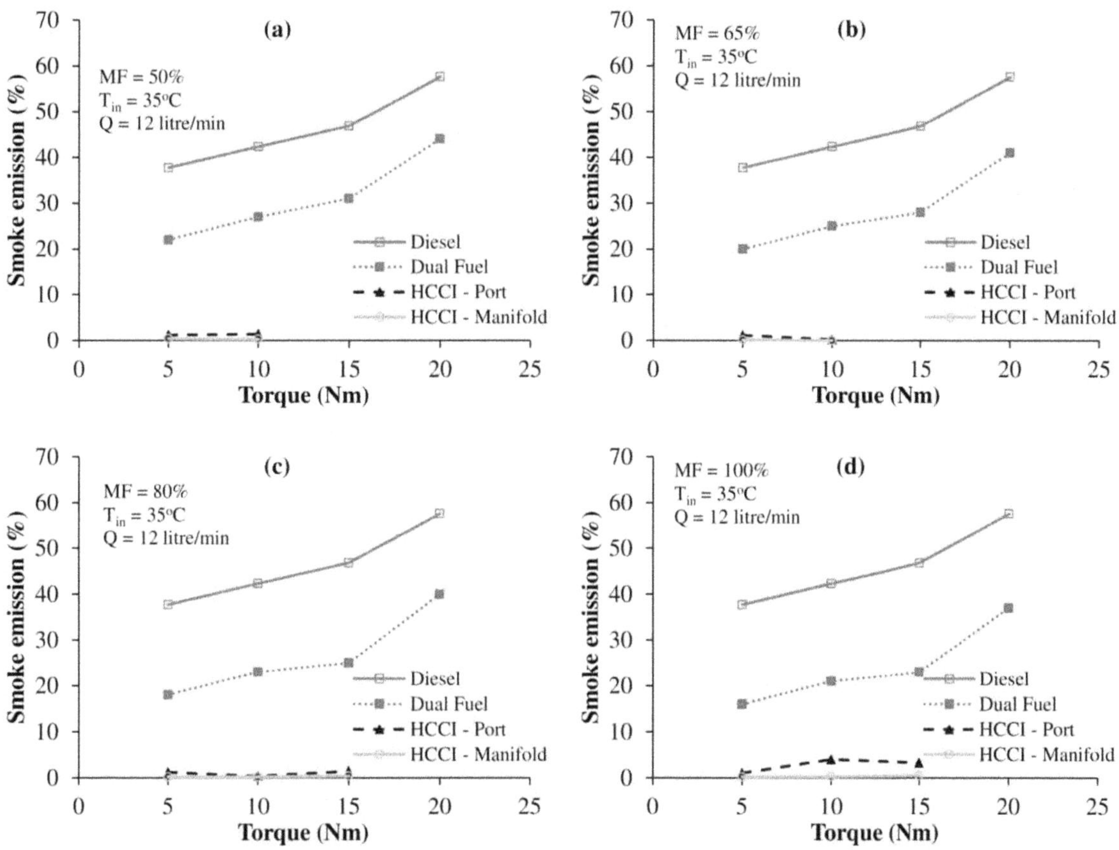

Figure 11. Net heat release rate histories: Dual fuel mode

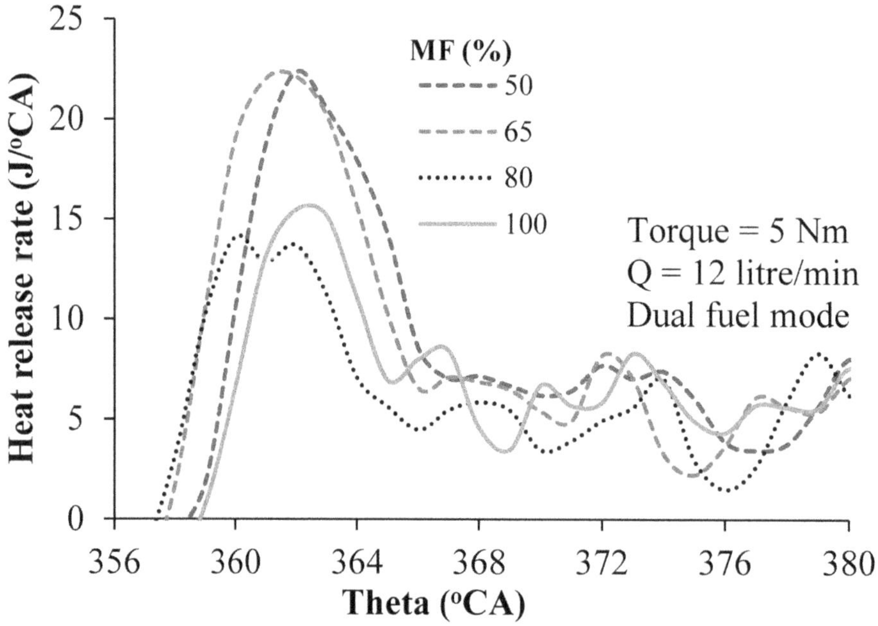

Figure 12. Comparison: MHRR

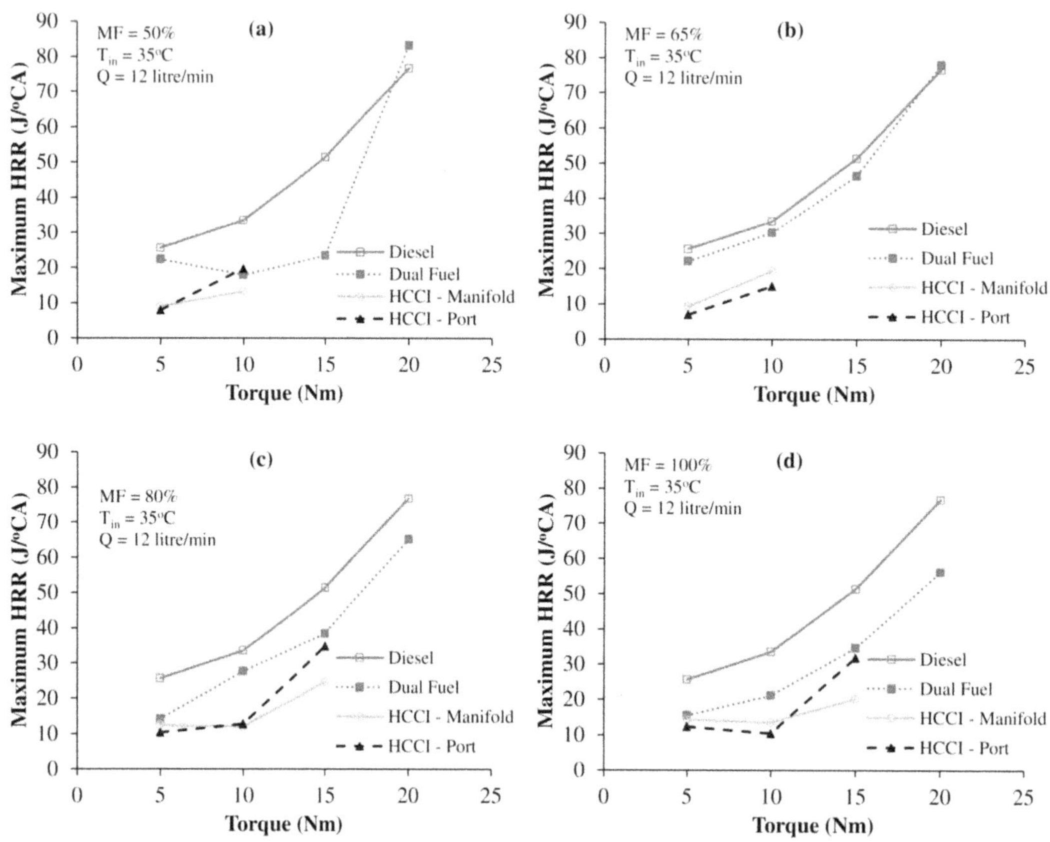

Ignition Delay

Figure 13 shows the comparison of ignition delay for diesel-only and dual fuel modes with various methane fractions and torques. Use of biogas in dual fuel mode shows high ignition delay value due to its low reactivity (measured in terms of cetane number). There is no significant variation in ignition delay at various methane fractions. Ignition delay could not be defined for HCCI mode, as there is no in-cylinder fuel injection in order to use the start of injection as the reference point.

Maximum Cylinder Pressure

Figure 14 shows representative cylinder pressure – crank angle (p-theta) diagrams of dual fuel mode for different methane fractions. Maximum cylinder pressure data is obtained from these pressure histories. Figure 15 depicts the effects of diesel-only, dual fuel and HCCI modes on maximum cylinder pressure at various torques and methane fractions. The trends of peak pressure correlate well with those of HRR. In general, the diesel-only mode is characterized by the fastest heat release and consequently greater peak pressures. Induction of biogas lowers the HRR as discussed earlier, reducing the peak pressure as reported

Figure 13. Comparison: Ignition delay

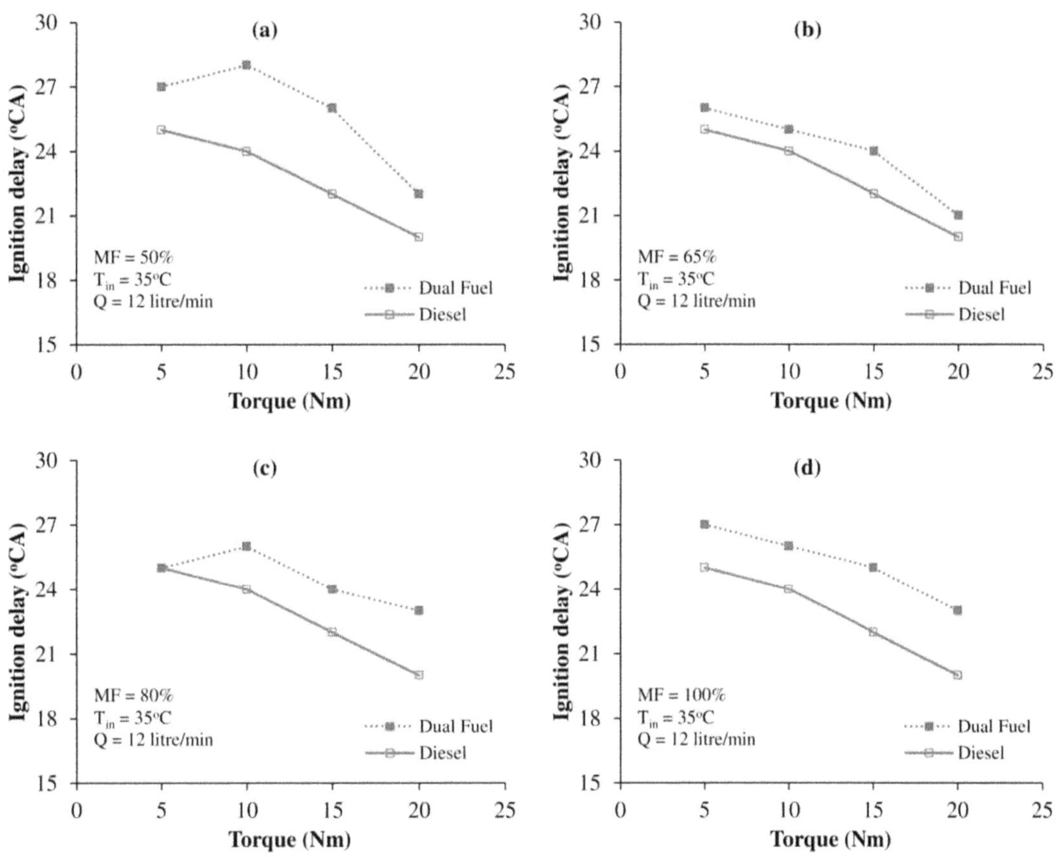

Figure 14. Cylinder pressure diagram: Dual fuel mode

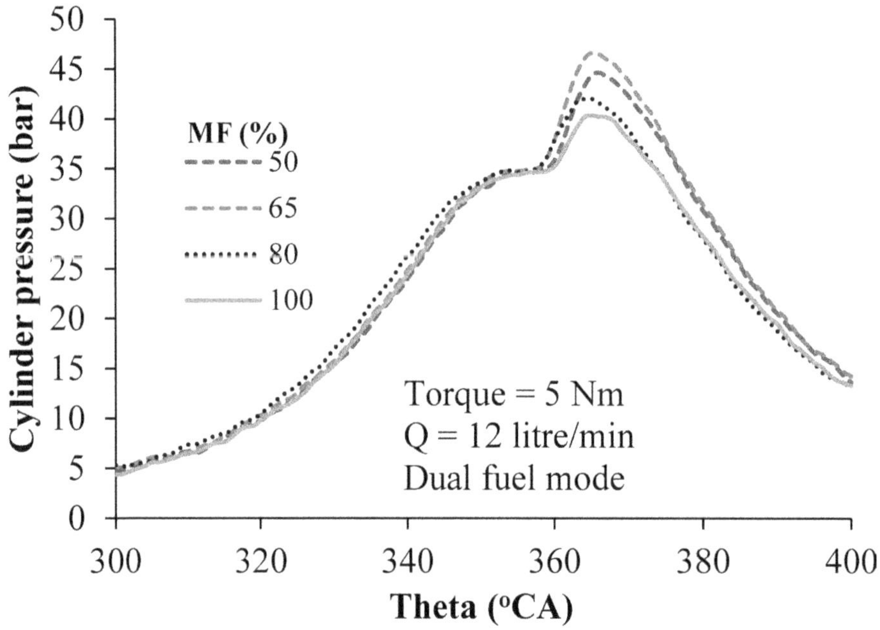

Figure 15. Comparison: Maximum cylinder pressure

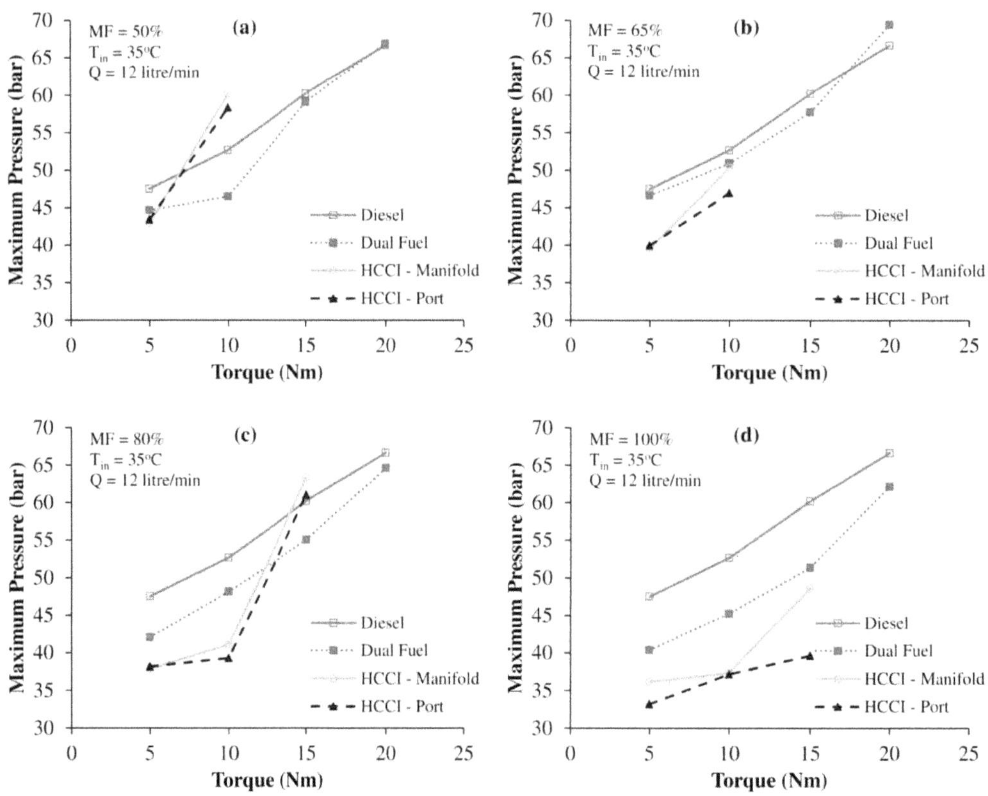

by Barik & Murugan(2014a), Königsson, Stalhammar, & Angstrom (2011), Yoon & Lee (2011b) etc. in the past. At low methane fractions, both HRR and peak pressures of HCCI mode are seen to be relatively high due to increased quantity of DEE injection, reflecting the findings of Rahman & Ramesh (2017).

FUTURE RESEARCH DIRECTIONS

In conjunction with biogas generation plant and purifier, a CI engine capable of multi-mode operation offers a viable energy solution, especially for rural communities whose access to conventional fuels like diesel may be limited. Typical applications of biogas fuelled stationary CI engines include pumping, irrigation and electric power generation.

Some of the topics which could be pursued for further research in this area include:

- Dual fuel operation with biogas and bio-diesels
- Emission reduction by addition of nano-particles to the pilot fuel
- Means of mitigating HC and CO emissions in HCCI operation.

CONCLUSION

- Addition of biogas minimises secondary fuel consumption, smoke and NO_x emissions.
- At high load, dual fuel mode provides better brake thermal efficiency than all other modes
- Ultra-low NO_x and smoke emissions can be achieved in HCCI mode.
- Increase in methane fraction extends the operating range of HCCI mode by reducing knocking tendency.
- Addition of biogas increases HC and CO emissions compared to diesel-only mode.
- Manifold injection provides higher thermal efficiency and lower emissions compared to port injection method.
- Simultaneous reduction of NO_x and smoke emissions are possible by using biogas in dual fuel and HCCI modes.

ACKNOWLEDGMENT

This research was supported by the Department of Science and Technology – Science and Engineering Research Board (DST-SERB), Government of India [project number EMR/2015/001379].

REFERENCES

Bailey, B., Eberhardt, J., Goguen, S., & Erwin, J. (1997). Diethyl Ether (DEE) as a Renewable Diesel Fuel. *SAE Transactions*. SAE Technical Paper 972978. doi:10.4271/972978

Barik, D., & Murugan, S. (2014a). Investigation on combustion performance and emission characteristics of a DI (direct injection) diesel engine fueled with biogas-diesel in dual fuel mode. *Energy*, *72*, 760–771. doi:10.1016/j.energy.2014.05.106

Barik, D., & Murugan, S. (2014b). Simultaneous reduction of NOx and smoke in a dual fuel DI diesel engine. *Energy Conversion and Management*, *84*, 217–226. doi:10.1016/j.enconman.2014.04.042

Barik, D., & Murugan, S. (2016). Effects of diethyl ether (DEE) injection on combustion performance and emission characteristics of Karanja methyl ester (KME)-biogas fueled dual fuel diesel engine. *Fuel*, *164*, 286–296. doi:10.1016/j.fuel.2015.09.094

Barik, D., & Sivalingam, M. (2013). Performance and emission characteristics of a biogas fueled DI diesel engine. Technical report, SAE Technical Paper 2013–01–2507. Retrieved from http://papers.sae.org/2013-01-2507/

Barik, D., & Sivalingam, M. (2014). Investigation on performance and exhaust emissions characteristics of a DI diesel engine fueled with Karanja methyl ester and biogas in dual fuel mode, Technical report, SAE Technical Paper 2014-01-1311 Retrieved from http://papers.sae.org/2014-01-1311/

Basu, S., Khan, A. L., Cano-Odena, A., Liu, C., & Vankelecom, I. F. J. (2010). Membrane-based technologies for biogas separations. *Chemical Society Reviews*, *39*(2), 750–768. doi:10.1039/B817050A PMID:20111791

Bedoya, I. D., Saxena, S., Cadavid, F. J., & Dibble, R. W. (2012). Exploring Strategies for Reducing High Intake Temperature Requirements and Allowing Optimal Operational Conditions in a Biogas Fueled HCCI Engine for Power Generation. *Journal of Engineering for Gas Turbines and Power*, *134*(7). doi:10.1115/1.4006075

Bedoya, I. D., Saxena, S., Cadavid, F. J., Dibble, R. W., & Wissink, M. (2012). Experimental evaluation of strategies to increase the operating range of a biogas-fueled HCCI engine for power generation. *Applied Energy*, *97*, 618–629. doi:10.1016/j.apenergy.2012.01.008

Bedoya, I. D., Saxena, S., Cadavid, F. J., Dibble, R. W., & Wissink, M. (2012). Experimental study of biogas combustion in an HCCI engine for power generation with high indicated efficiency and ultra-low NOx emissions. *Energy Conversion and Management*, *53*(1), 154–162. doi:10.1016/j.enconman.2011.08.016

Bhattacharya, T., Mishra, T., & Singh, B. (1988). Techniques for removal of CO2 and H2S from biogas. *XXIV Annual Convention of ISAE, Held at PKV, Akola*. Retrieved from https://scholar.google.co.in/scholar?hl=en&q=Techniques+for+removal+of+CO2+and+H2S+from+biogas&btnG=#0

Biswas, T., Kartha, A., & Pundarikakhadu, R. (1977). Removal of carbon dioxide from biogas. In *Proceedings of National Symposium on Biogas Technology and Uses*. New Delhi, India: IARI; Retrieved from https://scholar.google.co.in/scholar?q=Removal+of+carbon+dioxide+from+biogas+biswas&btnG=&hl=en&as_sdt=0%2C5#0

Bora, B. J., & Saha, U. K. (2015). Improving the Performance of a Biogas Powered Dual Fuel Diesel Engine Using Emulsified Rice Bran Biodiesel as Pilot Fuel Through Adjustment of Compression Ratio and Injection Timing. *Journal of Engineering for Gas Turbines and Power*, *137*(9). doi:10.1115/1.4029708

Bora, B. J., Saha, U. K., Chatterjee, S., & Veer, V. (2014). Effect of compression ratio on performance, combustion and emission characteristics of a dual fuel diesel engine run on raw biogas. *Energy Conversion and Management*, *87*, 1000–1009. doi:10.1016/j.enconman.2014.07.080

Cacua, K., Amell, A., & Cadavid, F. (2012). Effects of oxygen enriched air on the operation and performance of a diesel-biogas dual fuel engine. *Biomass and Bioenergy*, *45*, 159–167. doi:10.1016/j.biombioe.2012.06.003

Christensen, M., & Johansson, B. (2000). Supercharged Homogeneous Charge Compression Ignition (HCCI) with Exhaust Gas Recirculation and Pilot Fuel. *SAE Technical Papers*, (724), SAE 2000-01-1835. doi:10.4271/2000-01-1835

Chynoweth, D. P., Owens, J. M., & Legrand, R. (2000). Renewable methane from anaerobic digestion of biomass. *Renewable Energy*, *22*(1–3), 1–8. doi:10.1016/S0960-1481(00)00019-7

Demirbas, A. (2004). Combustion characteristics of different biomass fuels. *Progress in Energy and Combustion Science*, *30*(2), 219–230. doi:10.1016/j.pecs.2003.10.004

Demirbas, M., Balat, M., & Balat, H. (2011). Biowastes-to-biofuels. *Energy Conversion and Management*, *52*(4), 1815–1828. doi:10.1016/j.enconman.2010.10.041

Dept. of Animal Husbandry Dairying & Fisheries. (2012). *19th Livestock Census-2012: All India Report. Ministry of Agriculture*. Retrieved from http://dahd.nic.in/sites/default/files/Livestock 5.pdf

Dubey, A. (2000). Wet scrubbing of carbon dioxide. *Annual Report of CIAE, Bhopal (India)*. Retrieved from https://scholar.google.co.in/scholar?q=Wet+scrubbing+of+carbon+dioxide+dubey&btnG=&hl=en&as_sdt=0%2C5#0

Duc, P. M., & Wattanavichien, K. (2007). Study on biogas premixed charge diesel dual fuelled engine. *Energy Conversion and Management*, *48*(8), 2286–2308. doi:10.1016/j.enconman.2007.03.020

Feroskhan, M., & Ismail, S. (2016). Investigation of the effects of biogas composition on the performance of a biogas – diesel dual fuel CI engine. *Biofuels*, *7*(6), 593–601. doi:10.1080/17597269.2016.1168025

Feroskhan, M., & Ismail, S. (2017). A review on the purification and use of biogas in compression ignition engines. *International Journal of Automotive and Mechanical Engineering*, *14*(3), 4383–4400. doi:10.15282/ijame.14.3.2017.1.0348

Guha, A., Majumdar, S., & Sirkar, K. (1991). A larger-scale study of gas separation by hollow-fiber-contained liquid membrane permeator. *Journal of Membrane Science*, *62*(3), 293–307. doi:10.1016/0376-7388(91)80044-7

Gunaseelan, V. N. (1997). Anaerobic digestion of biomass for methane production: A review. *Biomass and Bioenergy*, *13*(1–2), 83–114. doi:10.1016/S0961-9534(97)00020-2

Henham, A., & Makkar, M. (1998). Combustion of simulated biogas in a dual-fuel diesel engine. *Energy Conversion and Management*, *39*(16–18), 2001–2009. doi:10.1016/S0196-8904(98)00071-5

Hullu, D. (2008). Biogas Upgrading: Comparing Different Techniques, Eindhoven, The Netherlands. Retrieved from students.chem.tue.nl/ifp24/Finalpresentation.pps%3E accessed 08.03.2013

Hyvönen, J., Haraldsson, G., & Johansson, B. (2003). Operating range in a Multi Cylinder HCCI engine using Variable Compression Ratio. *SAE Technical Paper 2003-01-1829*. doi:10.4271/2003-01-1829

Ibrahim, M. M., Varuna Narasimhan, J., & Ramesh, A. (2015). Comparison of the predominantly premixed charge compression ignition and the dual fuel modes of operation with biogas and diesel as fuels. *Energy*, *89*, 990–1000. doi:10.1016/j.energy.2015.06.033

Jun, D., Ishii, K., & Iida, N. (2003). Combustion Analysis of Natural Gas in a Four Stroke HCCI Engine Using Experiment and Elementary Reactions Calculation. *SAE Technical Papers, 2003*(724). doi:10.4271/2003-01-1089

Kalsi, S. S., & Subramanian, K. A. (2017). Effect of simulated biogas on performance, combustion and emissions characteristics of a bio-diesel fueled diesel engine. *Renewable Energy*, *106*, 78–90. doi:10.1016/j.renene.2017.01.006

Khapre, U. (1989). Studies on biogas utilization for domestic cooking. *XXV Annual Convention of ISAE, Held at CTAE, Udaipur*. Retrieved from https://scholar.google.co.in/scholar?q=Studies+on+biogas+utilization+for+domestic+cooking&btnG=&hl=en&as_sdt=0%2C5#0

Khoiyangbam, R. S., Gupta, N., & Kumar, S. (2011). *Biogas Technology Towards Sustainable Development*. Teri Publications.

Königsson, F., Stalhammar, P., & Angstrom, H.-E. (2011). Characterization and Potential of Dual Fuel Combustion in a Modern Diesel Engine. Technical report, SAE Technical Paper 2011–01–2223. doi:10.4271/2011-01-2223

Kozlov, V. E., Chechet, I. V., Matveev, S. G., Titova, N. S., & Starik, A. M. (2016). Modeling study of combustion and pollutant formation in HCCI engine operating on hydrogen rich fuel blends. *International Journal of Hydrogen Energy*, *41*(5), 3689–3700. doi:10.1016/j.ijhydene.2015.12.078

Kusworo, T. D., Johari, S., & Widiasa, I. N. (2012). The Uses of Carbon Nanotubes Mixed Matrix Membranes (MMM) for Biogas Purification. [IJWR]. *International Journal of Waste Resources*, *2*(1), 5–10. doi:10.12777/ijwr.2.1.2012.5-10

Li, K., & Teo, W. K. (1993). Use of an internally staged permeator in the enrichment of methane from biogas. *Journal of Membrane Science*, *78*(1–2), 181–190. doi:10.1016/0376-7388(93)85259-Y

Liu, H., Cooper, V. R., Dai, S., & Jiang, D. (2012). Windowed Carbon Nanotubes for Efficient CO 2 Removal from Natural Gas. *The Journal of Physical Chemistry Letters*, *3*(22), 3343–3347. doi:10.1021/jz301576s

Lounici, M. S., Loubar, K., Tazerout, M., Balistrou, M., & Tarabet, L. (2014). Experimental Investigation on the Performance and Exhaust Emission of Biogas-Diesel Dual-Fuel Combustion in a CI Engine. *Technical report, SAE Technical Paper 2014–01–2689*. doi:10.4271/2014-01-2689

Luijten, C. C. M., & Kerkhof, E. (2011). Jatropha oil and biogas in a dual fuel CI engine for rural electrification. *Energy Conversion and Management*, *52*(2), 1426–1438. doi:10.1016/j.enconman.2010.10.005

Moffat, R. J. (1988). Describing the uncertainties in experimental results. *Experimental Thermal and Fluid Science*, *1*(1), 3–17. doi:10.1016/0894-1777(88)90043-X

Mustafi, N. N., Raine, R. R., & Verhelst, S. (2013). Combustion and emissions characteristics of a dual fuel engine operated on alternative gaseous fuels. *Fuel*, *109*, 669–678. doi:10.1016/j.fuel.2013.03.007

Nathan, S. S., & Mallikarjuna, J. M. (2010). An experimental study of the biogas-diesel HCCI mode of engine operation. *Energy Conversion and Management*, *51*(7), 1347–1353. doi:10.1016/j.enconman.2009.09.008

Nathan, S. S., Mallikarjuna, J. M., & Ramesh, A. (2007). Effect of mixture preparation in a diesel HCCI engine using early in-cylinder injection during the suction stroke. *International Journal of Automotive Technology*, *8*(5), 543–553. Retrieved from http://www.scopus.com/scopus/inward/record.url?eid=2-s2.0-36849090822&partnerID=40&rel=R8.2.0

Nathan, S. S., Mallikrajuna, J. M., & Ramesh, A. (2009). Homogeneous charge compression ignition versus dual fuelling for utilizing biogas in compression ignition engines. *Proceedings of the Institution of Mechanical Engineers. Part D, Journal of Automobile Engineering*, *223*(3), 413–422. doi:10.1243/09544070JAUTO970

Nijaguna, B. (2006). Biogas technology. New Age International Retrieved from https://scholar.google.co.in/scholar?q=nijaguna&btnG=&hl=en&as_sdt=0%2C5#2

Pande, D. R., & Fabiani, C. (1989). Feasibility studies on the use of a naturally occurring molecular sieve for methane enrichment from biogas. *Gas Separation & Purification*, *3*(3), 143–147. doi:10.1016/0950-4214(89)80026-X

Park, S. H., & Yoon, S. H. (2016). Effect of dual-fuel combustion strategies on combustion and emission characteristics in reactivity-controlled compression ignition (RCCI) engine. *Fuel*, *181*, 310–318. doi:10.1016/j.fuel.2016.04.118

Park, S. H., Yoon, S. H., Cha, J., & Lee, C. S. (2014). Mixing effects of biogas and dimethyl ether (DME) on combustion andemission characteristics of DME fueled high-speed diesel engine. *Energy*, *66*, 413–422. doi:10.1016/j.energy.2014.02.007

Polk, A. C., Gibson, C. M., Shoemaker, N. T., Srinivasan, K. K., & Krishnan, S. R. (2013). Analysis of Ignition Behavior in a Turbocharged Direct Injection Dual Fuel Engine Using Propane and Methane as Primary Fuels. *Journal of Energy Resources Technology*, *135*(3). doi:10.1115/1.4023482

Rahman, A. K., & Ramesh, A. (2017). Effect of reducing the methane concentration on the combustion and performance of a biogas diesel predominantly premixed charge compression ignition engine. *Fuel*, *206*, 117–132. doi:10.1016/j.fuel.2017.05.100

Ray, N. H. S. (2013). Biogas as Alternate Fuel in Diesel Engines: A Literature Review. *IOSR Journal of Mechanical and Civil Engineering*, *9*(1), 23–28. doi:10.9790/1684-0912328

Ryckebosch, E., Drouillon, M., & Vervaeren, H. (2011). Techniques for transformation of biogas to biomethane. *Biomass and Bioenergy*, *35*(5), 1633–1645. doi:10.1016/j.biombioe.2011.02.033

Sahoo, B. B., Sahoo, N., & Saha, U. K. (2009). Effect of engine parameters and type of gaseous fuel on the performance of dual-fuel gas diesel engines-A critical review. *Renewable & Sustainable Energy Reviews*, *13*(6–7), 1151–1184. doi:10.1016/j.rser.2008.08.003

Sankaralingam, R. K., & Venugopal, (2016). Current trends on combustion control methodsusing fuel reactivities. *International Journal of Automotive & Mechanical Engineering, 13*(3), 3788–3811. Retrieved from http://10.0.59.178/ijame.13.3.2016.20.0310%0Ahttp://ezlibrary.technion.ac.il/login?url=http://search.ebscohost.com/login.aspx?direct=true&db=aci&AN=122893253&lang=he&site=eds-live&scope=site

Savery, C., & Cruzan, D. (1972). Methane recovery from chicken manure digestion. *Journal (Water Pollution Control Federation), 14*(2), 2349–2354. Retrieved from http://www.jstor.org/stable/25037689

Sorathia, H. S., & Yadav, H. J. (2012). Energy Analyses To a Ci-Engine Using Diesel and Bio-Gas Dual Fuel- a Review Study. *International Journal of Advanced Engineering Research and Studies, Vol.*, 212–217.

Sudheesh, K., & Mallikarjuna, J. M. (2010). Diethyl ether as an ignition improver for biogas homogeneous charge compression ignition (HCCI) operation - An experimental investigation. *Energy, 35*(9), 3614–3622. doi:10.1016/j.energy.2010.04.052

Tippayawong, N., & Thanompongchart, P. (2010). Biogas quality upgrade by simultaneous removal of CO2 and H2S in a packed column reactor. *Energy, 35*(12), 4531–4535. doi:10.1016/j.energy.2010.04.014

Verma, S., Das, L. M., & Kaushik, S. C. (2017). Effects of varying composition of biogas on performance and emission characteristics of compression ignition engine using exergy analysis. *Energy Conversion and Management, 138*, 346–359. doi:10.1016/j.enconman.2017.01.066

Vijay, V. K. (2007). Biogas refining for production of bio-methane and its bottling for automotive applications and holistic development. *Proc. of International Symposium on Eco Topia, 07*, 35–40. Retrieved from http://www.esi.nagoya-u.ac.jp/h/isets07/Contents/Session05/1018Vijay.pdf

Yao, M., Zheng, Z., & Liu, H. (2009). Progress and recent trends in homogeneous charge compression ignition (HCCI) engines. *Progress in Energy and Combustion Science, 35*(5), 398–437. doi:10.1016/j.pecs.2009.05.001

Yilmaz, I. T., & Gumus, M. (2017). Investigation of the effect of biogas on combustion and emissions of TBC diesel engine. *Fuel, 188*, 69–78. doi:10.1016/j.fuel.2016.10.034

Yoon, S. H., & Lee, C. S. (2011a). Effect of biofuels combustion on the nanoparticle and emission characteristics of a common-rail di diesel engine. *Fuel, 90*(10), 3071–3077. doi:10.1016/j.fuel.2011.05.007

Yoon, S. H., & Lee, C. S. (2011b). Experimental investigation on the combustion and exhaust emission characteristics of biogas–biodiesel dual-fuel combustion in a CI engine. *Fuel Processing Technology, 92*(5), 992–1000. doi:10.1016/j.fuproc.2010.12.021

KEY TERMS AND DEFINITIONS

Biogas: A combustible gas mixture formed by the decay of biomass in an oxygen deficient environment. Its primary constituents are methane and carbon dioxide in variable proportions.

Compression Ignition Engine: An IC engine where fuel is ignited by injecting it into hot compressed air, e.g. diesel engine.

Dual Fuel Engine: A CI engine where the primary fuel is inducted along with air and the secondary (pilot) fuel injected into the cylinder to ignite the compressed primary fuel-air mixture.

Homogeneous Charge Compression Ignition (HCCI) Engine: A CI engine where one or more fuels are homogeneously mixed with air in the intake, inducted into the cylinder and auto-ignited by compression.

Ignition Delay: The interval between the end of fuel injection and first instance of auto-ignition of the injected fuel. This definition is applicable to CI engines with in-cylinder injection.

Methane Enrichment: Purification of biogas by removing CO_2 in order to improve the combustion properties of biogas.

Methane Fraction: Fraction (or percentage) by volume of methane in biogas.

Secondary Fuel Energy Ratio (Sfer): Ratio of the energy released by combustion of the secondary fuel (diesel or DEE in the present case) to the overall energy release.

Simulated Biogas: A prepared mixture of methane and carbon dioxide.

Chapter 7
Comparative Assessment of Various Nanoadditives on the Characteristic Diesel Engine Powered by Novel Tamarind Seed-Methyl Ester Blend

V. Dhana Raju
Lakireddy Bali Reddy College of Engineering, India

Harish Venu
Vel Tech Rangarajan Dr. Sagunthala R&D Institute of Science and Technology, India

Lingesan Subramani
https://orcid.org/0000-0002-5785-4176
Madras Institute of Technology, India

S. Rami Reddy
Lakireddy Bali Reddy College of Engineering, India

ABSTRACT

This chapter focuses on enhancing the performance, combustion, and emission characteristics of a novel biodiesel blend-a mix of diesel (80%) and tamarind seed oil (20%), represented as tamarind seed methyl ester (TSME) with alumina oxide (Al2O3), Carbon nano tubes (CNT), and Cerium oxide(CeO2) considered as potential nanoparticles. These were added to TSME at concentration of 50 ppm and were uniformly dispersed in the biodiesel blend with the help of a magnetic stirrer as well as an Ultrasonicator to attain stable suspension. The immersed nanoparticles in the tamarind seed oil blend exhibit multiple advantages such as an enhanced air-fuel mixing, better oxidation process, larger surface area to volume ratio results in higher brake thermal efficiency, as well as a significant reduction in smoke opacity, hydrocarbon, and carbon monoxide emissions.

DOI: 10.4018/978-1-7998-2539-5.ch007

INTRODUCTION

The fast exhaustion of raw petroleum assets, rising natural pollution concerns and climb in fuel costs have required a more prominent attention on the need to exploitation of the biodiesel as an alluring sustainable feedstock for the diesel engine. Biodiesel is generally viewed as a domain inviting, prudent and plentifully accessible energy source. Over the past numerous years, researchers have analyzed distinctive feedstock of biodiesel got from linseed oil, palm oil, cotton seed oil, jatropha seed oil, pongamia oil, sunflower oil, coconut seed oil, lemon feel oil and so forth., and completed broad examinations on their impact on the characteristics of the compression ignition engine. Ramakrishnan et al. (2018) examined the comparative assessment of the diesel engine characteristics with and without preheated neem and pumpkin biodiesel blends. They noticed diesel had shown higher brake thermal efficiency than biodiesel blends. It was due to higher heating value of diesel. Preheated biodiesel blends neem oil and pumpkin oil have shown improved thermal efficiency and also greater reduction in engine tailpipe emissions like carbon monoxide, smoke and hydrocarbons when compared to the without preheated blends. Agarwal et al.(2008) studied the performance and emission characteristics of the various vegetable oils like mahua oil, linseed oil and rice bran oil as viable sources for the diesel fuel replacement. They found some problems such as operational and durability problems while engine was operated with the vegetable blends. Further, they suggested transesterification technique was highly effective approach for reducing the viscosity and the above-mentioned biodiesel problems. Venu and Madhavan (2017) performed experiments on diesel engine with different levels of alumina nano particles. They observed significant reductions in brake specific fuel consumption, oxides of nitrogen, and hydrocarbons. However, they were reported that a marginal increment in the carbon monoxide and smoke emissions at all load conditions. Basha and Anand (2014) studied the characteristics of the diesel engine fuelled with carbon nanotube and diethyl ether added biodiesel blends. They reported that additions fuel additives to the biodiesel fuel had shown promising results of the diesel engine.

Dhana Raju et al. (2018) investigated the viability of various biodiesel blends prepared from the tamarind seed oil through the transesterification process for the diesel engine applications. They found enhance performance and reduced engine tailpipe emissions for 20% blending of tamarind biodiesel with diesel fuel. Chen et al. (2018) studied the influence of three nanoparticles namely silicon oxide, alumina oxide and carbon nanotubes as fuel catalysts for the augmentation of combustion, performance and emission characteristics of the diesel engine. They used these nanoparticles at different levels like 25ppm, 50ppm and 100ppm with the diesel fuel and also they have done the ultraviolet-visible spectrophotometer for stability analysis of the nanoparticle blended diesel fuel. They noticed better stability for the silicon oxide and alumina oxide. However, the carbon nanotubes were least stable. They concluded that use of nanoparticles have shown significant reductions in NO_X emissions. Ramesh et al. (2018) explored the viability of the poultry litter biodiesel as an alternative fuel for diesel in diesel engines. They have tested 20% poultry litter biodiesel along with 30mg/l of alumina nanoparticle as fuel additive. They found greater reductions in hydrocarbon and carbon monoxide oxides of nitrogen emissions by the addition of nanoparticles when compared to the biodiesel blend without nanoparticles.

Nour et al. (2018) examined the effect of alumina nanoparticles at various concentrations such as 25, 50, 75 and 100mg/l into diesterol blended fuel (10% Jojoba biodiesel+ 20% ethanol +70% diesel) on the exploitation of diesel engine characteristics. They observed greater reduction in oxides of nitrogen and hydrocarbon emissions at dose level of 25mg/l when compared to the other additions of nanoparticles, but significant enhancement in peak cylinder pressure and considerable reduction in specific fuel consumption was found for 75mg/l. They suggested the dose level of 75mg/l alumina nanoparticle addition had the overall improvement of the characteristics of the diesel engine. Wu et al. (2018) investigated the influence of carbon coated aluminum nanoparticles as fuel catalysts for the biodiesel-diesel blends on the performance and emission characteristics of the diesel engine. The test results revealed that the dispersion nanoparticles have shown 14.5% reduction of hydrocarbon emissions, 6% reduction of brake specific fuel consumption and 10% reduction of carbon monoxide when compared to the 10% biodiesel blend.

El- Seessy et al. (2018) studied the influence of aluminum oxide nanoparticles on the characteristics of the diesel engine powered with 20% jojoba methyl ester-diesel blend at different loading conditions. They used the Al_2O_3nanoparticle at various concentrations of 10mg/l, 20mg/l, 30mg/l, 40mg/l and 50mg/l to the 20% biodiesel blend by the use of ultrasonic stabilization. They found promising results for 20mg/l additions to the biodiesel blend which revealed a significant reduction in exhaust emissions such as 60%by CO, 80% by HC and 70% by NO_X at full load. Selvan et al. (2009) analyzed the influence of cerium oxide nano additive in diesel-biodiesel-ethanol blends. They observed that the fuel consumption was increased and BTE was decreased of the ternary blends in comparison with diesel fuel. They also found that, addition of cerium oxide lowered the heat release rate for ternary blends. Shaafi and Velraj (2015) experimentally investigated the effect of alumina nano particles in biodiesel blends and they found higher heat release rate and cylinder pressure for alumina blended fuel, it was due to to higher surface area to volume ratio of alumina along with inherent oxygen present in soybean biodiesel which altogether makes rapid advances in combustion rate. Hosseini et al. (2017) experimentally investigated the effect of alumina nano particles as catalysts to various blend concentrations of waste cotton seed oil at levels of 30ppm, 60ppm and 90ppm and the test results they revealed improved performance and lowered exhaust emissions. Sivakumar et al. (2018) investigated the effect of 50ppm and 100ppm alumina nano additives for 25% pongamia biodiesel blend and they found 100ppm doped alumina –pongamia biodiesel blend 8.26% higher brake thermal efficiency and 16.16% reduced brake specific fuel consumption when compared to 25% pongamia blend. Prabu and Anand (2016) evaluated the combined effect of alumina and ceriua nano additives doped in jatropha biodiesel fuelled in a single cylinder diesel engine. They used nano particles at levels of 10ppm, 30ppm and 60ppm with the help of ultrasonicator. They found that jatropha biodiesel blended with 60ppm nano additives exhibits lowered BSFC of 0.283bkg/kWh and higher BTE of 31.52% which are closer to diesel fuel owing to higher surface area to volume ratio of nano additives improving the rate of combustion. Yilmaz and Atmanli (2017) used ternary blends (pentanol-diesel-biodiesel) in diesel engine. They noticed that use of ternary blends in diesel engine shown higher specific fuel consumption and reduction in brake thermal efficiency over the diesel. It was owing to superior evaporation of latent heat leads to produce cooling effect and subsequently reduction in combustion efficiency. Also, exhaust emissions were increased with increase in concentration of 1-pentanol blends in the biodiesel.

Further, the addition of nanoparticles to the diesel-biodiesel blends has shown larger surface area to volume ratio which leads to improved oxidation and evaporation [Ding et al. (2007); Yamin et al. (2013)] Nano-added substances blended with mineral diesel and biodiesel prompted advance flash point, kinematic viscosity and its extra properties inferable from higher volume nature of the nano added substances (Khan et al.2015). Sajeevan and Sajith et al. (2016) studied the feasibility of the jatropha biodiesel mixed with cerium-zirconium mixed oxides as nano additives in a diesel engine. Among the observation, the researcher revealed that the addition of cerium oxide nanoparticles had shown significant reduction in engine tailpipe emissions. They also proposed that BSFC is lowered linearly and marginal improvement in thermal efficiency due to faster combustion rate of biodiesel. Arockiasamy and Anand

(2015) explored the potential of alumina oxide and cerium oxide nanoparticle in jatropha fuel powered diesel engine. They found significant decline in the emission of nitrogen oxide by 9%, smoke by 17%, HC by 33% and CO by 20% due to alumina oxide addition and cerium oxide reduced NO_x, smoke, HC and CO by 7%, 20%, 28% and 20% respectively due to fuel catalyst behaviour of nano particles. Selvan et al. (2014) examined characteristics of VCR diesel engine powered by stable Diesterol- Ceria –Carbon Nano Tubes (CNT) blends. From the observation, the author stated that by addition of CNT and Ceria nanoparticles in Diesterol fuel, shows increased pressure of cylinder and lowered delay period due to the catalyst effect of nanoparticle addition.

Venu et al. (2019) investigated the effect of alumina nanoparticles doped ternary fuel operated diesel engine at various loading conditions. They reported that addition of nanopartciles have shown promising results in terms of higher efficiency and reduced emissions at all load conditions. Further, the authors also compared the test results with the simulation results generated by the Diesel-RK software and they concluded that closer values were found from the experimental tests and simulation software results. Raju et al. (2019) examined the characteristics of the diesel engine powered with novel tamarind biodiesel with nanoparticles at alumina oxide at various concentrations of 30ppm, 60ppm and 90ppm and blends were homogeneously achieved by the application of ultrasonicator. The test results revealed that the addition of nanoparticles have shown improved brake thermal efficiency and significant reduction in engine harmful exhaust emissions. Saibabu et al. (2019) explored the application of butylated hydroxyl anisole (BHA) as promising antioxidant additives for the waste tamarind biodiesel fuelled compression ignition engine for analysis of engine characteristics. They presented that addition of 2000ppm BHA have shown 1.79% enhancement in thermal efficiency and 23.32% reduction in oxides of nitrogen emissions at full operations when compared to the 20% blend of tamarind biodiesel. Hawi et al. (2019) conducted tests on diesel engine with iron-doped cerium oxide nanopartciles added waste cooking oil methyl ester at different concentrations. All the tests were performed at rated speed of 2000rpm with varying load of 0 to 12N. The experimental test results revealed that CO emissions were reduced by 24.6%, NOX emissions reduced by 15.7% when compared to the waste cooking methyl ester at full load.

Even though many authors contemplated the numerous biodiesel feedstock's for diesel engine applications. A very limited work was explored in the literature by the use of tamarind seed methyl ester as biodiesel for diesel engine. Furthermore, the addition of various nanoparticles in tamarind biodiesel blend is a methodology proposed here that examines the important characteristics as well as generates credible technical data on tamarind seed oil as a viable renewable fuel for effective use in biodiesel engines. For the experimental purpose, three different fuels are prepared by using alumina oxide (Al_2O_3), Carbon nanotubes (CNT) and Cerium oxide (CeO_2) at same concentration addition of 50 ppm with TSME (80% of diesel and 20% of tamarind biodiesel) biodiesel blend to examine the performance, combustion and emissions characteristics of the diesel engine.

MATERIALS AND METHODS

Tamarind Seed Oil as an Alternative Fuel

The origin of Tamarind tree or "*Tamarindus indica*" may historically be traced to the African continent. However, it is found abundantly across the length and breadth of the Indian subcontinent from the distant past. Presently, India enjoys the status of the largest supplier of tamarind fruit in the world, which is

Figure 1.

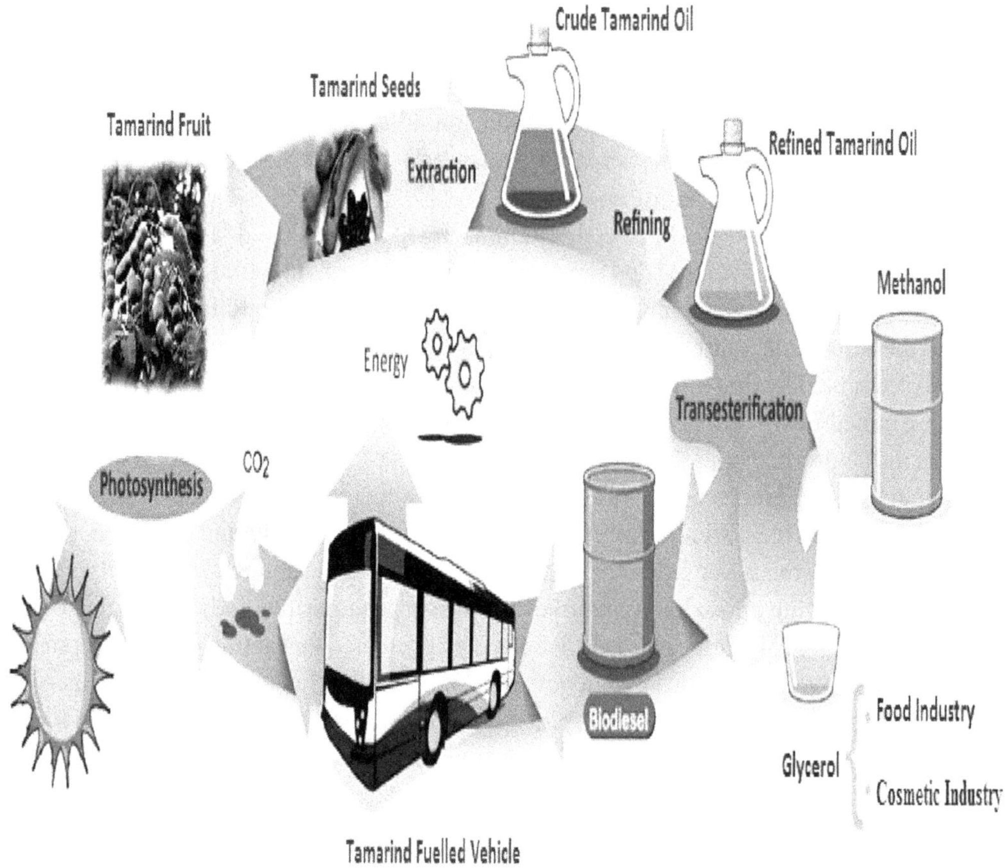

popularly used in various cuisines or preparations worldwide. It is largely found in the states like Madhya Pradesh, Kerala, Andhra Pradesh, Karnataka, West Bengal, and Tamilnadu. The tamarind seeds used in the present study for the production of biodiesel were collected from locally available trees. It offers high potential as a dry land crop since its investment is one-time and the harvest is lifelong. The seeds extracted from the tamarind fruit were dried to remove moisture content and then, mechanically pressed to produce crude tamarind seed oil. Since the crude bio fuel has higher viscosity than diesel, tamarind biodiesel was produced through transesterification process from the crude oil. The physio-chemical properties of tamarind seed methyl ester were evaluated experimentally and it showed encouraging biodiesel parameters such as higher cetane number, higher inherent oxygen content, low sulfur content and higher flash and fire point than diesel. Another environmental benefit of biodiesel is less green house effect and less air pollution. Tamarind biodiesel cycle is presented in Figure 1. The CO_2 emissions released by the transportation vehicles when fuelled with tamarind biodiesel is again absorbed by the plants for photosynthesis process for its growth and it reduces the global warming and green house gas emissions.

Nano Additives as Fuel Catalyst

Nanoparticles were considered as fuel catalyst and it has many merits like higher energy density, enhanced combustion phenomenon and reduced engine tailpipe emissions. Novel nano additive is renowned as a better fuel catalyst material in these recent centuries.In the present work, tamarind seed biodiesel blend (TSME) is brought under the influence alumina oxide (Al_2O_3), Carbon nanotubes (CNT) and Cerium oxide (CeO_2) as nano fuel catalysts for studying engine characteristics that is further analyzed with diesel fuel. Then, the nano additives of alumina, Carbon nanotubes and Cerium oxide have been synthesized by the application of sol-gel method leading to crystallite grain size characterization that uses a Scanning Electron Microscope (SEM). Table 1 represents the properties of Al_2O_3, CNT and CeO_2 nanoparticles.

The prepared aluminium, carbon nanotubes and cerium oxide nanoparticles were then characterized for grain size by using a Scanning Electron Microscope (SEM) as presented in Figure 2 (a), Figure 3 (a) and Figure 4 (a). The SEM morphology of Al_2O_3, CNT and CeO_2 nanoparticles maintain a definite crystalline nature with minimal agglomeration as well as aggregate formation. The average grain size of alumina oxide is measured at about 20 nm, whereas it is measured at 33 nm for carbon nanotubes and 50nm for CeO_2. The X-ray diffraction (XRD) pattern of synthesized Al_2O_3, CNT and CeO_2 nanoparticles is depicted in Figure 2 (b), Figure 3 (b) and Figure 4 (b) respectively. The aluminium oxide (Al_2O_3), Carbon nanotubes (CNT) and Cerium oxide (CeO_2) nano particles of 50 ppm accurately weighed mass fractions that are disseminated in the tamarind seed methyl ester biodiesel blend by using Ultrasonicator at a frequency of 40 kHz, 120 W for 30 minutes, in order to produce homogeneous mixture.

The nano additives were blended homogeneously with biodiesel by using a magnetic stirrer as well as an ultrasonicator. The physio-chemical properties of the tested fuels are determined experimentally and analyzed with the base fuel as listed in Table 2. The immersed metallic nanoparticles in the tamarind seed oil blends exhibit multiple advantages such as enhanced air-fuel mixing, better oxidation process, larger surface area to volume ratio etc.

Figure 2.

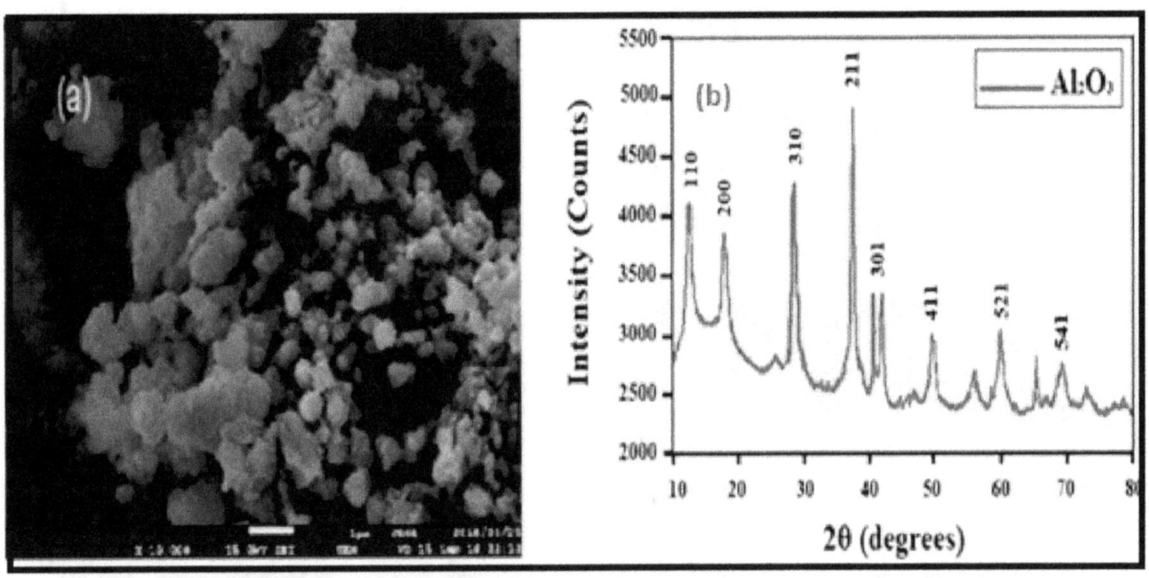

Figure 3.

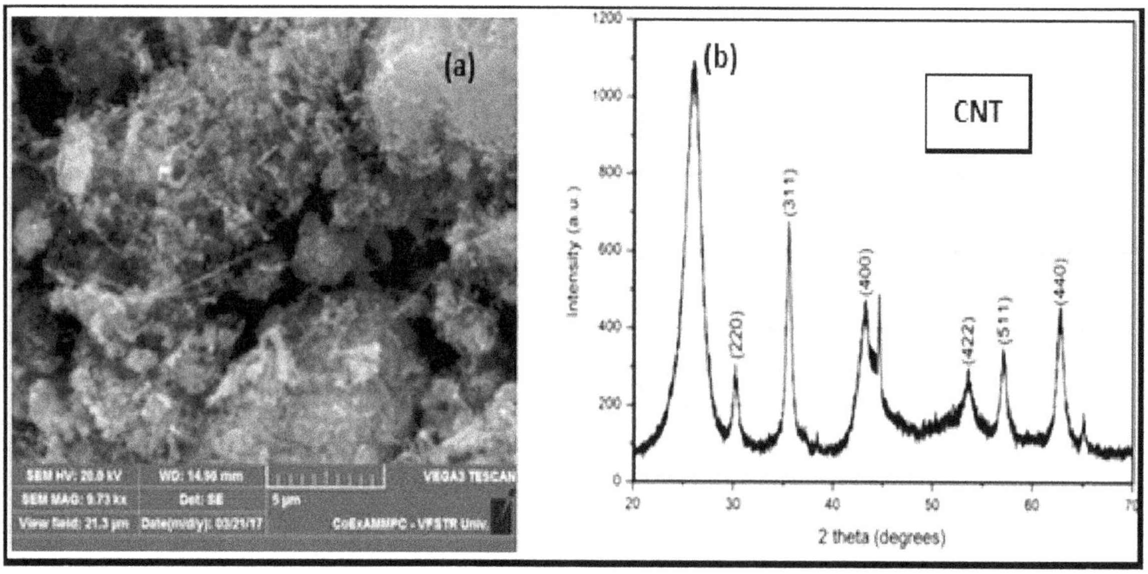

Figure 4.

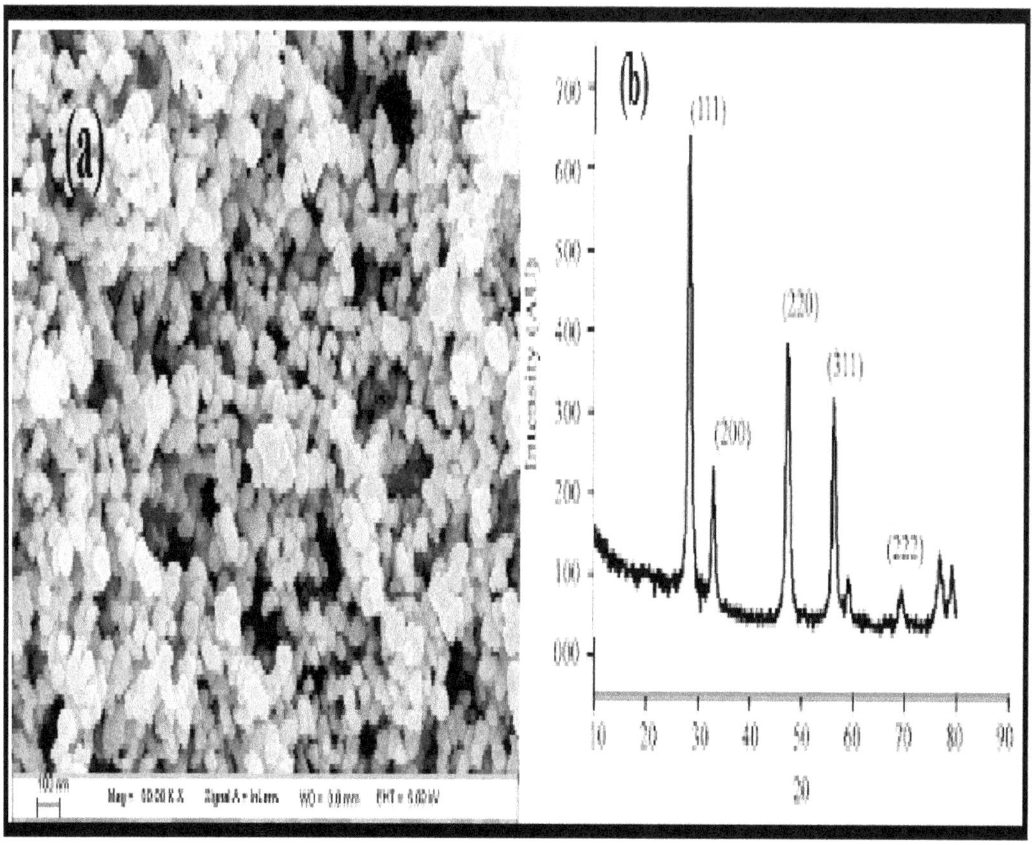

Table 1 Properties of Al_2O_3, CNT and CeO_2 nanoparticles

Nano additive	Alumina oxide	Carbon nanotubes	Cerium oxide
Physical appearance	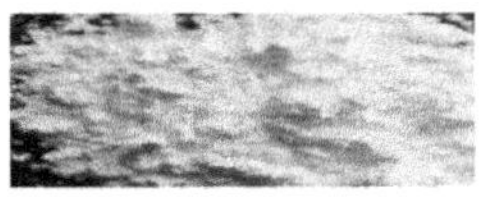		
Chemical Formula	Al_2O_3	CNT	CeO_2
Crystallite size	20 nm	33.77nm	50nm
surface area	20 m^2/g	39 m^2/g	18 m^2/g
Shape	spherical	tubes	octahedrons/cubes
Purity	99.9%	93%	99%
Form	powder	powder	Pale yellow powder
APS	50 - 200 nm	Outer Diameter 10-30 nm, Length 1-10 μm	50-100nm

Table 2. Fuel properties of diesel and biodiesel feedstock

Properties	Test method ASTM 6751	Diesel	TSME	TSME Al_2O_3 50ppm	TSME CNT 50ppm	TSME CeO_2 50ppm
Density (kg/m3) @15°C	830	844	844.3	846.4	847.9	848
Kinematic Viscosity(Cst)	3.05	3.84	3.8	3.88	3.91	3.93
Calorific Value(MJ/kg)	42.5	42.5	41.72	41.86	41.74	41.76
Flash point (o C)	56	73	74	77	79	81
Cetane index	43	47	51	53	52	52

EXPERIMENTAL SETUP

Experimental investigations were conducted on a four stroke, single cylinder, natural aspirated, water cooled direct injection compression ignition engine. Schematic arrangement of the experimental test setup is as shown in Figure 5. These diesel engines are popularly used in irrigation applications as portable generators in India. The required instruments have been arranged after inspection and calibration to estimate engine parameters as well as exhaust emissions. The diesel engine is utilized for examination which has the displacement volume of 661 cc with compression ratio of 17.5:1. The diesel engine develops 5.2 kW at an appraised speed of 1500 rpm.

The tail pipe exhaust of direct injection diesel engine contains of different elements such as hydrocarbons (HC), carbon monoxide (CO), oxygen ($O_{2)}$, carbon dioxide ($CO_{2)}$ and nitrogen oxides (NO_X) emissions. The concentrations of exhaust emissions (CO, CO_2, HC, O_2 and NO_X) were measured with an AVL 444N five gas analyzer. An AVL 437C smoke meter was used to measure the concentration of smoke present in the engine exhaust. The intensity of smoke opacity was measured with the help of an. It works on the principle of light extinction and the smoke opacity is measured in percentage. The technical details of the experimental setup are given in Table 3.

Figure 5.

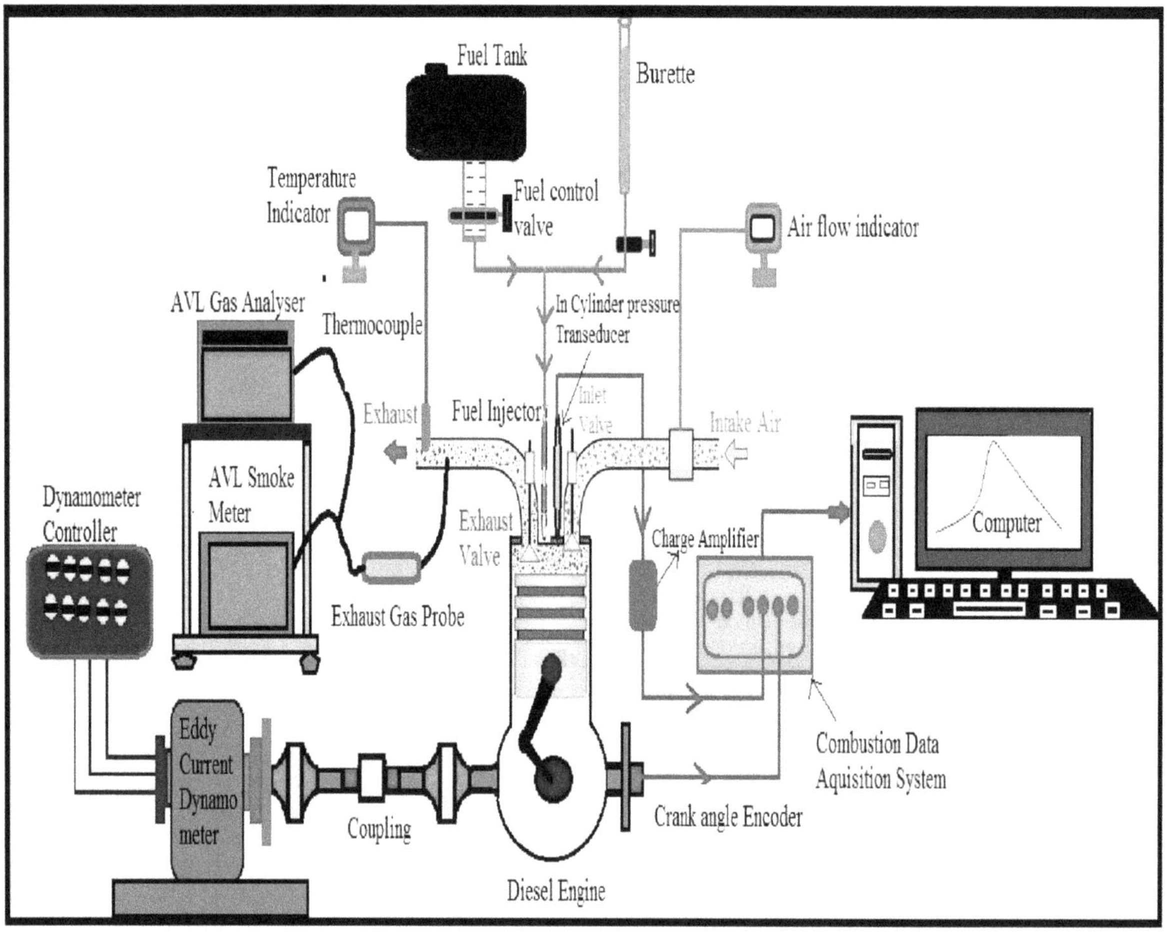

AVL pressure transducer (GH14D/AH01) and AVL crank angle encoder (602-TI0602) are used to measure the maximum in-cylinder pressure variation with respect to crank angles respectively. The obtained pressure variation data with respect to crank angle is fed to the computer with the help of the data acquisition system (AVL INDMICRA). The data acquisition systems software is used to process the pressure variation data with respect to crank angle and deduce the variation of heat release rate with the principle of conservation of energy.

RESULTS AND DISCUSSION

Brake Thermal Efficiency (BTE)

The variation of brake thermal efficiency with brake mean effective pressure for all the fuels used in this examination is delineated in Figure 6. Thermal efficiency of any engine is mainly depends on the net energy available in the fuel. It represents the efficient translation of energy present in given fuel

Table 3. Technical specifications of the diesel engine setup

Parameter	Specification
Make	Kirloskar TV1
Engine Type	Diesel engine
Type of injection	Direct injection
Type of cooling	Water cooled
Rated Power	5.2 kW
Rated Speed	1500 rpm
Bore	87.5 mm
Stroke	110 mm
Compression ratio	17.5:1
Displacement	661 cm^3
Injection timing	23° BTDC
Injection pressure	200 bar
Inlet valve open before TDC	4.5°
Inlet valve closes after BDC	35.5°
Exhaust valve open before BDC	35.5°
Exhaust valve closes after TDC	4.5°

to the mechanical power generated at the engine crank shaft. From the graph, it is observed that brake thermal efficiency for all fuels are shown similar pattern of increasing nature with the engine load. BTE is found higher for diesel when compared to other tested fuels, due to high energy content. The brake thermal efficiency of diesel, TSME, TSME Al_2O_3 50ppm, TSME CNT 50ppm and CeO_2 50ppm are 36.12%, 32.39%, 34.91%,33.12% and 32.83% respectively at full load condition. The BTE of TSME Al_2O_3 50ppm is 7.78% higher than the TSME biodiesel blend. The considerable enhancement in BTE is fundamentally credited to improved atomization, upgraded evaporation rate and better air-fuel blending within the sight of nano metal particles and furthermore the bigger surface zone to volume proportion prompting total burning. The nanoparticles present in biodiesel acts as a fuel catalyst for ignition and oxygen support for the total consuming of the air-fuel mixture (Sajeevan and Sajith 2016)

Brake Specific Fuel Consumption (BSFC)

It is a measure of efficiency of any fuel supplied to the heat engine and it is evaluated as quantity of fuel consumed per unit power development. The variations of brake specific fuel consumption of the various fuels with respect to the brake mean effective pressure is shown in Figure 7. It is highly influenced by several parameters such as calorific value, viscosity, density and cetane number. It is found that brake specific fuel consumption is decreased with increase in engine load for all the tested fuels. Further, all the experimented fuels are followed same nature of fuel consumption throughout the engine operation. The brake specific fuel consumption of diesel, TSME, TSME Al_2O_3 50ppm, TSME CNT 50ppm and CeO_2 50ppm are 0.233 kg/kWh, 0.256 kg/kWh, 0.242 kg/kWh, 0.245 kg/kWh and 0.252 kg/kWh respectively at full load condition. The BSFC is found minimum for diesel followed by TSME Al_2O_3 50ppm at full

Figure 6.

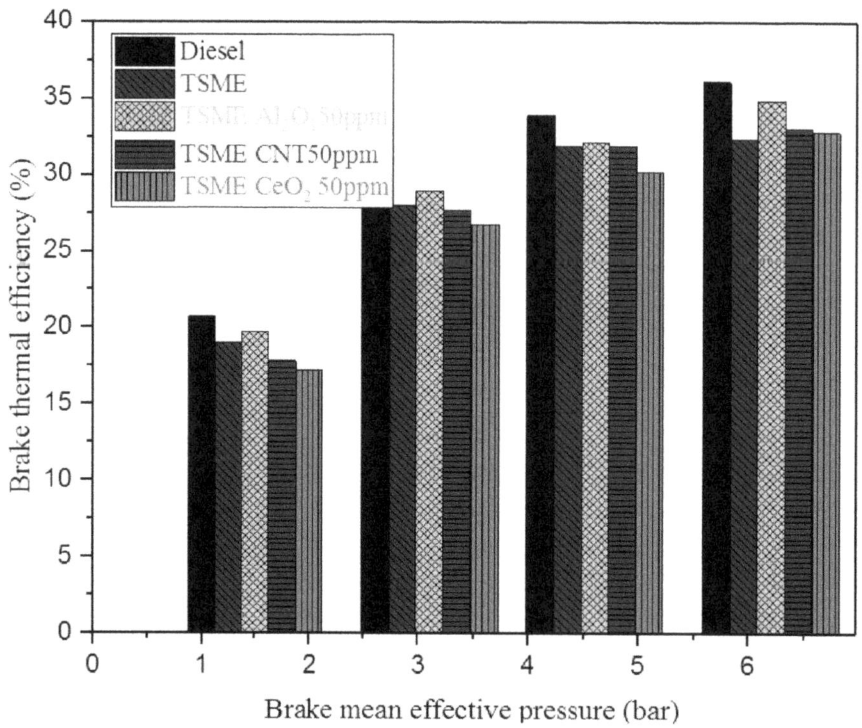

load condition. It is noticed that BSFC for TSME Al_2O_3 biodiesel blend is shown 5.5% lower than the TSME at full load. Also, the BSFC for other nanoparticles like CNT and CeO_2 are shown reduced values over TSME at all load operations. This is owing to the fuel catalyst behaviour of nanoparticles results in better combustion phenomenon in the engine cylinder (Basha and Anand 2014)

Cylinder Pressure (CP)

The variations of cylinder pressure with respect to the crank angle for all fuels used in this experimentation are delineated in Figure 8. Cylinder pressure played a vital role in combustion process and it is generated by the equivalent effect in combustion analysis, power output of the engine followed by the exhaust tailpipe emissions. The cylinder pressure is mainly influenced by ignition delay, fuel viscosity and cetane number and fuel concurrence effect. From the figure, the cylinder pressure for the fuels of diesel TSME, TSME Al_2O_3 50ppm, TSME CNT 50ppm and CeO_2 50ppm are 67.39 bar, 65.86 bar, 66.45 bar, 66.01 bar and 66.21 bar respectively at peak load. The addition of nanoparticles, it is observed minor increase in cylinder pressure owing to improved ignition lag and availability of oxygen in air-fuel mixture, rapid evaporation and fine atomization which advances in the overall combustion phase. At peak load, diesel fuel blend exhibits higher cylinder pressure of 67.39 bar which is attributed to improving delay period, more heating value with shortened diffusion combustion.

Figure 7.

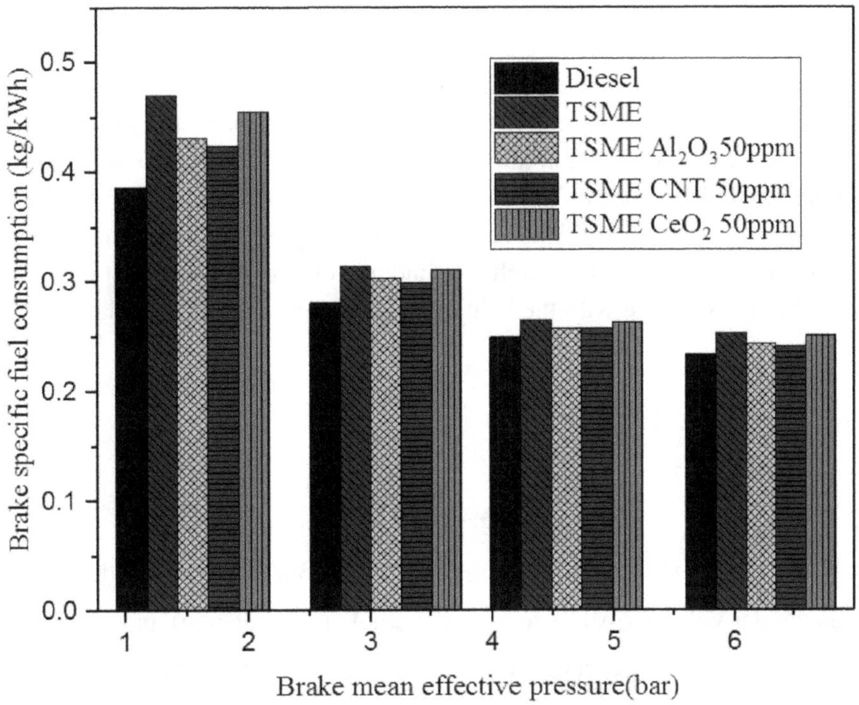

Figure 8.

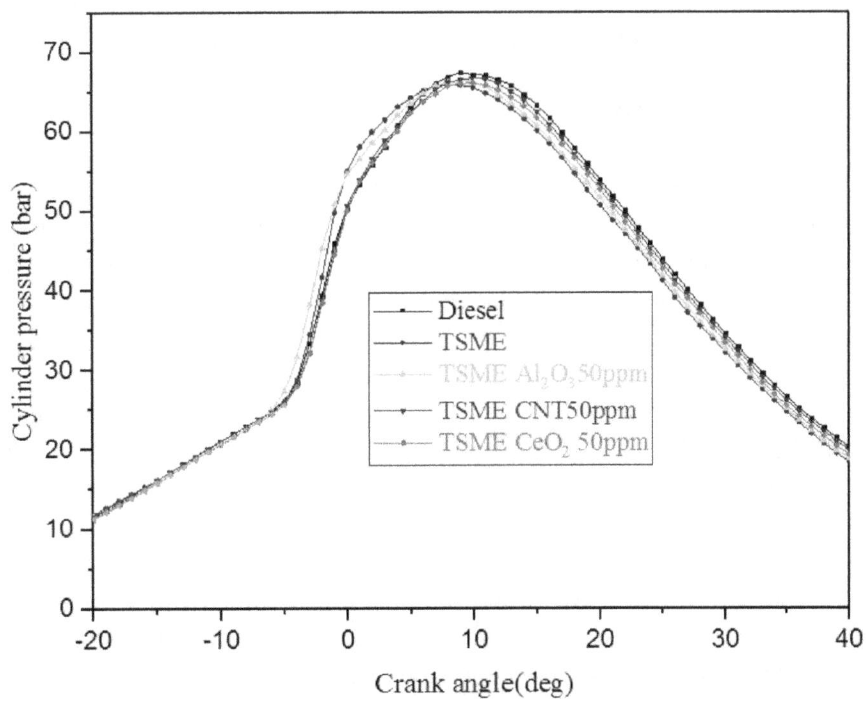

Heat Release Rate (HRR)

The variation of heat release rate at various crank angles for the used fuel samples are depicted in Figure 9. Heat release rate represents the quantity of thermal energy generated in the combustion chamber due to burning of fuel/air mixture. It is noticed from the figure that all the tested fuels are followed similar pattern throughout engine operation. Also, the maximum heat release rate is found for diesel followed by TSME Al_2O_3 50ppm fuel blend at peak load are 68.51 J/degree and 66.63 J/degree crank angle (CA) at -2° before top dead centre respectively. This could be attributed to excess amounts of premixed mixture which induces longer delay followed by higher cylinder pressure. The HRR at every crank angle for different fuels used were evaluated with the help of equation derived from first law of thermodynamics as follows

$$\left(\frac{dQn}{d\theta}\right)=\left(\frac{\gamma}{\gamma-1}P\frac{dV}{d\theta}\right)+\left(\frac{1}{\gamma-1}V\frac{dP}{d\theta}\right)+Q_{lw} \quad (1)$$

where, $\frac{dQ_n}{d_,}$ indicates the net HRR (J/°CA), P represents the instantaneous in-cylinder pressure (N/m^2), V stands for the cylinder volume (m^3), θ denotes the crank angle (degree) and γ is the adiabatic index and is equal to C_p/C_v (kJ/kgK). It depends on temperature and influences the intensity of $\frac{dQ_n}{d_,}$ and Q''_{lw}. Q''_{lw} Indicates blow by losses which were evaluated by Rakopoulas (2012).

From the figure, the heat release rate values for the fuels of diesel TSME, TSME Al_2O_3 50ppm, TSME CNT 50ppm and CeO_2 50ppm are 68.51 J/CA, 60.38 J/CA, 66.63 J/CA, 62.52 J/CA and 60.91 J/CA respectively at full load condition. With nanoparticle addition, there is an enhancement in HRR which can be corroborated with improved ignition delay and ability of nanoparticles in accelerating early pre-mixed combustion which therefore results in increased heat release rate.

Carbon Monoxide Emissions (CO)

Carbon monoxide emissions variation with brake mean effective pressure for different fuels used in this current investigation are presented in Figure 10. It is for the most part generated owing to improper mixing of fuel/air mixture in combustion chamber. It is seen that CO emissions of biodiesel blend with fuel added substances and diesel diminishes with increment in load and least at maximum load. The higher percentage of carbon monoxide is present in engine tailpipe emission is due to inferior combustion, moderate combustion, air/fuel mixture, higher engine load condition. The carbon monoxide emissions for the fuels of diesel TSME, TSME Al_2O_3 50ppm, TSME CNT 50ppm and CeO_2 50ppm are 5.75 g/kWh, 5.17 g/kWh, 3.2089 g/kWh, 3.45 g/kWh and 4.36 g/kWh respectively at full load. From the figure, it is observed that CO emissions are significantly reduced with the applications of these nanoparticles in the tamarind seed biodiesel blend when compared to diesel fuel. Further, it is found the CO reduction for TSME with 50 ppm addition of Al_2O_3, CNT and CeO_2 nanparticles are 37.91%, 33.26% and 15.67% respectively when compared to the TSME at maximum load condition. Hence, the test results revealed that nanoparticles acts as fuel catalysts in the tamarind biodiesel blend which promotes the combustion process results in significant reduction of CO emissions throughout engine operation.

Figure 9.

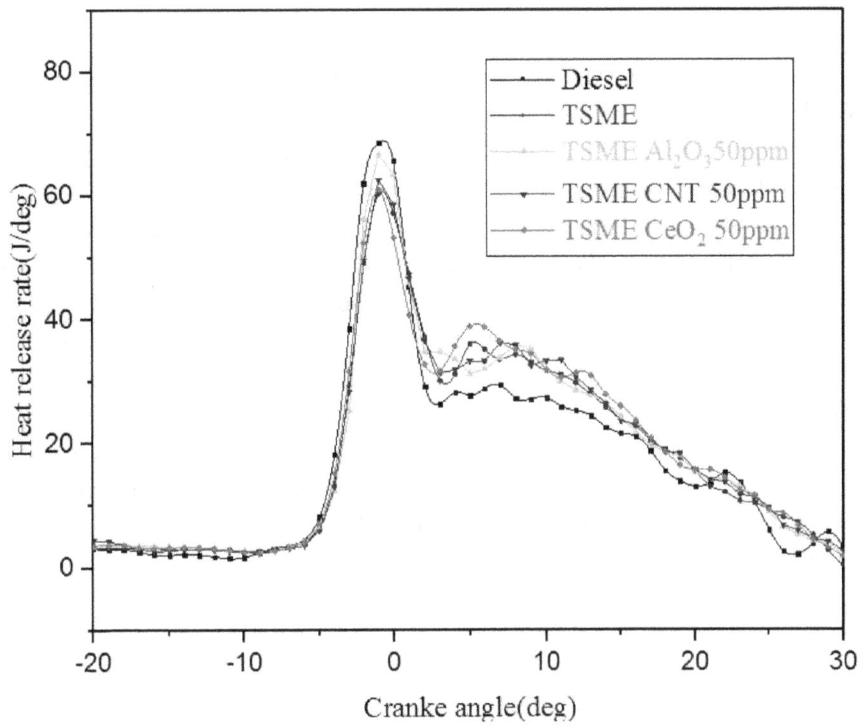

Figure 10.

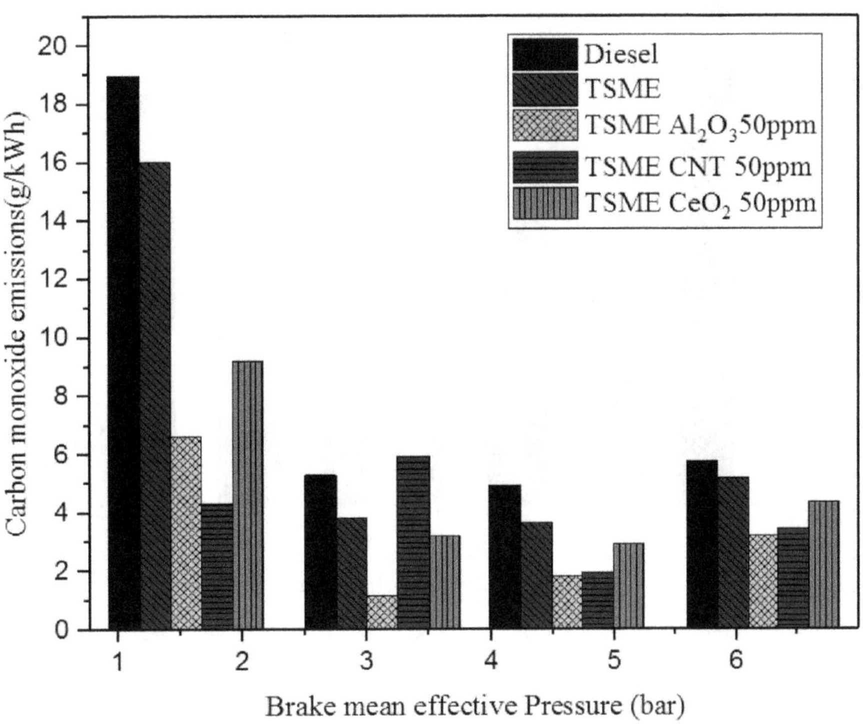

Hydrocarbons Emissions (HC)

Hydrocarbon emissions are generated owing to incomplete burning phenomena. The variations of HC emissions with respect to brake mean effective pressure for the tested fuels are represented in Figure 11. Higher levels of HC emission shows deteriorate engine efficiency and this could be overcome by the addition of nanoparticles as fuel additives. The lower efficiency of any engine indicates higher quantity of HC and CO engine tailpipe exhausts. The HC emissions are reduced by means of dispersion of nano additives in fuel samples which is delineated in the Fig.11. The experimental results reveal that the use of nanoadditives to TSME blend fundamentally decreased the hydrocarbon emissions when contrasted with diesel fuel as well as TSME throughout the engine operating condition. For the entire engine load the diesel fuel got higher UBHC emission when compared to other fuel test samples. The hydrocarbon emissions for the fuels of diesel TSME, TSME Al_2O_3 50ppm, TSME CNT 50ppm and CeO_2 50ppm are 0.197 g/kWh, 0.172 g/kWh, 0.068 g/kWh, 0.081 g/kWh and 0.142 g/kWh respectively for the tested fuels at full load. The reason for nanoparticles blend exhibiting lowered HC emission owing to availability of more oxygen content and activation energy released by the nanoparticles. The HC emissions are reduced significantly by 60.41% for Al_2O_3, 52.91% CNT and 17.45% CeO_2 when compared to the TSME at full load.

Figure 11.

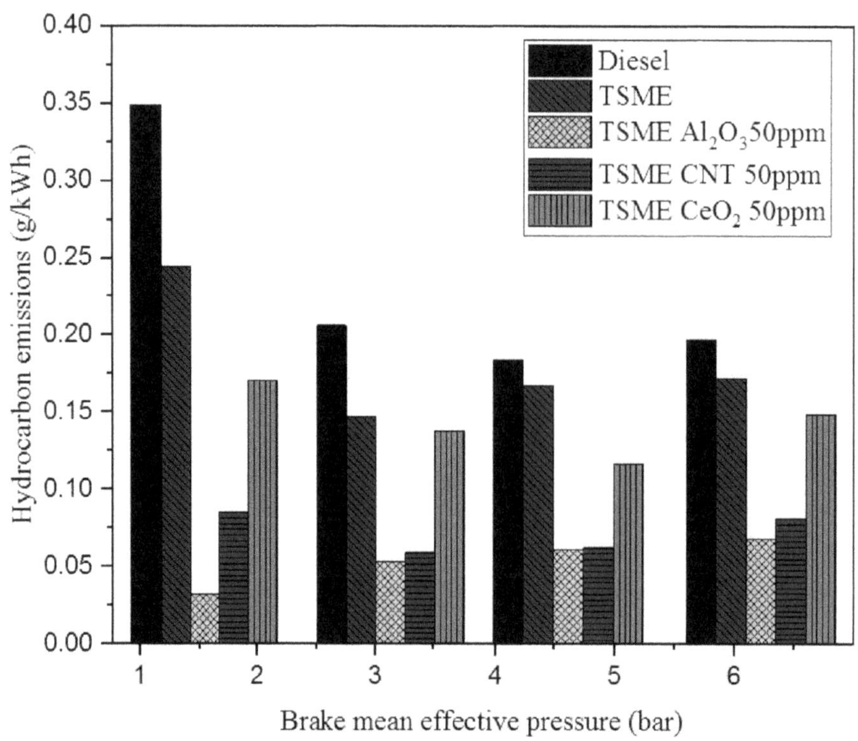

Nitrogen Oxide Emissions (NO_x)

The two common ingredients of nitrogen oxide (NO_x) are nitrogen dioxide and nitric oxide. NO_x is named as toxic and noxious gases for all the environment and human life. Nitrogen oxides development fundamentally relies upon the availability of oxygen and higher temperature while combustion of air/fuel mixture in the combustion chamber. Figure 12 presents the deviation of emissions of nitrogen oxide for the used fuel samples in this investigation at different loads. The three prime stages of NO_x production were derived by Zeldovich Mechanism as displayed below:

$$N_2 + (O) = NO + (N) \tag{2}$$

$$(N) + O_2 = NO + O \tag{3}$$

$$(N) + (OH) = NO + (H) \tag{4}$$

The hydrocarbon emissions for the fuels of diesel TSME, TSME Al_2O_3 50ppm, TSME CNT 50ppm and CeO_2 50ppm are 14.66 g/kWh, 15.69 g/kWh, 1.19 g/kWh, 13.27 g/kWh and 13.44 g/kWh respectively at full load. It is found that higher NO_X generated for TSME when compared to all other fuels used in this investigation. The use of nanoparticles are lowered the oxides of nitrogen formation marginally

Figure 12.

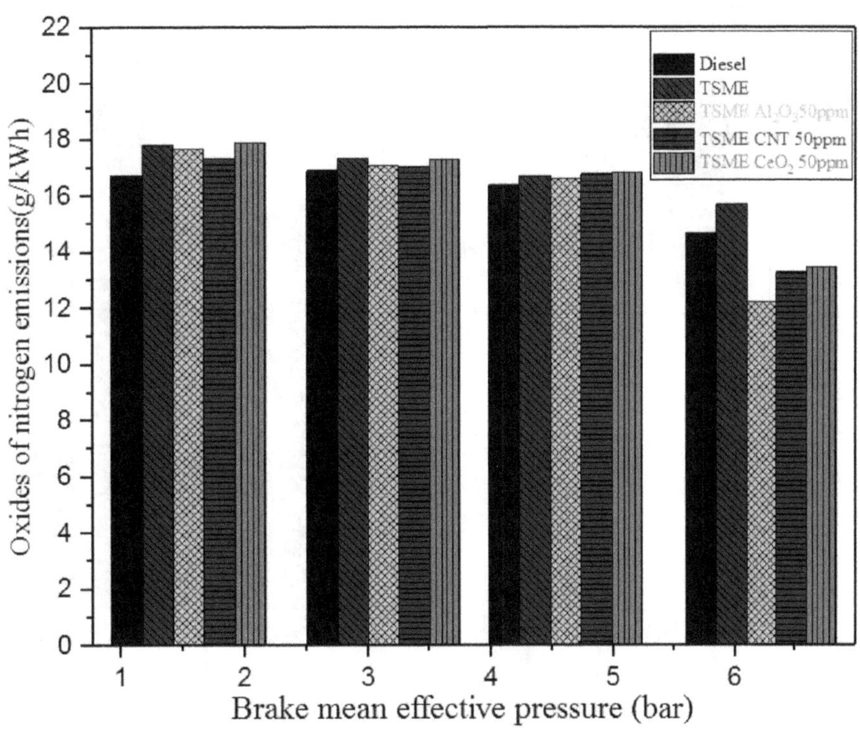

when compared to the diesel as well as the tamarind biodiesel blend at all load conditions. Nanoparticles act as a reducing agent and absorb oxygen during the combustion followed by improved heat transfer between the cylinder walls, lowered peak temperature followed by lowered NO_x formation. The conclusions made by Basha and Anand (2013) about NO_X emissions are also in accordance with the present experimental result.

Smoke Opacity (SO)

The deviation of smoke opacity with respect to brake mean effective pressure for the tested fuel samples like diesel, tamarind biodiesel blend and tamarind biodiesel blend with nanoadditives of Al_2O_3, CNT and CeO_2 at 50ppm additions are delineated in Figure 13. In general, fewer oxygen levels in the fuel deliver improper combustion and it causes higher smoke emission in the tailpipe. It represents the intensity of smoke concentration in engine exhaust. Smoke opacity is measured in percentage. Smoke opacity is reduced when the fuel/air mixture is completely burned in the combustion chamber. The smoke opacity for the fuels of diesel TSME, TSME Al_2O_3 50ppm, TSME CNT 50ppm and CeO_2 50ppm are 68.78%, 66%, 53.4%, 56.7% and 58.7% respectively at full load. The reduction of smoke emissions for Al_2O_3, CNT and CeO_2 of 50ppm addition to the TSME biodiesel blend are 19.1%, 14% and 11.2% respectively when compared to the TSME without nano additives. It is also found; the smoke emissions of diesel fuel are maximum at all load operations when compared to the all tested fuels. The lowered smoke with nano additives can be corroborated with improved evaporation rate, exhaust fuel-air mixing and improved

Figure 13.

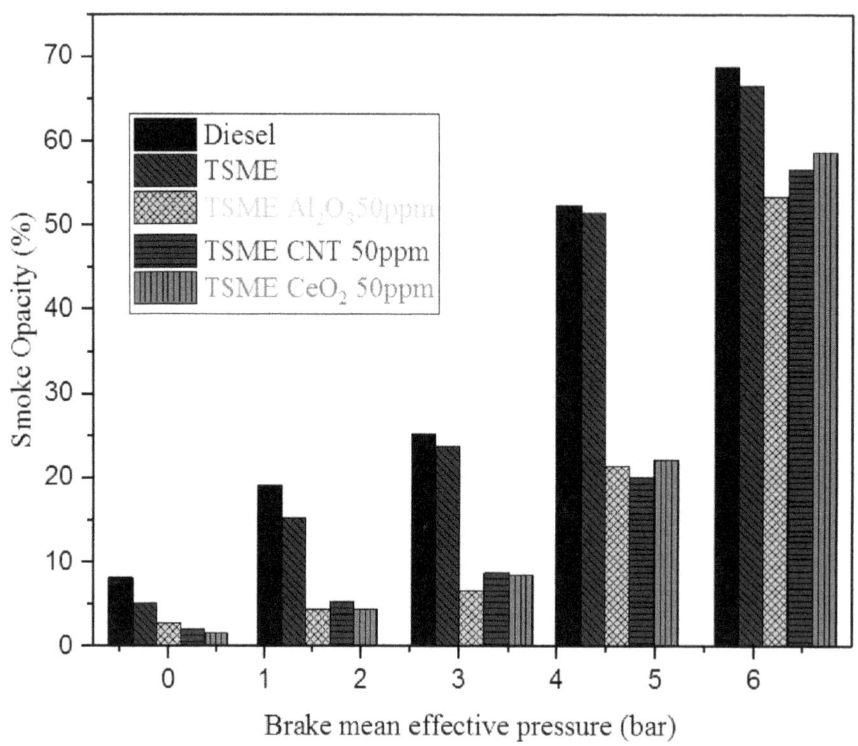

ignition characteristics. All the test fuels were shown similar nature of smoke opacity trend throughout the engine load operation reported by Qi et al. (2010). With increasing additive concentration in blends, there is a reduction in smoke opacity, due to the presence of inbuilt oxygen in alcohols which assists combustion thereby lowering the fuel-rich zones, enhanced mixing rates and lesser probability of soot formation, especially at diffusion combustion phase reported by Ashok et al. (2017).

CONCLUSION

The present research work focused on the effect of various nanoadditives on the diesel engine combustion, performance and emission characteristics powered with tamarind biodiesel blend was investigated and compared with diesel. Different nanoadditives like Al_2O_3, CNT and CeO_2 of 50ppm addition to the TSME biodiesel blend. The major conclusions were drawn from the present investigation are as follows.

- The test results revealed that ternary blends formed with diesel-biodiesel-nanoparticles on the diesel engine were performed successfully.
- The use of 50ppm Al_2O_3 addition to TSME is shown considerable enhancement in BTE when compared to nanoparticles at all load operation. BTE for 50ppm Al_2O_3 addition to TSME20 is 7.78% higher than the TSME biodiesel blend at full load.
- It is also found for TSME with Al_2O_3, CNT and CeO_2 of 50ppm addition have shown significant reductions in engine tailpipe exhaust such as CO, HC and smoke when contrasted to diesel fuel as well as TSME at maximum load.
- Combustion characteristics such as heat release rate and cylinder pressure values are found slightly higher for Al_2O_3 50ppm addition to TSME of 66.63 J/CA and 66.45 bar respectively when compared to other nanoadditives of TSME biodiesel blend.

Therefore, it is recommending that the use of nanopaticles to the waste tamarind biodiesel blend of TSME at smaller volumes to generate promising results of diesel engine.

REFERENCES

Agarwal, D., Kumar, L., & Agarwal, A. K. (2008). Performance evaluation of a vegetable oil fuelled compression ignition engine. *Renewable Energy*, *33*(6), 1147–1156. doi:10.1016/j.renene.2007.06.017

Arockiasamy, P., & Anand, R. B. (2015). Performance, combustion and emission characteristics of a D.I. diesel engine fuelled with nanoparticle blended jatropha biodiesel. *Period. Polytech. Mech. Eng.*, *59*(2), 88–93. doi:10.3311/PPme.7766

Ashok, B., Nanthagopal, K., Jeevanatham, A. K., Bhowmick, P., Malhotra, D., & Agarwal, P. (2017). An assessment of Calophyllum Inophyllum biodiesel fuelled diesel engine characteristics using novel antioxidant additives. *Energy Conversion and Management*, *148*, 953–943. doi:10.1016/j.enconman.2017.06.049

Basha, J. S., & Anand, R. B. (2013). The influence of nano additive blended biodiesel fuels on the working characteristics of a diesel engine. *Journal of the Brazilian Society of Mechanical Sciences and Engineering*, *35*(3), 257–264. doi:10.100740430-013-0023-0

Basha, J. S., & Anand, R. B. (2014). Performance, combustion, and emission characteristics of diesel engine using carbon nano tubes blended jatropha methyl ester emulsions. *Alexandria Engineering Journal*, *53*, 259–273.

Chen, A. F., Adzmi, M. A., Adam, A., Othman, M. F., Kamaruzzaman, M. K., & Mrwan, A. G. (2018). Combustion characteristics, engine performances and emissions of a diesel engine using nanoparticle-diesel fuel blends with aluminium oxide, carbon nanotubes and silicon oxide. *Energy Conversion and Management*, *171*, 461–477. doi:10.1016/j.enconman.2018.06.004

Ding, Y., Chen, H., Wang, L., & Yang, C. (2007).. . *Heat Transfer Intensification Using Nanofluids*, *25*(25), 23–38.

El-Seessy, A. I. (2018). Atiat AMA and El-Batsh HM. The effect of aluminum oxide nanoparticles addition with jojoba methyl ester –diesel fuel blends on a diesel engine performance, combustion, and emission characteristics. *Fuel*, *224*, 147–166. doi:10.1016/j.fuel.2018.03.076

Hawi, M., Elwardany, A., Ismail, M., & Ahmed, M. (2019). Experimental investigation on performance of a compression ignition engine fuelled with waste cooking oil biodiesel-diesel blend enhanced with Iron-doped cerium oxide nanoparticles. *Energies*, *12*(5), 798. doi:10.3390/en12050798

Hosseini, S. H., Taghizadeh-Alisaraei, A., Ghobadian, B., & Abbaszadeh-Mayvan, A. (2017). Effect of added alumina as nano-catalyst to diesel-biodiesel blends on performance and emission characteristics of CI engine. *Energy*, *124*, 543–552. doi:10.1016/j.energy.2017.02.109

Nour, M., EL-Seesy, A. I., Abdel-Rahman, A. K., & Bady, M. (2018). Influence of adding aluminum oxide nanoparticles to diesterol blends on the combustion and exhaust emission characteristics of a diesel Engine. *Experimental Thermal and Fluid Science*, *98*, 634–644. doi:10.1016/j.expthermflusci.2018.07.009

Prabu, A., & Anand, R. B. (2016). Emission control strategy by adding alumina and cerium oxide nano particle in biodiesel. *Journal of the Energy Institute*, *89*(3), 366–372. doi:10.1016/j.joei.2015.03.003

Qi, D. H., Chen, H., Geng, L. M., & Bian, Y. Z. (2010). Experimental studies on the combustion characteristics and performance of a direct injection engine fueled with biodiesel /diesel blends. *Energy Conversion and Management*, *51*(12), 2985–2992. doi:10.1016/j.enconman.2010.06.042

Raju, V. D., Kishore, P. S., & Yamini, K. (2018). Experimental studies on four stroke diesel engine fuelled with tamarind seed oil as potential alternative fuel sustainable green environment. *European Journal of Sustainable Development Research*, *2*. doi:10.20897/ejosdr/78489

Raju, V. D. (2019). V., Kishore, P. S., Kumar, M. H., & Reddy, S. R. Experimental investigation of alumina nanoparticles effects on the performance and emission characteristics of the tamarind seed biodiesel fuelled diesel engine, Materials today. *Proceedings*, *18*(03), 1229–1242. doi:10.1016/j.matpr.2019.06.585

Rakopoulos, D. C. (2012). Heat release analysis of combustion in heavy-duty turbocharged diesel engine operating on blends of diesel fuel with cottonseed or sunflower oils and their bio-diesel. *Fuel*, *96*, 524–534. doi:10.1016/j.fuel.2011.12.063

Ramakrishnan. M, Thansekhar, M. R., & Viswanathan, K. (n.d.). Comparative studies on the performance and emissions of a direct injection diesel engine fuelled with neem oil and pumpkin seed oil biodiesel with and without fuel preheater, Environmental Science and Pollution Control, 25(13), 1-11. Doi:10.100711356-017-0838-9

Ramesh, D. K., Dhanajaya Kumar, J. L., Hemanth Kumar, S. G., Namith, V., & Parashuram, B. J. (2018). And Sharath S. Study on effects of alumina nanoparticles as additive with poultry litter biodiesel on performance, combustion and emission characteristics of diesel engine. Materials today. *Proceedings*, *5*, 1114–1120. doi:10.1016/j.matpr.2017.11.190

Saibabu, K., Raju, V. D., Rao, K. A., Reddy, S. R., & Sai, P. T. Experimental studies on the influence of antioxidant additives with waste tamarind biodiesel on the diverse characteristics of diesel engine, International Journal of Ambient Energy, Published online July 9, 2019. . doi:10.1080/01430750.2019.1636871

Sajeevan, A. C., & Sajith, V. (2016). Synthesis of stable cerium zirconium oxide nanoparticle - Diesel suspension and investigation of its effects on diesel properties and smoke. *Fuel*, *183*, 155–163. doi:10.1016/j.fuel.2016.06.048

Selvan, V. A. M., Anand, R. B., & Udayakumar, M. (2009). Effects of cerium oxide nanoparticle addition in diesel and diesel-biodiesel-ethanol blends on the performance and emission characteristics of a CI engine. *Journal of Engineering and Applied Sciences (Asian Research Publishing Network)*, *4*(7), 1819–6608.

Selvan, V. A. M., Anand, R. B., & Udayakumar, M. (2014). Effect of Cerium Oxide Nanoparticles and Carbon Nanotubes as fuel-borne additives in Diesterol blends on the performance, combustion and emission characteristics of a variable compression ratio engine. *Fuel*, *130*, 160–167. doi:10.1016/j.fuel.2014.04.034

Shaafi, T., & Velraj, R. (2015). Influence of alumina nanoparticles, ethanol and isopropanol blend as additive with diesel–soybean biodiesel blend fuel: Combustion, engine performance and emissions. *Renewable Energy*, *80*, 655–663. doi:10.1016/j.renene.2015.02.042

Sivakumar, M., Sundaram, N. S., & Thasthagir, M. H. S. (2018). Effect of aluminium oxide nanoparticles blended pongamia methyl ester on performance, combustion, and emission characteristics of diesel engine. *Renewable Energy*, *116*, 518–526. doi:10.1016/j.renene.2017.10.002

Venu, H., & Madhavan, V. (2017). Effect of diethyl ether and Al_2O_3 nano additives in diesel-biodiesel-ethanol blends: Performance, combustion and emission characteristics. *Journal of Mechanical Science and Technology*, *31*(1), 409–420. doi:10.100712206-016-1243-x

Venu, H., Raju, V. D., & Subramani, L. (2019). Combined effect of influence of nano additives, combustion geometry and injection timing in a diesel engine fuelled with (diesel-biodiesel-ethanol) blends. *Energy*, *174*, 386–406. doi:10.1016/j.energy.2019.02.163

Wu, Q., Xie, X., Wang, Y., & Roskilly, T. (2018). Effect of carbon coated aluminum nanoparticles ass additive to biodiesel-diesel blends on performance and emission characteristics of diesel engine. *Applied Energy*, *221*, 597–604. doi:10.1016/j.apenergy.2018.03.157

Yamin, J. A., Sakhnini, N., Sakhrieh, A., & Hamdan, M. A. (2013). Environmental and performance study of a 4-Stroke CI engine powered with waste oil biodiesel. *Sustain. Cities Soc.*, *9*, 32–38. doi:10.1016/j.scs.2013.02.002

Yilmaz, N., & Atmanli, A. (2017). Experimental assessment of a diesel engine fuelled with diesel-biodiesel-1-pentanol blends. *Fuel*, *191*, 190–197. doi:10.1016/j.fuel.2016.11.065

Yunus Khan, T. M., Atabani, A. E., Badruddin, I. A., Ankalgi, R. F., Mainuddin Khan, T. K., & Badarudin, A. (2015). Ceiba pentandra, Nigella sativa and their blend as prospective feedstocks for biodiesel. *Industrial Crops and Products*, *65*, 367–373. doi:10.1016/j.indcrop.2014.11.013

Chapter 8
Reduction of NOx on a Single Cylinder CI Engine Running on Diesel–Biodiesel Blends by New Approach

Mohamed Musthafa M.
https://orcid.org/0000-0001-5714-7183
SASTRA University, India

ABSTRACT

Diesel-water emulsion has been used in diesel engine combustion for a long time with encouraging results, but the point of efficiency and NOx trade-off represent a highly challenging task for diesel engines. A new approach was used in this study. The new blends which were obtained by mixing diesel-neem oil biodiesel blend (70:30 by volume) with water (5% by volume), span-80 surfactant (1% by volume), and cetane enhancing additive of Di-tertiary butyl peroxide (0.5% by volume). The blend is designated as B3. This chapter investigates performance and emission characteristics of a single cylinder diesel engine running on B3 fuel. Performance and emission of the engine fueled by B3 fuel results were compared with diesel (D), diesel-biodiesel blend (B1), and diesel-biodiesel with water emulsion through surfactant (B2). B3 fuel had better performance and improved emissions than B1 fuel and diesel fuel, with NOx emission especially reduced by up to 35%.

INTRODUCTION

Biodiesel derived from plant oils and animal fats is considered a promising substitute for petroleum diesel fuel because of its advantages, such as renewability, biodegradability, less environmental toxicity, and superior combustion efficiency (MingHuo et al. 2014). Consistent scientific investigations are performed to find alternate to petroleum oil. Vegetable oils are a viable substitute for diesel fuel, and short-term tests using pure vegetable oils did not show any variation on performance and the results are as good as to diesel fuel (Dwivedi et al.2013). However, the problems arise due to the high oil viscosity

DOI: 10.4018/978-1-7998-2539-5.ch008

after long-term usage. Direct use of vegetable oil blends has been considered to be not agreeable and unrealistic for both direct and indirect diesel engines (Baskar and Senthil Kumar 2017). To reduce the viscosity, transesterification reactions on the triacylgycerol are carried out to convert it into esters by converting each fat oil molecule into one molecule of glycerine and three molecules of ester. Separation of glycerine from the ester produces biodiesel. The prevalent method followed by the researcher for biodiesel production is the transesterification process. Transesterification is the reaction of a fat or oil with an alcohol to form esters and glycerol. The catalyst (Alkalis and acid) and enzymes are used to improve the reaction rate and yield. An alkali-catalyzed process is much faster than the acid-catalyzed process and is used commercially. A diesel-biodiesel blends up to 80:20 by volume is widespread in the worldwide for any biodiesel blends (Nabi et al.2009). The main drawback of commercial use of B20 blend operation is higher NO_x emission (Subramanian 2011; Suresh and Amirthagadeswaran 2016). Hence research has been a quest toward reduction of NOx emission for biodiesel blend fuels. Simple and easiest method might be carried out by many studies in water emulsion with a surfactant. Based on the literature (Alahmer et al.2010; Fahd et al. 2013; Noge et al 2015; Selim and Ghannam 2010; Varatharaju Perumal, Ilangkumaran 2018), the methods for admitting water into the combustion zone are (i) direct injection of water into the engine through separate injectors; (ii) modification to be made on nozzle to inject both diesel and water known as hybrid injection (iii) fumigate the water into the engine intake air along with air (iv) Diesel-water emulsions through surfactant. Among the four methods of water addition, diesel-water emulsion with a surfactant is most widely used (Suresh and Amirthagadeswaran 2016). Water addition to the blended fuel influences on reducing the peak flame temperature thereby NO_x emission is reduced (Koc and Abdullah 2013; Basha and Anand 2014). An emulsion is a finely dispersed mixture of two liquids that cannot normally be mixed without visible separation. A surfactant can reduce the oil and water surface tension, activate their surfaces, and maximize their superficial contact areas to make oil-in-water or water-in-oil two-phase emulsions (Mura et al. 2010; Ochoterena et al 2010). Based on the past literature, the many studies dealing with performance and emission study on diesel engines fueled with diesel-water emulsions. Their conclusion reveals both convergent and divergent results (Attia and Kulchitskiy, 2014; Noor El-Din et al., 2013; Sachin et al., 2014; Selim and Ghannam, 2010). In general, engine performance decreases with water content, due to the lower heating value of emulsion fuel compared to pure diesel fuel however a significant reduction in NO_x emission. The trade-off between the improved engine performances and reduction in NOx emission for blended fuel with water emulsion was obtained through the addition of cetane number enhancing additive(Basha and Anand 2012: Mohamed Musthafa et al.2018; Yang et al. 2013). Study on diesel engine performance fuelled by diesel-biodiesel-water- Di-tertiary butyl peroxide blend is unexploded. Hence the study has focused on the engine performance and emission characteristics test running on the new blend (B3) so that to bring research works in this area under one platform to research community and to further update the possible area of intervention for researchers.

MATERIALS AND METHODS

Preparation of Biodiesel

The raw neem oil was obtained from the local market. In the transesterification method, the raw oil was first filtered and then heated to a temperature of 110^{0}C for the removal of water content and then fed to the base catalyst esterification process. The potassium hydroxide (KOH) 1% and 30% methanol (by volume of raw oil) solution was prepared in order to maintain the molar ratio of 6:1. The prepared methanolic solution was added to pre processed oil in a reaction flask with continuous stirring by a magnetic stirrer. The mixture was kept for one hour at 50^0 C then left settling for 24 h to form the two distinct layers. The upper layer was the biodiesel and the bottom layer was glycerin. The upper layer of ester was separated out. The separated ester was washed with warm water (around 10% volume of ester) in thrice to remove the catalyst present in the ester and then the washed biodiesel was heated to 110^0C for removal of moisture.

Blend Preparation

The prepared biodiesel is mixed with diesel in the proportion of 30:70 by volume and is assigned as B1. Water (5%) and span80 surfactant (1%) by volume is added to blend 1 with vigorous stirring and is designated as B2. Di-Tert-Butyl Peroxide (DTBP) (0.5% by volume) is mixed to blend 2 and is designated as B3. The colour of B3 has appears to be white. Trials are made by the different proportion of water addition (5%, 10% and 20%) and DTBP addition (0.5%, 1% and 1.5%) based on fuel stability. Physico-chemical properties of fuels are essential for proper combustion of diesel fuel. The prepared blends properties are tested as per ASTM and listed in Table 1.

Experimental Setup

Experiments were carried out on a test engine and the detailed engine specifications are given in Table 2. The experimental setup is shown in Figure 2. An eddy current dynamometer is coupled to the engine for loading. The measurement of applied load, air and fuel flow rate, instantaneous cylinder pressure, injection pressure, the position of crank angle are made through online using the various sensors and instruments integrated with computerized data acquisition system. The exhaust emissions (HC, CO and NO_x) from the engine were measured by Delta 1600L exhaust gas analyser. The test fuels included as conventional diesel fuel and three different fuel blends B1, B2, & B3.

The percentage of uncertainty of the measured parameter can be estimated using the following relations.

Table 1. Physicochemical properties of test fuels (B1, B2 & B3)

Properties	Diesel ASTM D 975	Biodiesel (B100) ASTM D6751	Blend1	Blend2	Blend3
			(B6- B40) ASTM D6467		
Density (Kg/m^3)	0.85	0.87-0.89	0.747	0.722	0.736
Viscosity @ 40°C (cSt)	1.3-4.1	1.9-6.0	5.26	3.19	4.39
Flash Point (°C)	40-50	110-150	50	57	46
Fire Point (°C)	55-60	170-200	55	65	51
Calorific Value (kJ/kg)	43500	42000-43000	42646.7	42390.46	42949.2

Figure 1. Photographic view of engine setup

Uncertainty of this experimental setup =

$$\sqrt{uncertainty\ of\left[(load)^2+(speed)^2+(pressure)^2+(temperature)^2+(fuel\ consumed)^2\right]}$$
$$=\sqrt{[(0.2)^2+(0.2)^2+(0.9)^2+(0.25)^2+(0.2)^2}$$
$$=\mp 0.996\%$$

Table 2. Engine specifications

Make	Kirloskar-TV1
Engine loading device	Eddy current dynamometer
Rated output	3.5 kW
Rated speed	1500 rpm
Bore	87.5mm
Stroke	110mm
Compression ratio	12:1–18:1
Injection operating pressure	210 bar
Injection timing	20-40 ^{0}CA

It can be observed that uncertainty for the experimental setup is ±0.996% which may not affect the accuracy of the results.

RESULTS AND DISCUSSIONS

The performance and emission characteristics are an essential review of fuel modification other than conventional fuel. The enhancement in performance and emission of an engine by modified fuel leads to economic viability to replace the existing diesel fuel. The diesel and prepared blends have been tested in a variable compression ratio (VCR) engine at a fixed compression ratio of 18. The performance and emission characteristics of the engine were obtained for each test fuel at different load conditions. The results were compared through graphs from Figure 2 to Figure 7 as discussed below.

Comparison of Brake Thermal Efficiency vs. Load

Brake thermal efficiency against load for the test fuels was compared and presented in the Figure 2. B3 fuel showed efficiency close to the diesel operation and higher than B1 and B2 test fuels. B2 fuel result illustrated a slight reduction in efficiency than B1 fuel the reason might be the engine power decreases with water content due to lower calorific. However, the addition of DTBP to the B2 fuel increases the efficiency owed to increase in calorific value and enhancing cetane number which results in complete combustion.

Figure 2. Comparison of brake thermal efficiency vs. load for the test fuels

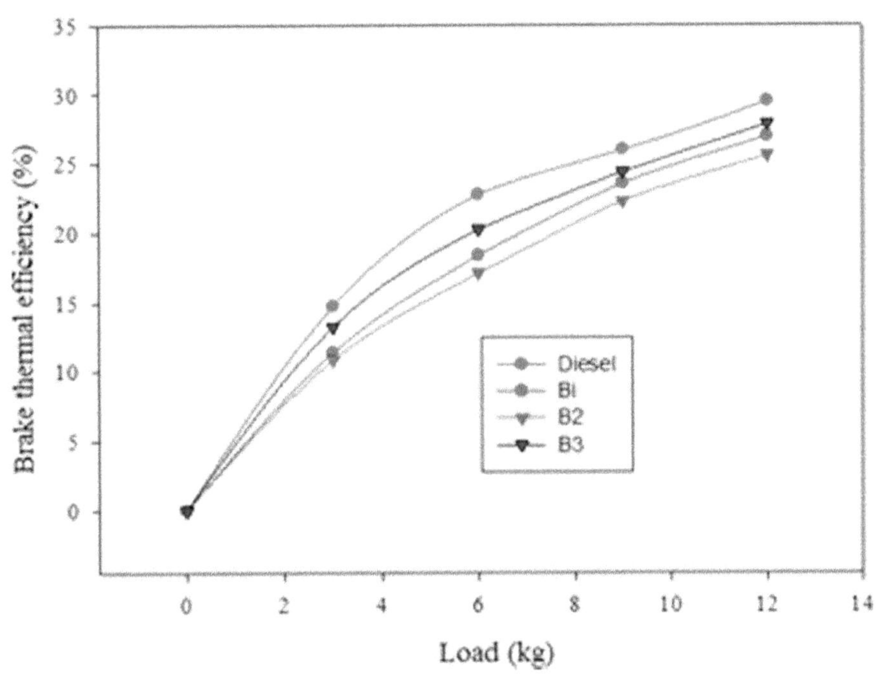

Comparison of Brake Specific Fuel Consumption vs. Load

Figure 3. depict the comparison of brake specific fuel consumption (BSFC) with the load for the test fuels. More deviation was observed among the test fuels at low load up to 1.5 kW of load. Beyond1.5 kW of load, there is not much variation was noted. BSFC for B3 fuel is slightly higher than diesel. B2 and B3 fuels BSFC trends are close to each other. The water content in the B2fuel is gets evaporated rapidly during combustion at higher loads due to the higher cylinder wall and gas temperature. Lower energy content is compensated by the enthalpy of vaporization of water. B3 fuel showed lower BSFC than B1 and B2 fuels and close to diesel operation. The DTBP additive enhances cetane number of B3 fuel which reduces the ignition delay promotes the rate of burning matches with the rate of fuel injection and results in better combustion.

Comparison of Exhaust Gas Temperature vs. Load

Figure 4 illustrates the comparison of exhaust gas temperature versus load. Increasing fashion of exhaust gas temperature with increase in load as the fuel consumption increased. It was noted that the decrease in exhaust gas temperature for B2 and B3 fuel at entire load spectrum than diesel and B1 fuels. The reason might be water content in the B2 and B3 fuel gets evaporated results in a reduction in an exhaust gas temperature. B1 fuel showed higher exhaust gas temperature over entire spectrum than diesel due to lower calorific value, and high viscosity leads to poor atomization.

Figure 3. Comparison of brake specific fuel consumption vs. load for the test fuels

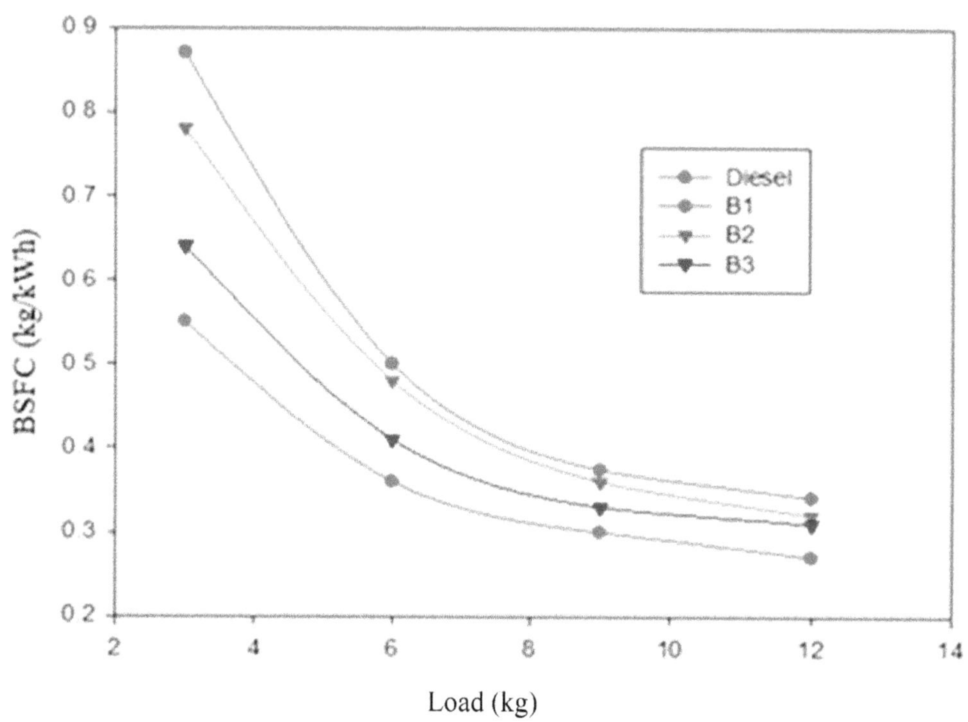

Figure 4. Comparison of exhaust gas temperature vs. load for the test fuels

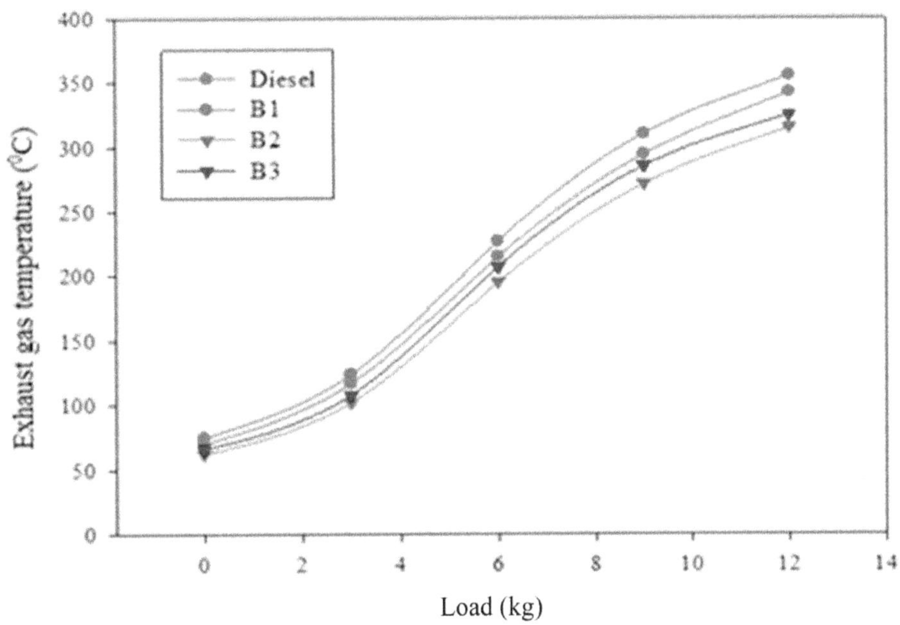

Hydrocarbon Emission

Figure 5 illustrates the comparison of hydrocarbon against the load for the test fuels. Incomplete combustion and lack of homogeneity are the main causes to form un-burnt HC in the diesel engine. B3 fuel demonstrates less HC emission than B1 and higher than B2 fuel due to micro explosion process. At higher combustion chamber temperature, the water hubs in the blend expand as in an explosion, in sequence the adjacent diesel drop is worn out into countless parts, an even finer spray is created in the combustion chamber this phenomenon is called a micro-explosion(Mura et al. 2012; Morozumi and Saito 2010). Owing to this micro explosion process the effectiveness of the engine increases slightly which reflects the considerable HC emission reduction.

Comparison of Carbon Monoxide Emission vs. Load

Carbon monoxide emission in the exhaust indicates how far efficient burning the fuel occurs during the combustion. Higher the CO emission designates incomplete combustion. The variation of CO emission for the test fuel at different loads is illustrated in Figure 6. It was found that B3 fuel has lower CO emission among other test fuels due to additive present in the fuel enhances oxygen content results the complete combustion. The decrease in HC emission upgrades the efficiency of the engine. This point reflects increased efficiency for B3 fuel.

Figure 5. Comparison of hydrocarbon emission vs. load for the test fuels

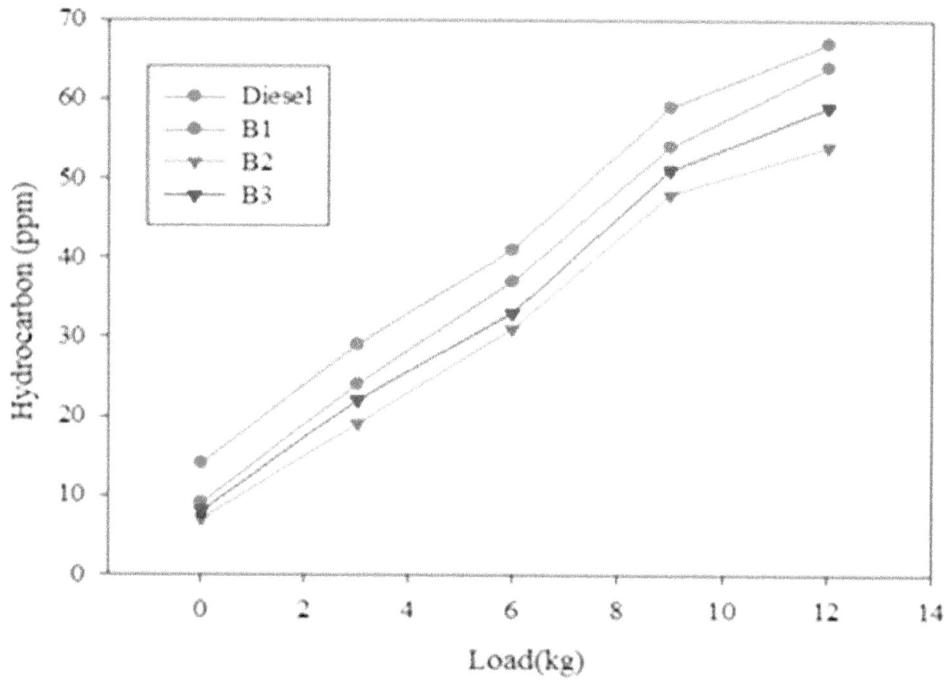

Comparison of NO_x Emission vs. Load

In general, NOx formation mechanisms are described as thermal NO_x, prompt NO_x and fuel NO_x. Among them thermal NO_x is most predominant in diesel engine combustion, At higher combustion temperatures, N_2 and O_2 can react through a series of chemical steps known as the Zeldovich mechanism. The rate of NO_x formations increases with an increase in combustion temperature and it takes place at temperatures above 1500°C (Ni peiyong and Wang Xiangli 2012). Figure.7 shows the variation of NO_x emission against the load for the test fuels. It was prominent that B2 fuel showed a significant reduction in NOx emission among the test fuels. The reason might be the combustion chamber heat is removed through the water evaporation process. This also results in a dropping of the combustion chamber peak temperatures. As the formation of nitrogen oxides increases exponentially with the combustion temperature, NOx formation decreases accordingly at lower peak combustion temperature water content in the blend reduces the combustion temperature results in NOx emission reduction(Babu and Sendilvelan, 2011). However, the addition of DTBP to B2 fuel enhances O_2 in the B3 fuel provides better combustion with a slight increase in NO_x emission than B2 fuel but less than D and B1 fuel.

COMBUSTION CHARACTERISTICS

The combustion characteristics of fuel can be analyzed through cylinder pressure-crank angle data and heat release rate-crank angle data.

Figure 6. Comparison of carbon monoxide emission vs. load for the test fuels

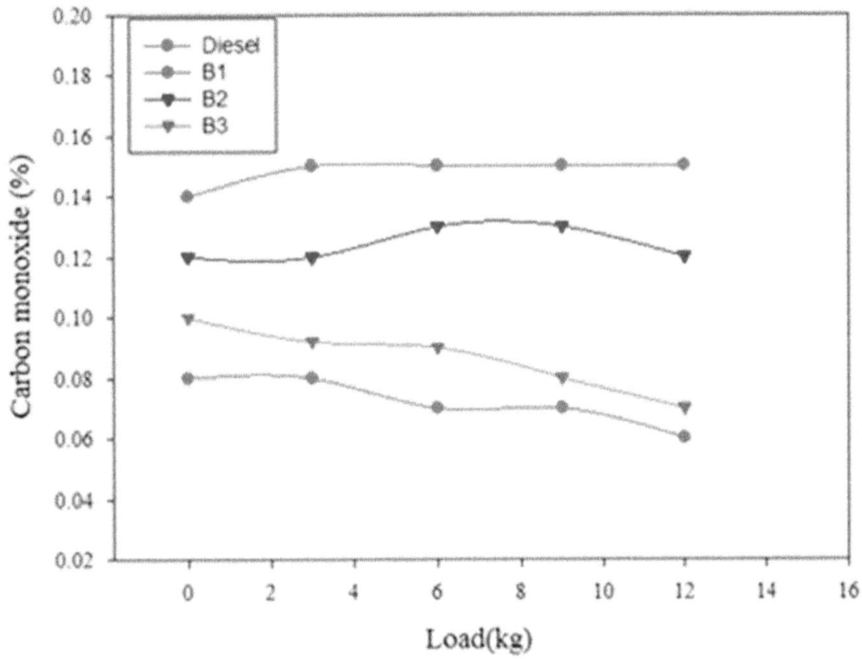

Figure 7. Comparison of NOx emission versus load for the test fuels

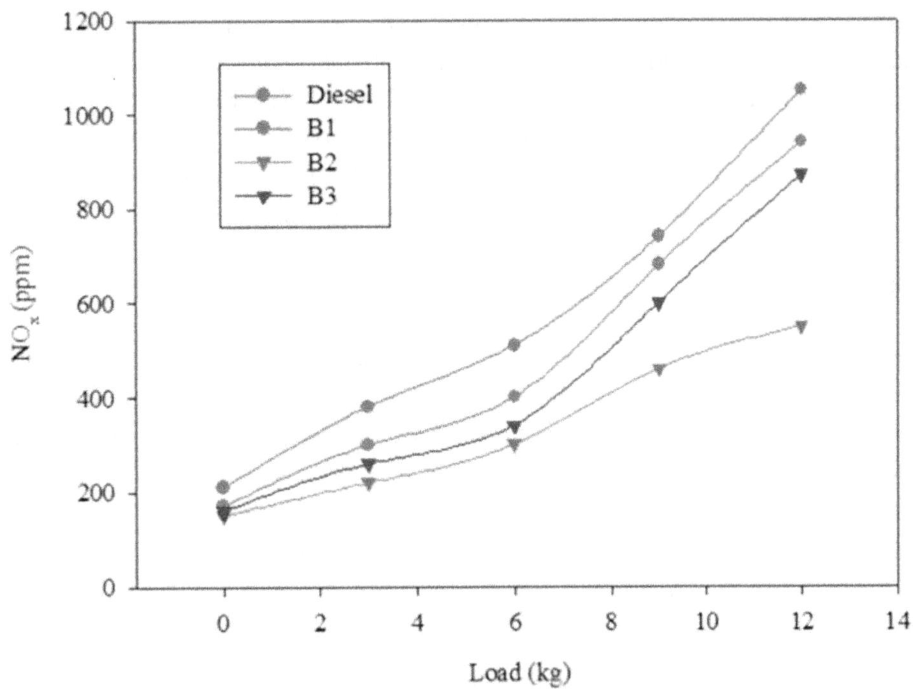

Figure 8. Comparison of cylinder pressure vs. crank angle at peak load for the test fuels

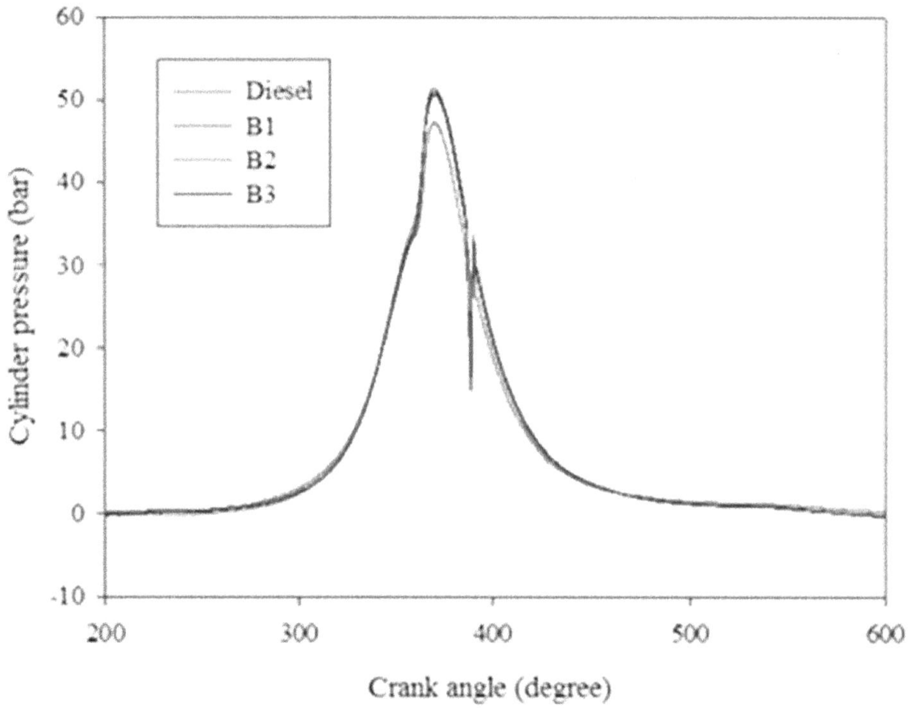

Rate of Pressure Rise

Figure 8 describes the cylinder pressure variation with respect to crank angle at rated load for the test fuels. It can be seen that the in-cylinder pressure increases with increasing load for the test fuels. As seen in the Figure, the peak cylinder pressures for B2 and B3 fuels are close to the diesel fuel due to the latent heat of evaporation of water during combustion along fuel combustion (Babu and Sendilvelan 2010). It is observed that B1 fuel had lower peak pressure at TDC. The reason might be due to high viscosity and lower calorific value which resulting in lower thermal efficiency.

Rate of Heat Release

Figure 9 depicts the comparison of cumulative heat release rate versus crank angle at peak load for the test fuels. It is observed that the maximum heat release rate is recorded for B2 and B3 fuels than B1 fuel. This phenomenon can be explained on the basis of the heat is added in the combustion chamber during latent heat of vaporisation of water along normal combustion of diesel and biodiesel blends. (Khan et al 2014). B1 fuel showed less heat release rate than B2, B3 & diesel fuels due to lower calorific value and higher viscosity results in the poor air-mixed fuel in the cylinder.

Figure 9. Comparison of cumulative heat release rate vs. crank angle at peak load

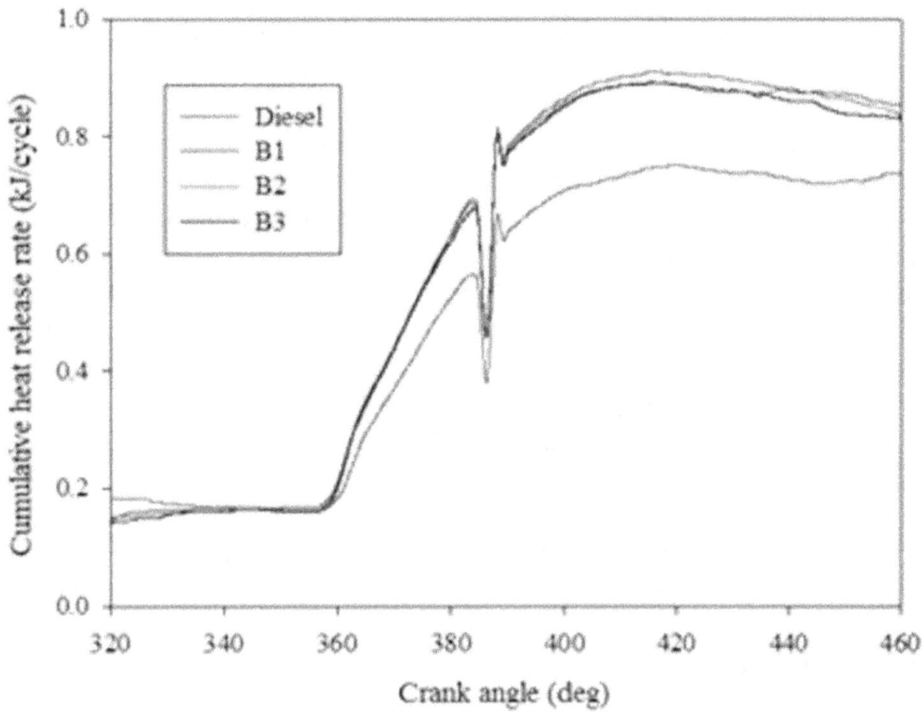

Water Dissociation

During the combustion of water in the blend, water dissociates into hydrogen and oxygen at high combustion temperature. Hydrogen takes part in combustion and the elemental oxygen atom seeks to find reaction associates like partially combusted fuel components, soot particles and hydrocarbons in the combustion chamber(Watanabe et al. 2010). This will direct to better combustion and lower soot emission. The homogeneous combustion process provides an obvious improvement in the engine operation to become smooth.

Di-Tertiary-Butyl Peroxide Dissociation

Di-tertiary-butyl peroxide in the blend (B3) get decomposed into acetone and ethane in the presence of oxygen during the combustion of a diesel engine in turn which results in two-stage ignition. In the first stage, DTBP is oxidized to acetone at lower combustion temperature which leads to the final combustion products due to additional oxidation. In succession with the first stage, combustion of acetone and ethane takes place as a chain branching reaction at high temperature in the second stage (Sebbar et al. 2017). The two-stage ignition provides complete burning of surrounded fuel spray

CONCLUSION

In the present study, the preparation and testing of diesel-biodiesel blends with water emulsion and additive were investigated. The conclusion of the results are summarized as follows

- Water emulsion technology used in diesel-biodiesel blends is simple and no modifications to the engine or the injection system are required.
- Higher NOx emission was observed by 40% in the diesel-biodiesel blend (B1) than diesel operation for the entire load spectrum.
- As expected, water emulsion (5% by volume) with a surfactant to B1 fuel reduces NOx emission up to 42%, but unexpected with a reduction in performance in terms decreasing thermal efficiency and increasing specific fuel consumption.
- A trade-off between better performance and improved emission, an addition of DTBP (0.5% by volume) to B2 fuel showed the increasing brake thermal efficiency by 3.5% without any deviation of emissions level compared to B2 fuel.

Based on the above conclusion, the new blend B3 can be used for future alternative fuel with less cost by replacement of 30% biodiesel and 5% water in their volume of the fossil diesel fuel.

Scope for Future Work

This study can be extended to evaluate the performance, combustion and emission characteristics of the engine by ANN modeling for varying the compression ratio, blends proportion ratio and can find optimum value for conducting experiment.

REFERENCES

Alahmer, A., Yamin, J., Sakhrich, A., & Hamdan, M. A. (2010). Engine performance using emulsified diesel fuel. *Energy Conversion and Management, 51*(8), 1708–1713. doi:10.1016/j.enconman.2009.11.044

Attia, A. M. A., & Kulchitskiy, A. R. (2014). Influence of the structure of water-in-fuel emulsion on diesel engine performance. *Fuel, 116*, 703–708. doi:10.1016/j.fuel.2013.08.057

Babu, R. V., & Sendilvelan, S. (2011). Reduction of pollutants in CI engine using emulsion fuels to reduce overall traffic-induced emissions. *Global Journal of Research Engineering, 11*(2).

Basha, J. S., & Anand, R. B. (2012). Effects of nanoparticle additive in the water-diesel emulsion on the performance, emission and combustion characteristics of a diesel engine. *International Journal of Vehicle Design, 59*(2/3), 164–181. doi:10.1504/IJVD.2012.048692

Basha, J. S., & Anand, R. B. (2014). An experimental study in a CI engine using nano additive blended water-diesel emulsion. *International Journal of Green Energy, 8*(3), 332–348. doi:10.1080/15435075.2011.557844

Baskar, P., & Senthil Kumar, A. (2017). Experimental investigation on performance characteristics of a diesel engine using diesel-water emulsion with oxygen enriched air. *Alexandria Engineering Journal, 56*(1), 137–146. doi:10.1016/j.aej.2016.09.014

Dwivedi, G., Jain, S., & Sharma, M. P. (2013). Diesel engine performance and emission analysis using biodiesel from various oil sources. *J Mater Environ Sci, 4*, 434–447.

Fahd, M. E. A., Wenming, Y., Lee, P. S., Chou, S. K., & Yap, C. R. (2013). Experimental investigation of the performance and emission characteristics of the direct injection diesel engine by water emulsion diesel under varying load condition. *Applied Energy, 102*, 1042–1049. doi:10.1016/j.apenergy.2012.06.041

Ghojel, J. I., & Tran, X. (2010). Ignition characteristics of diesel-water emulsion sprays in a constant-volume vessel: Effect of injection pressure and water content. *Energy & Fuels, 24*(7), 3860–3866. doi:10.1021/ef1001875

Huo, M., Lin, S., Liu, H., & Lee, C. F. (2014). Study on the spray and combustion characteristics of water–emulsified diesel. *Fuel, 123*, 218–229. doi:10.1016/j.fuel.2013.12.035

Khan, M. Y., Abdul Karim, Z. A., Hagos, F. Y., Aziz, A. R. A., & Tan, I. M. (2014). Current trends in the water-in-diesel emulsion as a fuel. *The Scientific World Journal*, 1–15. PMID:24563631

Koc, A. B., & Abdullah, M. (2013). Performance and NOX emissions of a diesel engine fueled with biodiesel-diesel-water nanoemulsions. *Fuel Processing Technology, 109*, 70–77. doi:10.1016/j.fuproc.2012.09.039

Mohamed Musthafa, M., Ashok Kumar, T., Mohanraj, T., & Chandramouli, R. (2018). A comparative study on performance, combustion and emission characteristics of a diesel engine fuelled by biodiesel blends with and without an additive. *Fuel, 225*, 343–348. doi:10.1016/j.fuel.2018.03.147

Morozumi, Y., & Saito, Y. (2010). Effect of physical properties on micro explosion occurrence in water-in-oil emulsion droplets. *Energy & Fuels, 24*(3), 1854–1859. doi:10.1021/ef9014026

Mura, E., Josset, C., Loubar, K., Huchet, G., & Bellettre, J. (2010). Effect of dispersed water droplet size in micro explosion phenomenon for water in oil emulsion. *Atomization and sprays, 20*(9), 791–799.

Mura, E., Massoli, P., Josset, C., Loubar, K., & Bellettre, J. (2012). Study of the micro-explosion temperature of water in oil emulsion droplets during the leidenfrost effect. *Experimental Thermal and Fluid Science, 43*, 63–70. doi:10.1016/j.expthermflusci.2012.03.027

Nabi, M. N., Najmul Hoque, S. M., & Akhter, M. S. (2009). Karanja (Pongamia Pinnata) biodiesel production in Bangladesh, characterization of Karanja biodiesel and its effect on diesel emissions. *Fuel Processing Technology, 90*(9), 1080–1086. doi:10.1016/j.fuproc.2009.04.014

(2012). Ni, P., & Wang, X. Modelling the formation of NOx and Soot emissions in a diesel engine at different humidity. *International Journal of Green Energy, 9*(8), 815–828. doi:10.1080/15435075.2011.641701

Noge, H., Kidoguchi, Y., Yahya, W. J., Imai, Y., & Tajima, K. (2015). An investigation into the relationship between the formation of thermal cracked components and PM reduction during diesel combustion using water emulsified fuel. *J. Therm. Sci. Technol, 10*(2), 1–11. doi:10.1299/jtst.2015jtst0024

Noor El-Din, M. R., & Sabrnal, H., El-Hamouly, M. H. M., Mishrif, M. R., & Ragab, A. M. (2013). Water-in-diesel fuel nanoemulsions: Preparation, stability and physical properties Egyptian. *Journal of Petrology*, *22*(4), 517–530.

Ochoterena, R., Lif, A., Nyden, M., Anderson, S., & Denbratt, I. (2010). Optical studies of spray development and combustion of water in- diesel emulsion and microemulsion fuels. *Fuel*, *89*(1), 122–132. doi:10.1016/j.fuel.2009.06.039

Perumal, V., & Ilangkumaran, M. (2018). Water emulsified hybrid pongamia biodiesel as a modified fuel for the experimental analysis of performance, combustion and emission characteristics of a direct injection diesel engine. *Renewable Energy*, *121*, 623–631. doi:10.1016/j.renene.2018.01.060

Sachin, Z., Tuti, M., & Durgun, O. (2014). Experimental investigation of the effects of water adding to the intake air on the engine performance and exhaust emission in a DI automotive diesel Engine. *Fuel*, *115*, 884–895. doi:10.1016/j.fuel.2012.10.080

Sebbar, N., Habisreuther, P., Bockhorn, H., Auzmendi-Murua, I., & Bozzelli, J. W. (2017). Di-tertiary-butyl peroxide decomposition and combustion with air: Reaction mechanism, Ignition, flame structure, and laminar flame velocities. *Energy & Fuels*, *31*(3), 2660–2273. doi:10.1021/acs.energyfuels.6b02201

Selim, M. Y. E., & Ghannam, M. (2010). Combustion study of stabilized water-in-diesel fuel emulsion. Energy Sources A Recov. *Utiliz. Environ. Effects*, *32*(3), 256–274. doi:10.1080/15567030802467621

Selim, M. Y., & Ghannam, M. T. (2009). Combustion study of stabilized water-in-diesel fuel emulsion. *Energy Sources, Part A: Recovery, Utilization, and Environmental Effects, 32*(3), 256-274.

Subramanian, K. A. (2011). A comparison of water-diesel emulsion and timed injection of water into the intake manifold of a diesel engine for simultaneous control of NO and smoke emissions. *Energy Conversion and Management*, *52*(2), 849–857. doi:10.1016/j.enconman.2010.08.010

Suresh, V., & Amirthagadeswaran, K. S. (2015). Combustion and performance characteristics of water-in- diesel emulsion fuel. Energy Sources A Recov. *Utiliz. Environ. Effects*, *37*(18), 2020–2028. doi:10.1080/15567036.2015.1072605

Suresh, V., & Amirthagadeswaran, K. S. (2016). The role of water-in-diesel emulsion and its additives on diesel engine performance and emission levels: A retrospective review. *Alexandria Engineering Journal*, *55*(3), 2463–2472. doi:10.1016/j.aej.2016.07.021

Watanabe, H., Suzuki, Y., Harada, T., Matsushita, Y., Aoki, H., & Miura, T. (2010). An experimental investigation of the breakup characteristics of secondary atomization of emulsified fuel Droplet. *Energy*, *35*(2), 806–813. doi:10.1016/j.energy.2009.08.021

Yang, W. M., An, H., Chou, S. K., Chua, K. J., Mohan, V., Sivasankaralingma, V., ... Li Maghbouli, J. (2013). Impact of emulsion fuel with nano-organic additives on the performance of diesel engine. *Applied Energy*, *112*, 1206–1212. doi:10.1016/j.apenergy.2013.02.027

Chapter 9
Impact of Diesel–Butanol–Waste Cooking Oil Biodiesel Blends on Stationary Diesel Engine Performance and Emission Characteristics

H. Sharon
Karunya Institute of Technology and Sciences, India

Joel Jackson R.
Karunya Institute of Technology and Sciences, India

Prabha C.
Karunya Institute of Technology and Sciences, India

ABSTRACT

Feed stock cost and NO_X emission are the major barriers for commercialization of biodiesel. Waste cooking oil is well identified as one of the cheapest feed stocks for biodiesel production. This chapter reduces NO_X emission of waste cooking oil biodiesel. Test fuel blends are prepared by mixing diesel (20 to 50 v/v%), butanol (5 v/v%), and waste cooking oil biodiesel (45 to 75 v/v%). Fuel properties of waste cooking oil biodiesel are enhanced due to addition of diesel and butanol. Brake specific energy consumption of the blends is higher than diesel fuel. Harmful emissions like carbon monoxide, nitrous oxide, and smoke opacity are lower for blends than diesel fuel. Increasing biodiesel concentration in blend also reduces hydrocarbon emission to a significant extent. The obtained results justify the suitability of proposed cheap blends for diesel engine emission reduction.

DOI: 10.4018/978-1-7998-2539-5.ch009

INTRODUCTION

Diesel engines occupy an important role in industrial, agricultural and transportation sectors (Ceramicrotaryengines, (2019)). However, continuous rise in diesel fuel price along with harmful gas emissions during its combustion have forced nations around the globe to look for alternate clean fuels (Rahman et al. (2014)). Emissions from diesel engines can be reduced by fuelling them with vegetables oils instead of fossil diesel fuel. Long term operation of diesel engines with raw vegetable oils reduces durability of engine components (Ramadhas et al., (2004); Shahid and Jamal (2008); Basinger et al. (2010)). Durability problems can be avoided if viscosity of vegetable oils is reduced by converting them into their ester forms/bio-diesel (Meher et al. (2006); Agarwal et al., (2007)). Biodiesel production cost can be reduced by utilizing non-edible oils, animal fats and used oils as feed stocks instead of edible oils (Ma and Hanna (1999); Bousbaa et al., (2012)). Among these feed stocks, used/waste oils are cheap, readily available and in addition their storage stability properties are very well studied (Knothe and Steidley (2009)). In biodiesel-based fuels NO_X emissions will be higher. In this chapter, the effort taken to reduce NO_X emission is presented.

The major objectives of the work are listed below:

1. Conversion of waste cooking oil into biodiesel
2. Preparation of diesel-butanol-waste cooking oil biodiesel blends
3. Assessment of diesel-butanol-waste cooking oil biodiesel fuel properties
4. Impact assessment of fuel blends on diesel engine performance and emissions

Background

Feedstock for biodiesel production is sustainable only if it is non-edible and has low water footprint. Waste cooking oil falls under this category (Tabatabaei et al., (2019)). Trans-esterification is widely used to produce biodiesel. Different technologies have been used for trans-esterification of waste cooking oils. **Table 1** lists out different procedures along with their reaction parameters for biodiesel production from waste cooking oils. Detailed review on types of reactors used for biodiesel production can be found in (Tabatabaei et al., (2019)). Waste cooking oil biodiesel is gaining momentum throughout the globe. Biodiesel production from waste oils of oil and soap industries is found to be economically viable in Egypt (El-Galad et al. (2015)). Biodiesel produced from locally available waste cooking oil can replace 17.8% and 16.0% of diesel fuel in power generation and transport sectors of eco-tourism island Langkawi, Malaysia (Kumaran et al., (2011)). Government of India has issued an order to food business operators whose edible oil consumption is more than 50 L/d to handover the used /waste cooking oil to authorized collection agencies of biodiesel manufacturers (FSSAI. (2018)). Use of waste cooking oil as diesel engine fuel not only conserves fossil fuel but also prevents environmental pollution and human health degradation.

A lot of research has been done around the globe to visualize the emissions of diesel engine fuelled by waste cooking oil bio-diesel.

Gracia-Martin et al., (2019) have successfully predicted cetane number of waste cooking biodiesel from raw feed stock using Near Infra-red spectroscopy there by leading a new path to select suitable feed stock for producing biodiesel with superior fuel qualities. Emissions during waste cooking biodiesel blend combustion are dependent on type of engine and biodiesel concentration. Biodiesel blends up to 60%

Table 1. Comparison of technologies available for trans-esterification of used/waste cooking oils

Trans-esterification Technology	Catalyst	Molar Ratio	Reaction Temperature (° C)	Reaction Time (h)	Yield (%)	Ref
Base Catalyzed Trans-esterification (single step process)	1.0wt% KOH	12:1 (Ethanol)	78.0	2.0	74.2	(Encinar et al., 2007)
Base Catalyzed Trans-eterification (Single step process) (steam treated feed)	1.0wt% KOH	6:1(Methanol)	60.0	1.0	83.5	(Supple et al., (2002))
Base Catalyzed Trans-esterification (Two step process)	4.2 g of NaOH for the first step and 1.8 g for second step	6:1 (Methanol (140 ml for the first step; 60 ml for second step))	25.0	1.0	86.0	(Cayli and Kusefoglu (2008))
Enzyme Catalysed Trans-esterification	10.0% Novozym 435	25:1 (Methanol)	50.0	4.0	89.1	(Maceiras et al.,(2009))
Solid Catalyzed Trans-esterification	Mo-Mn/γ-Al2O3-15.0wt%Mgo	27:1(Methanol)	100.0	4.0	91.4	(Farooq et al., (2013))
Solid Catalyzed Trans-esterification	3.0 wt% of Calcinated clamshell	6.03:1(Methanol)	60.0	3.0	89.0	(Nair et al., (2012))

waste cooking oil biodiesel is very effective in reducing particulate matter and toxic organic pollutants (Cheruiyot et al., (2019)). (Hwang et al., (2016)) observed long injection delay, poor-air fuel mixing, longer ignition delay and increased NO_X emission for waste cooking oil biodiesel in comparison to diesel fuel. Plamondon and Sheers (2019) observed the influence of fuel injection strategies on performance and emission of waste cooking biodiesel-fuel blends. (Dhanasekaran et al., (2017)) made an attempt to replace 50% diesel with 50% waste cooking oil. These blends decreased NO_X & smoke and increased HC emission relative to diesel fuel. Smoke, hydrocarbon and carbon mono-oxide emission of waste cooking oil biodiesel is lower than jatropha and karanja biodiesel (Hirner et al., (2019)). NO_X emission of waste cooking oil biodiesel- diesel blends is higher than jatropha and soyabean biodiesel-diesel blends (Chaurasiya et al., (2019)). At high loads, NO_X emission of waste cooking oil biodiesel blends is lower than algal oil biodiesel due to increased O_2 content in algal biodiesel in comparison to waste cooking oil biodiesel (Nirmala et al., (2019)). Advancing injection timing and increasing compression ratio reduces smoke and hydro carbon emissions and increase NO_X emission during waste cooking oil biodiesel blend combustion (Shivakumar et al., (2011)).

Increased NO_X emissions from waste cooking oil biodiesel- diesel blends are widely observed in comparison to diesel fuel (Abed et al., (2018) ; Enweremadu and Rutto (2010); Kumar et al., (2016)). Exhaust gas re-circulation (Agarwal et al., (2006); Praveen et al., (2018)) and blending of fuels with ethanol (Rakopoulos et al., (2008)) has been practiced to reduce NO_X emissions. On the other hand, implementation of exhaust gas re-circulation and blending of fuels with ethanol has increased CO and HC emissions in diesel engines (Praveen et al., (2018); Rakopoulos et al., (2008)). Blending of waste cooking biodiesel with higher alcohols like butanol can enhance fuel properties (Atabani et al., (2019)). Mixing coconut oil-ethanol blend with 5 v/v% butanol has reduced NO_X emission from diesel engines (Singh et al.,(2010)). Mehta et al., (2010) also observed reduced NO_X emission with jatropha biodiesel-diesel blends mixed with 5 v/v% of butanol. Increasing butanol concentration beyond 5 v/v%

in blends has favored increased NO_X emission (Singh et al.,(2010) ; Altun et al., (2011)) and aldehyde emissions (He et al., (2013)) hence minimum concentration of butanol is recommended in biodiesel blend preparation. Butanol can also be produced from agricultural wastes and has no negative impacts on engine components (Jin et al., (2011)). Moreover, fuel properties, emission reduction potential and phase stability of butanol-based fuel blends are more superior to ethanol based fuel blends (Sukjit et al., (2012); Chotwichien et al., (2009)).

MAIN FOCUS OF THE CHAPTER

Suitability of waste cooking oil for viable biodiesel production and potential of butanol in reducing NO_X emission can be seen clearly from above cited literatures. Impact of butanol addition to waste cooking biodiesel on engine performance and emissions has not been reported so far by researchers. Hence, in the current work, an effort has been made to understand the performance and emission profile of stationary diesel engine fueled by waste cooking oil biodiesel, diesel and 5 v/v% butanol blends. Butanol content in produced fuel blends is kept to a minimum of 5 v/v% to prevent increased HC and CO emissions, while diesel content and waste cooking oil biodiesel content in the blends is varied from 20 to 50 v/v% and 45 to 75 v/v%, respectively. Engine performance and emission results with fuel blends have also been compared with diesel fuel. The observations from the study are encouraging and worthwhile.

MATERIALS AND METHODS

In this section, waste cooking oil biodiesel production, fuel blend preparation and engine testing procedures will be discussed.

Fuel Blends Preparation

It can be observed from **Table 1**, that base catalyzed trans-esterification requires less temperature, less molar ratio of alcohol to oil and less reaction time. Acid and enzyme catalyzed trans-esterification are suitable for feeds with high free fatty acids (Morais et al., (2010); Kara et al., (2018)). Heterogeneous bio-catalyst (El-Gendy et al., (2015)) and ionic liquids (Fathy et al., (2015)) have also been tried for biodiesel production from waste vegetable oils. In this work, waste cooking oil collected from Tirunelveli region of Tamil Nadu, India is converted to biodiesel using base catalyzed trans-esterification with NaOH (0.6 wt%) and methanol (6:1 molar ratio) at a process temperature of 65 °C for 3 hrs heating in a batch type reactor at National Institute for Interdisciplinary Science and Technology (NIIST), Trivandrum. Methyl esters (biodiesel) thus obtained is distilled at reduced pressure and is subjected to gas chromatography analysis to identify its fatty acid composition. Fuel blends are prepared by blending diesel, butanol and waste cooking oil biodiesel in required proportions using magnetic stirring arrangement. Biodiesel, diesel and butanol are non-polar therefore they are easily miscible (Solubility of alcohols (2019)). This makes the fuel blends more stable as a result no phase separation is noticed during experimentation. Composition of fuel blends are displayed in **Table 2.** Properties of the fuel blends are determined using standard methods (Knothe, (2006)) at ETA lab, Chennai.

Table 2. Fuel Blend Composition

Fuel Type	Composition		
	Diesel	Butanol	Biodiesel
100% Diesel	100	--	--
50%D-5%B-45%BD	50	5	45
40%D -5%B-55%BD	40	5	55
30%D -5%B-65%BD	30	5	65
20%D- 5%B-75%BD	20	5	75
100% Biodiesel	-	-	100

Experimental Setup

The schematic representation of Constant speed (1500 rpm) stationary DI diesel engine coupled to an eddy current dynamometer used for experimentation is shown in **Figure 1.** The specifications of the test engine are displayed in **Table 3.** Experiments are conducted by varying engine load from 0 to 100% with a step size of 25%. Fuels used are diesel fuel and pre-mixed diesel-butanol-biodiesel blends. Standard

Figure 1. Test engine setup

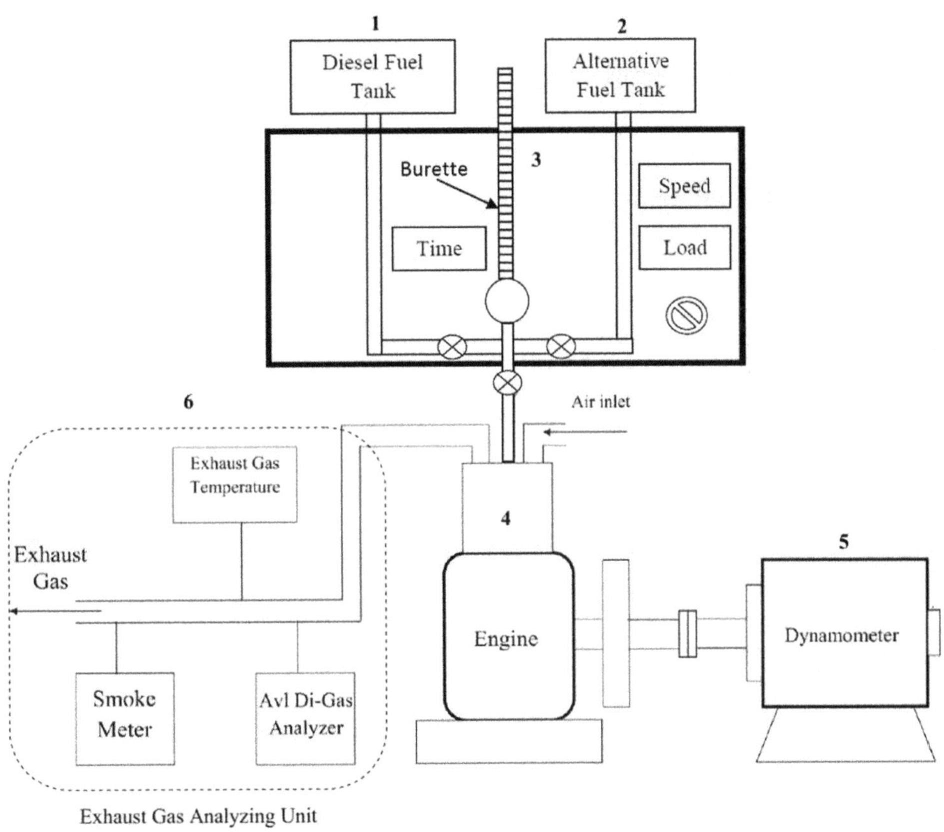

burette (accuracy ± 0.02 ml) and stop watch (accuracy 0.003%) arrangement are used to measure time taken for 10 cc fuel consumption. AVL digas 444 (five gas analyzer) is used to measure engine emissions NO_X, CO, HC and CO_2 emissions of engine (Accuracy of HC and NO_X measurement ± 10 ppm; Accuracy of CO, CO_2 measurement ± 0.01% of indicated value). AVL 437C smoke meter is used to measure smoke opacity (Accuracy & reproducibility ± 1.0% of full scale reading). K-type thermocouple is used to measure exhaust gas temperature (Accuracy ± 0.4°C, Sheath diameter: 0.13 inch, Probe length: 3.0 inch; Junction type: grounded).

Brake specific energy consumption (BSEC) is estimated by,

$$\text{BSEC} = \text{BSFC} \times \text{Calorific value of fuel blend} \tag{1}$$

$$\text{Brake specific fuel consumption} = \frac{\text{Total fuel consumption (TFC)}}{\text{Brake power}} \tag{2}$$

where,

$$TFC = \frac{x \times 3600 \times s}{t \times 1000} \tag{3}$$

't' is the time taken for consumption of 'x' ml of fuel blend and 's' is the specific gravity of the fuel blend. TFC in kg/h.

Brake thermal efficiency (BTE) is estimated by,

$$\text{BTE} = \frac{\text{Brake Power}}{\text{Heat energy supplied by fuel blend}} \tag{4}$$

$$\text{BTE} = \frac{\text{Brake power} \times 3600}{\text{TFC} \times \text{Calorific value of fuel blend}} \tag{5}$$

Total fuel consumption represents quantity of fuel blend consumed per hour during engine operation at desired loads.

Total uncertainty (%) = ((uncertainty of HC)^2 + (uncertainty of NO_X)^2 + (uncertainty of smoke)^2 + (uncertainty of CO)^2 + (uncertainty of exhaust gas temperature)^2)$^{1/2}$

Total uncertainty = ± 8.18%

Table 3. Specification of the test engine

Manufacturer	Kirloskar
Model	TAF 1
Type	Direct Injection, Air Cooled, 4 stroke
Lubrication	Forced Type
Bore x Stroke (mm)	87.5 x 110 mm
Compression ratio	17.5:1
Swept volume	0.661 L
Rated power	4.7 kW
Rated speed	1500 rpm
Start of injection	24° bTDC
Type of fuel pump	Mechanical
Fuel injection pressure	21.0 MPa
Fuel Injector	One, 3 holes of 0.25 mm diameter

RESULTS AND DISCUSSIONS

In this section, the properties of test fuel, gas chromatography results along with engine performance and emission characteristics will be discussed in a detailed manner.

Test Fuel Properties

Fatty acid composition of waste cooking oil biodiesel obtained by trans-esterifcation is displayed in **Table 4.** Fatty acid profile confirms the presence of large amount of saturates. Presence of large amount of palmitic acid and oleic acid confirms that the waste cooking oil is mainly used palm oil which is widely used for cooking in southern part of Tamil Nadu, India. Fuel properties of blends are displayed in **Table 5.** Flash point, viscosity and density of waste cooking oil biodiesel are higher than diesel fuel. Kinematic viscosity of blends is measured at 30° C due to the presence of butanol which has lower flash point. Blends containing butanol and diesel reported reduced viscosity, density and flash point in comparison to waste cooking oil biodiesel. Lower flash point of blends in comparison to diesel fuel is due to the presence of butanol which has lower flash point (36 to 38 °C). Increased viscosity is noticed with increase in biodiesel content in blends. Kinematic viscosity and flash point of waste cooking oil biodiesel is within ASTM standards (Rahman et al., 2014). Calorific value of fuel blends is higher than biodiesel but it is lower than diesel fuel. Pour point and cloud point for biodiesel occurred at higher temperatures because of the presence of saturated fatty acids. Low temperature properties of waste cooking oil biodiesel are enhanced by blending it with diesel and butanol. However, lower temperature properties of the blends are not within the limits set by ASTM standards (Rahman et al., (2014)). However, it won't be a problem to use these fuel blends in southern part of Tamil Nadu where ambient temperature will always be above 25 °C. Correlation given by Lapuerta et al., (2010) is utilized to find the cetane index of waste cooking oil biodiesel. Addition of butanol has made the blends more miscible and hence no phase separation is observed during experimental investigations.

Table 4. Fatty acid content of waste cooking oil biodiesel

Fatty acid	Carbon structure	Waste cooking oil biodiesel (wt %)
Lauric acid	C12:0	1.805
Myristic acid	C14:0	1.739
Palmitic acid	C16:0	43.480
Stearic acid	C18:0	4.640
Oleic acid	C18:1	41.126
Linoleic acid	C18:2	7.210
CY:n: Y carbon atoms and n double bond If n = 0; saturated fatty acid; if n=1,2,3,....(unsaturated fatty acid with n double bonds)		

INVESTIGATIONS ON DIESEL ENGINE PERFORMANCE AND EMISSION

Brake Specific Energy Consumption and Brake Thermal Efficiency

Ratio of fuel energy consumed to brake power produced is defined as brake specific energy consumption (BSEC). It is more effective to compare fuels with different properties in terms of BSEC rather than brake specific fuel consumption. BSEC profile for blends and diesel fuel is calculated using Eqn(1) and is displayed in **Figure 2.** Blends showed increased BSEC compared to diesel fuel due to their low calorific value and high viscous nature (causes poor spray characteristics) (Domkundwar (2009); Sanjid et al., (2014); Ahmed et al., (2014)). BSEC tends to decrease with increase in load and it increases with increase in biodiesel content for loads due to its lower calorific value compared to diesel fuel. Presence of 5 v/v% butanol of lower calorific value in fuel blends might also have contributed to the increased BSEC.

Brake thermal efficiency (BTE) profile for the test fuels is displayed in **Figure 3.** BTE of diesel fuel is higher than the blends while BTE of the blends are closer to each other. BTE is related to the energy conversion, rather than energy content. It depends mainly on the nature of combustion which in turn depends of ignition delay, viscosity and spray structure of the fuel. The reduced BTE of biodiesel blends

Table 5. Fuel Blends Properties

Fuel Type	Flash Point	Kinematic viscosity at 30°C	Calorific value	Density at 15 °C	Cetane number	Cloud Point	Pour Point
	(°C)	(mm^2/s)	(kJ/kg)	(g/cm^3)		(°C)	(°C)
100% Diesel	50	7.18	42,650	0.822	47.0	0.0	-3.0
50%D-5%B-45%BD	36	8.89	40,237	0.846	-	8.0	4.0
40%D -5%B-55%BD	39	9.56	39,805	0.854	-	9.0	5.0
30%D -5%B-65%BD	41	9.88	39,380	0.858	-	13.0	10.0
20%D- 5%B-75%BD	42	10.21	38,950	0.863	-	14.0	10.0
100% Biodiesel	161	12.46	38,350	0.876	66.7*	16.0	12.0
* Biodiesel Cetane index							

Figure 2. Variation of brake specific energy consumption with load for test fuels

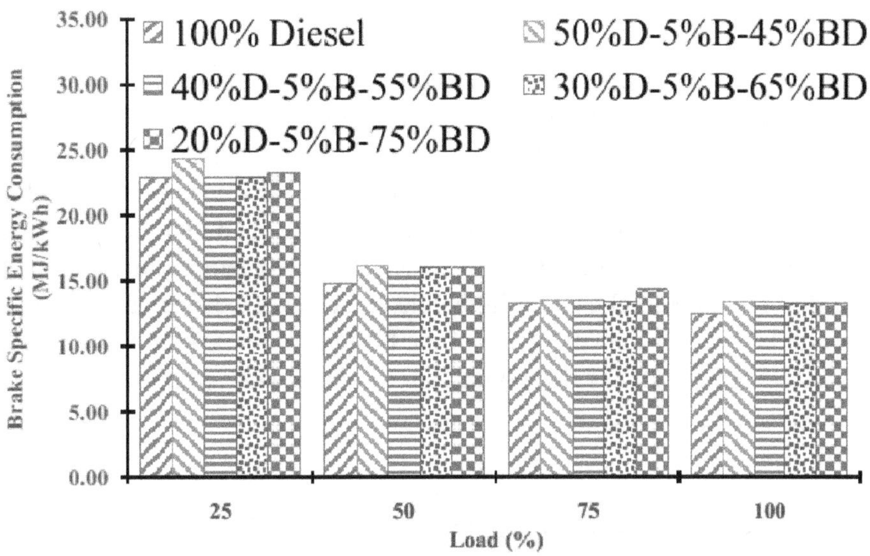

may be due to its high viscous nature and consequent poor spray structure. In addition blends contain butanol which has high latent heat of vaporization which quenches the combustion chamber there by more fuel is consumed to produce same power. Among the blends, BTE increases with increase in biodiesel content in the blend this may be due to reduced ignition delay (high cetane number of biodiesel) and reduced heat loss due to quenching effect caused by butanol in fuel.

Figure 3. Variation of brake thermal efficiency with load for test fuels

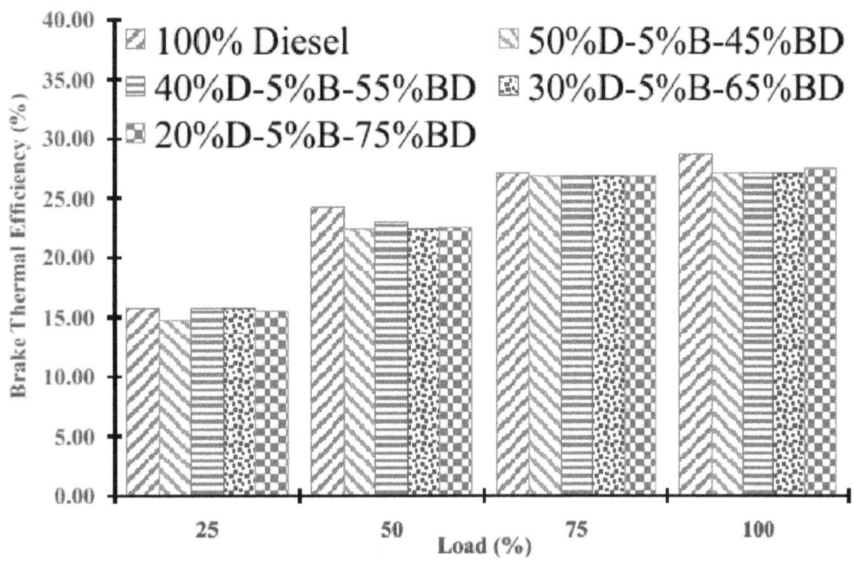

Exhaust Gas Temperature

Exhaust gas temperature profile for blends and diesel fuel is displayed in **Figure 4.** Higher exhaust gas temperature is observed at high loads due to combustion of more fuel at high loads to meet the required load demand (Gad et al., (2018)). Generally, exhaust gas temperature of biodiesel based fuels will be higher than conventional diesel fuel (Ramadhas et al., (2005); Buyukkaya (2010); Enweremadu and Rutto (2010)). The reduced exhaust gas temperature for the prepared blends is due to high latent heat of vapourization of butanol in addition to its lower calorific value and ignition delay (Dogan (2011)). Lower calorific value and lower ignition delay represents low peak temperature thereby heat losses to exhaust is reduced. As the biodiesel content in the blend increases exhaust gas temperature increases slightly due to slight increase in calorific value of the blend (**Table 5**). Karabetkas and Hosoz (2009) and Rakopoulos D.C et al., (2010) has also noticed similar exhaust gas profiles.

NO_X Emission

Zeldovich mechanism is mainly responsible for the formation of NO_X (Ban-Weiss et al., (2007)). NO_X emission profile for the test fuels is displayed in **Figure 5.** At high loads, engine in-cylinder temperature increases which could be seen from exhaust gas temperature profile shown in **Figure 4.** The high in-cylinder temperature induces reaction of inert nitrogen gas with oxygen to form NO_X (Ahmed et al.,(2014)). NO_X emission for the blends 50%D-5%B-45%BD, 40%D-5%B-55%BD, 30%D-5%B-65%BD and 20%D-5%B-75%BD are lower than diesel fuel by 7.04%, 4.74%, 3.74% and 2.85%, respectively. NO_X emissions are generally higher for ester based blends (Ban-Weiss et al., (2007); Nabi et al., (2009)). In the present experimental investigation, it could be seen that NO_X emission of blends are lower than the diesel fuel at all loads due to the presence of butanol which lowers in-cylinder temperature (Karabetkas and Hosoz (2009)) which can be justified from the exhaust gas profile shown in **Figure4.** Among blends,

Figure 4. Variation of exhaust gas temperature with load for test fuels

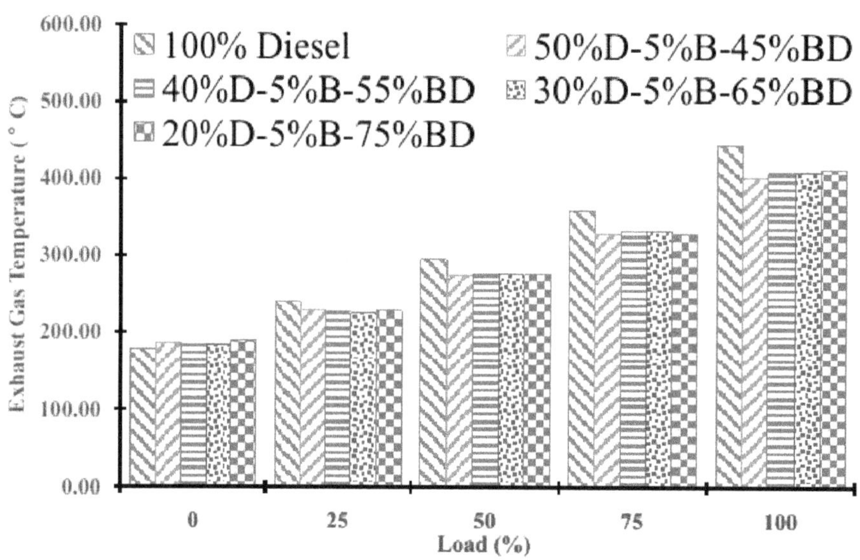

Figure 5. Variation of NO_X emission with load for test fuels

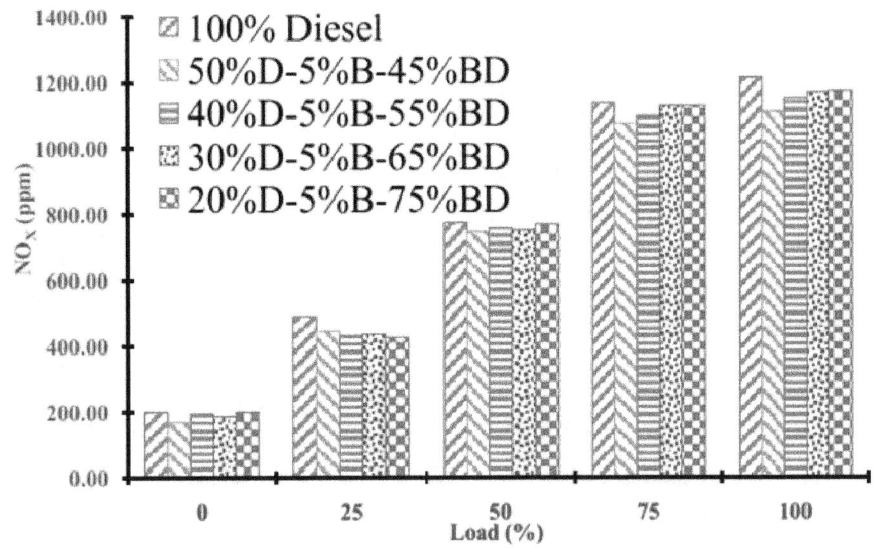

NO_X emission increases with increasing biodiesel content this may be due to rise of calorific value which favors high in cylinder temperature leading to enhanced NO_X formation.

Smoke Opacity

Smoke is formed when local temperature is high enough to cause decomposition of fuel and there is inadequate oxygen to burn the carbon (Domkundwar (2009)). Smoke opacity profile for fuel blends and diesel fuel is displayed in **Figure 6.** Smoke opacity increases with load due to reduced residence time of combustion gases inside the cylinder at high loads there by combustion products are left outside partially unburned (Sharon et al., (2013)). Mean smoke opacity of diesel fuel is around 23.10%, 29.10%, 36.10% and 38.10% higher than 50%D-5%B-45%BD, 40%D-5%B-55%BD, 30%D-5%B-65%BD and 20%D-5%B-75%BD fuel blends, respectively. Reduced smoke opacity of blends is due to the presence of oxygen rich butanol and biodiesel which makes these blends to burn at lean side (Mehta et al., (2010); Ramadhas et al., (2005); Dogan (2011)).

Hydrocarbon Emission

Hydro Carbon (HC) emission profile for the test fuels is displayed in **Figure 7.** HC emission profile of diesel increases with load due to reduced oxygen content under fuel rich conditions (Gad et al., (2018)). HC emission of blends drops considerably with increase in load due to the presence of oxygen in the fuel blends which helps in effective combustion. Mean HC emission of 50%D-5%B-45%BD, 40%D-5%B-55%BD and 30%D-5%B-65%BD fuels are higher than diesel fuel by 85.24%, 48.36%, and 9.01%, respectively. Blend 20%D-5%B-75%BD shows an average HC reduction of 1.60% in comparison with diesel fuel. Increased HC emissions of the blends are mainly because of the presence of butanol which lowers the engine cylinder temperature thereby causing poor combustion of the supplied fuel (Dogan (2011); Karabetkas and Hosoz (2009); Rakopoulos D.C et al., (2010)).

Figure 6. Variation of smoke opacity with load for test fuels

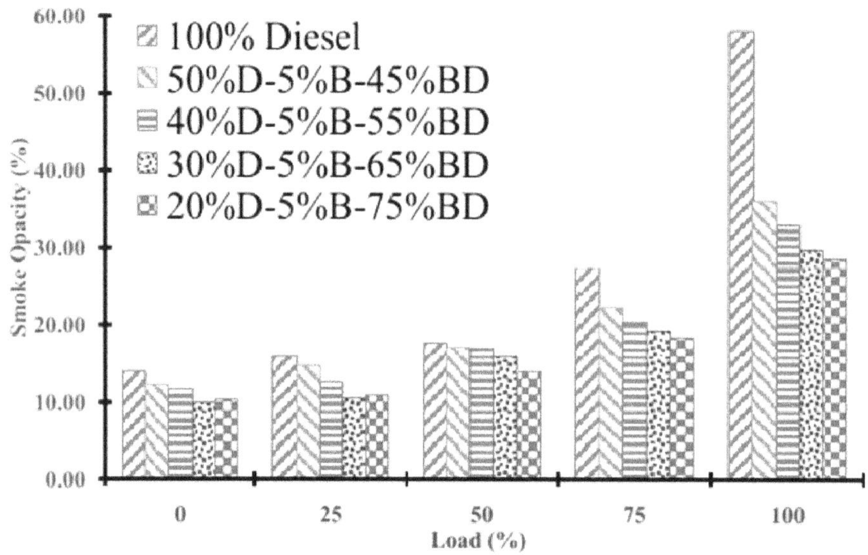

On contrary, Altun et al., (2011) obtained reduced HC emission with increasing butanol percentage in the blends, they explained that the reduction is due to enhanced combustion caused by the presence of inbuilt oxygen molecule. Sukjit et al., (2012) noticed increased HC emissions with long chain saturated fatty acid - butanol blends and reduced HC emissions with short chain fatty acid - butanol blends and unsatuared fatty acid – butanol blends. Biodiesel used in the current study is produced from used palm oil which contains 48.12 wt% of long chain saturated fatty acid (palmitic acid and stearic acid) and 48.336 wt% of unsaturated fatty acid (oleic acid). So, it can be justified that as biodiesel volume percentage in

Figure 7. Variation of hydrocarbon emission with load for test fuels

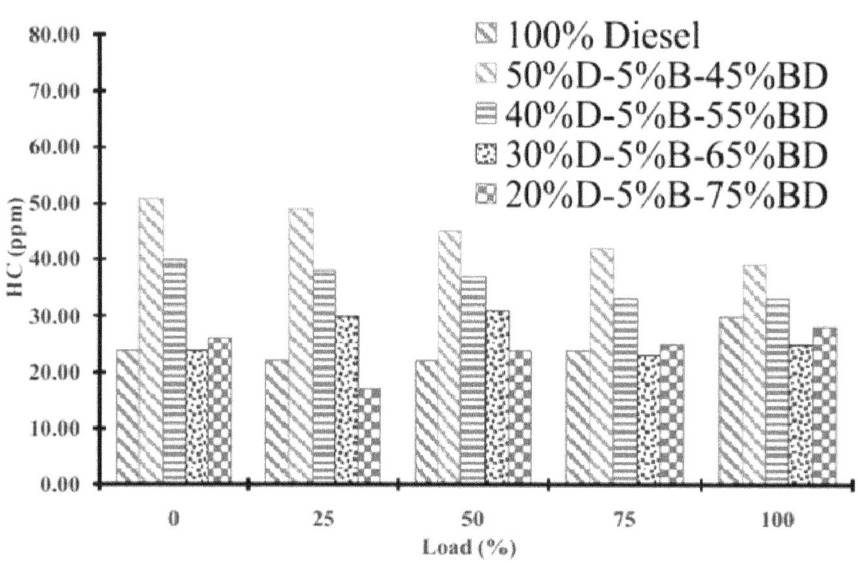

Figure 8. Variation of carbon monoxide emission with load for test fuels

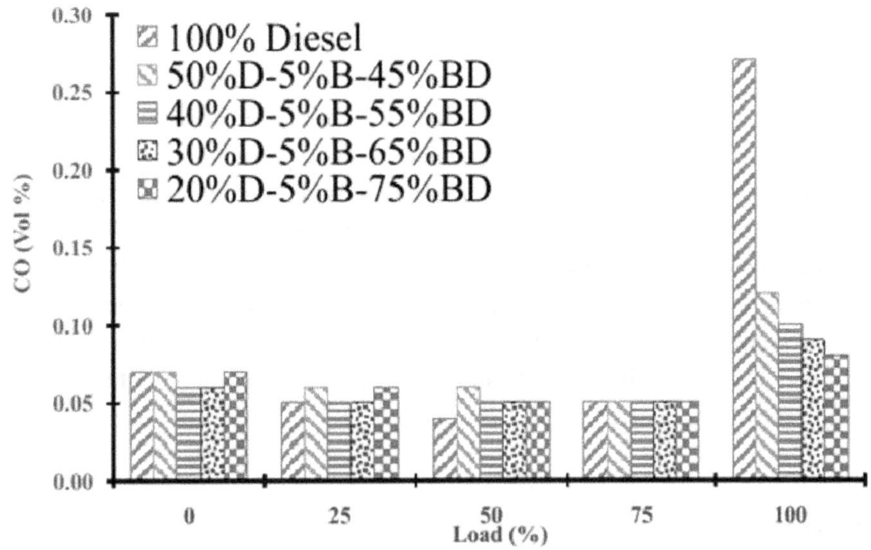

the blend increases, unsaturated fatty acid content also increases thereby HC emission reduces. Hence, from **Figure 7**., it can be seen clearly that HC emission of the blends containing high v/v% of biodiesel is more or less closer to diesel fuel.

Carbon Monoxide Emission

Insufficient oxygen to oxidize the fuel during combustion is responsible for the formation of carbon monoxide (CO) (Domkundwar (2009)). CO emission profile for the blends and diesel fuel is displayed in **Figure 8.** CO emission during low loads is found to be closer to each other. Reduced CO emission of the blends at 100% load is mainly due to inbuilt oxygen molecule in biodiesel and butanol (Karabetkas and Hosoz (2009)) in combination with high combustion temperature (Yadav et al., (2018)). CO emission for blends at low loads might be due to the quenching effect caused by butanol over engine cylinder which reduces in-cylinder temperature.

FUTURE RESEARCH DIRECTIONS

Stringent emission regulations in upcoming years will be a challenge for researchers. There is a lot of scope available for further reduction of emissions and improvement of combustion process in biodiesel fueled engines. Future research must address the problem of phase instability of blends and identification/development of new and cheap biodiesel production methodology. Researches relating to engine modification to incorporate bio-fuels directly will be more fruitful.

CONCLUSION

Collected waste cooking oil is successfully trans-esterified with methanol in the presence of sodium hydroxide. Gas chromatography analysis of produced waste cooking oil biodiesel indicated the presence of saturated and unsaturated fatty acids. Waste cooking oil biodiesel blended with diesel and butanol showed improved fuel properties, increased brake specific energy consumption and brake thermal efficiency nearly closer to diesel fuel. Presence of oxygenated fuels (butanol and biodiesel) in the blends reduced CO emission and smoke opacity. Average smoke opacity of the blend 20%D-5%B-75%BD is nearly 38.1% lower than diesel fuel. Blends reported reduced NO_X emission and increased HC emission in comparison to diesel fuel. Increasing concentration of biodiesel reduced HC emission. From experimental results, it can be confirmed that diesel-butanol-biodiesel blends are capable of mitigating harmful emissions and the blend 20%D-5%B-75%BD can be considered as a suitable alternate fuel because of its reduced emission profile (reduced CO, HC, Smoke and NO_X) in comparison to diesel fuel.

ACKNOWLEDGMENT

The authors wish to express their appreciation to D.R.Soban Kumar and Dr.A.Sundaresan of NIIST, Trivandrum. Thanks are extended to Mr.Laxmanan of I.C Engine Laboratory, Chennai.

Funding: This research received no specific grant from any funding agency in the public, commercial or non-profit sectors.

REFERENCES

Abed, K. A., El Morsi, A. K., Sayed, M. M., El Shaib, A. A., & Gad, M. S. (2018). Effect of waste cooking oil biodiesel on performance and exhaust emissions of a diesel engine. *Egyptian Journal of Petroleum*, *27*(4), 985–989. doi:10.1016/j.ejpe.2018.02.008

Agarwal, D., Kumar, L., & Agarwal, A. K. (2007). Performance evaluation of a vegetable oil fuelled compression ignition engine. *Renewable Energy*, *33*(6), 1147–1156. doi:10.1016/j.renene.2007.06.017

Agarwal, D., Sinha, S., & Agarwal, A. K. (2006). Experiental investigation of control of NO_X emissions in biodiesel-fueled compression ignition engine. *Renewable Energy*, *31*(14), 2356–2369. doi:10.1016/j.renene.2005.12.003

Ahmed, S., Hassan, M. H., Kalam, M. A., Rahman, S. M. A., Abedin, M. J., & Shahir, A. (2014). An experimental investigation of biodiesel production, characterization, engine performance, emission and noise of Brassica juncea methyl ester and its blends. *Journal of Cleaner Production*, *79*, 74–81. doi:10.1016/j.jclepro.2014.05.019

Altun, S., Oner, C., Yasar, F., & Adin, H. (2011). Effect of n-Butanol blending with a blend of diesel and biodiesel on performance and exhaust emissions of a diesel engine. *Industrial & Engineering Chemistry Research*, *50*(15), 9425–9430. doi:10.1021/ie201023f

Atabani, A. E., Shobana, S., Mohammed, M. N., Uğuzd, G., Kumar, G., Arvindnarayan, S., ... Al-Muhtase, A. H. (2019). Integrated valorization of waste cooking oil and spent coffee grounds for biodiesel production: Blending with higher alcohols, FT–IR, TGA, DSC and NMR characterizations. *Fuel*, *244*, 419–430. doi:10.1016/j.fuel.2019.01.169

Ban-Weiss, G. A., Chen, J. Y., Buchholz, B. A., & Dibble, R. W. (2007). A numerical investigation into the anomalous slight NO_X increase when burning biodiesel; A new (old) theory. *Fuel Processing Technology*, *88*(7), 659–667. doi:10.1016/j.fuproc.2007.01.007

Basinger, M., Reding, T., Rodriguez-Sanchez, F. S., Lackner, K. S., & Modi, V. (2010). Durability testing modified compression ignition engines fueled with straight plant oil. *Energy*, *35*(8), 3204–3220. doi:10.1016/j.energy.2010.04.004

Bousbaa, H., Sary, A., Tazerout, M., & Liazid, A. (2012). Investigations on a compression ignition engine using animal fats and vegetable oil as fuels. *Journal of Energy Resources Technology*, *134*(2), 1–11. doi:10.1115/1.4005660

Buyukkaya, E. (2010). Effects of biodiesel on a DI diesel engine performance, emission, and combustion characteristics. *Fuel*, *89*(10), 3099–3105. doi:10.1016/j.fuel.2010.05.034

Cayli, G., & Kusefoglu, S. (2008). Increased yields in biodiesel production from used cooking oils by a two step process: Comparison with one step process by using TGA. *Fuel Processing Technology*, *89*(2), 118–122. doi:10.1016/j.fuproc.2007.06.020

Ceramicrotaryengines. (2019). Retrieved from www.ceramicrotaryengines.com

Chaurasiya, P. K., Singh, S. K., Dwivedi, R., & Choudri, R. V. (2019). Combustion and emission characteristics of diesel fuel blended with raw jatropha, soybean and waste cooking oils. *Heliyon (London)*, *5*(5). doi:10.1016/j.heliyon.2019.e01564 PMID:31193108

Cheruiyot, N. K., Hou, W.-C., Wang, L.-C., & Chen, C.-Y. (2019). The impact of low to high waste cooking oil-based biodiesel blends on toxic organic pollutant emissions from heavy-duty diesel engines. *Chemosphere*, *235*, 726–733. doi:10.1016/j.chemosphere.2019.06.233 PMID:31279123

Chotwichien, A., Luengnaruemitchai, A., & Jai-In, S. (2009). Utilization of palm oil alkyl esters as an additive in ethanol-diesel and butanol-diesel blends. *Fuel*, *88*(9), 1618–1624. doi:10.1016/j.fuel.2009.02.047

Dhanasekaran, R., Krishnamoorthy, V., Rana, D., Saravanan, S., Nagendran, A., & Kumar, B. R. (2017). A sustainable and eco-friendly fueling approach for direct-injection diesel engines using restaurant yellow grease and n-pentanol in blends with diesel fuel. *Fuel*, *193*, 419–431. doi:10.1016/j.fuel.2016.12.030

Dogan, O. (2011). The influence of n-butanol/diesel fuel blends utilization on a small diesel engine performance and emissions. *Fuel*, *90*(7), 2467–2472. doi:10.1016/j.fuel.2011.02.033

Domkundwar, V. M. (2009). *A course in internal combustion engines*. Dhanpat Rai & Co.

El-Galad, M. I., El-Khatib, K. M., & Zaher, F. A. (2015). Economic feasibility study of biodiesel production by direct esterification of fatty acids from the oil and soap industrial sector. *Egyptian Journal of Petroleum*, *24*(4), 455–460. doi:10.1016/j.ejpe.2015.06.002

El-Gendy, N. Sh., Deriase, S. F., Hamdy, A., & Abdallah, R. I. (2015). Statistical optimization of biodiesel production from sunflower waste cooking oil using basic heterogeneous biocatalyst prepared from egg shells. *Egyptian Journal of Petroleum*, *24*(1), 37–48. doi:10.1016/j.ejpe.2015.02.004

Encinar, J. M., Gonzalez, J. F., & Rodriguez-Reinares, A. (2007). Ethanolysis of used frying oil. Biodiesel preparation and characterization. *Fuel Processing Technology*, *88*(5), 513–522. doi:10.1016/j.fuproc.2007.01.002

Enweremadu, C. C., & Rutto, H. L. (2010). Combustion, emission and engine performance characteristics of used cooking oil biodiesel-A review. *Renewable & Sustainable Energy Reviews*, *14*(9), 2863–2873. doi:10.1016/j.rser.2010.07.036

Farooq, M., Ramli, A., & Subbarao, D. (2013). Biodiesel production from waste cooking oil using bifunctional heterogeneous solid catalysts. *Journal of Cleaner Production*, *59*, 131–140. doi:10.1016/j.jclepro.2013.06.015

FSSAI. (2019). Retrieved from https://fssai.gov.in/upload/advisories/2019/06/5cf8fe9068fc1Order_Used_Cooking_Oil_06_06_2019.pdf

Gad, M. S., El-Araby, R., Abed, K. A., El-Ibiari, N. N., El Morsi, A. K., & El-Diwani, G. I. (2018). Performance and emission characteristics of C.I. engine fueled with palm oil/palm oil methyl ester blended with diesel fuel. *Egyptian Journal of Petroleum*, *27*(2), 215–219. doi:10.1016/j.ejpe.2017.05.009

García-Martín, J. F., Alés-Álvarez, F. J., López-Barrera, M. C., Martín-Domínguez, I., & Álvarez-Mateos, P. (2019). Cetane number prediction of waste cooking oil-derived biodiesel prior to transesterification reaction using near infrared spectroscopy. *Fuel*, *240*, 10–15. doi:10.1016/j.fuel.2018.11.142

He, B.-Q., Liu, M.-B., Yuan, J., & Zhao, H. (2013). Combustion and emission characteristics of a HCCI engine fuelled with n-butanol-gasoline blends. *Fuel*, *108*, 668–674. doi:10.1016/j.fuel.2013.02.026

Hirner, F. S., Hwang, J., Bae, C., Patel, C., Gupta, T., & Agarwal, A. K. (2019). Performance and emission evaluation of a small-bore biodiesel compression-ignition engine. *Energy*, *183*, 971–982. doi:10.1016/j.energy.2019.07.015

Hwang, J., Bae, C., & Gupta, T. (2016). Application of waste cooking oil (WCO) biodiesel in a compression ignition engine. *Fuel*, *176*, 20–31. doi:10.1016/j.fuel.2016.02.058

Jin, C., Yao, M., Liu, H., Lee, C. F., & Ji, J. (2011). Progress in the production and application of n-butanol as a biofuel. *Renewable & Sustainable Energy Reviews*, *15*(8), 4080–4106. doi:10.1016/j.rser.2011.06.001

Kara, K., Ouanji, F., & El, M. (2018). Biodiesel production from waste fish oil with high free fatty acid content from Morroccan fish processing industries. *Egyptian Journal of Petroleum*, *27*(2), 249–255. doi:10.1016/j.ejpe.2017.07.010

Karabetkas, M., & Hosoz, M. (2009). Performance and emission characteristics of a diesel engine using isobutanol-diesel fuel blends. *Renewable Energy*, *34*(6), 1554–1559. doi:10.1016/j.renene.2008.11.003

Knothe, G. (2006). Analyzing biodiesel: Standards and other methods. *Journal of the American Oil Chemists' Society*, *83*(10), 823–833. doi:10.100711746-006-5033-y

Knothe, G., & Steidley, K. R. (2009). A comparison of used cooking oils; A very heterogeneous feedstock for biodiesel. *Bioresource Technology*, *100*(23), 5796–5801. doi:10.1016/j.biortech.2008.11.064 PMID:19592239

Kumar, P., Sharma, M. P., & Dwivedi, G. (2016). Impact of ternary blends of biodiesel on diesel engine performance. *Egyptian Journal of Petroleum*, *25*(2), 255–261. doi:10.1016/j.ejpe.2015.06.010

Kumaran, P., Mazlini, N., Hussein, I., Nazrain, M., & Khairul, M. (2011). Technical feasibility studies for Langkawi WCO (waste cooking oil) derived – biodiesel. *Energy*, *36*(3), 1386–1393. doi:10.1016/j.energy.2011.02.002

Lapuerta, M., Rodriguez-Fernandez, J., & Armas, O. (2010). Correlation for the estimation of the density of fatty acid esters fuels and its implications. A proposed Biodiesel Cetane Index. *Chemistry and Physics of Lipids*, *163*(7), 720–727. doi:10.1016/j.chemphyslip.2010.06.004 PMID:20599853

Ma, F., & Hanna, M. A. (1999). Biodiesel production: A review. *Bioresource Technology*, *70*(1), 1–15. doi:10.1016/S0960-8524(99)00025-5

Maceiras, R., Vega, M., Costa, C., Ramos, P., & Marquez, M. C. (2009). Effect of methanol content on enzymatic production of biodiesel from waste frying oil. *Fuel*, *88*(11), 2130–2134. doi:10.1016/j.fuel.2009.05.007

Meher, L. C., Sagar, V. D., & Naik, S. N. (2006). Technical aspects of biodiesel production by transesterification-a review. *Renewable & Sustainable Energy Reviews*, *10*(3), 248–268. doi:10.1016/j.rser.2004.09.002

Mehta, R. N., Chakaraborty, M., Mahanta, P., & Parikh, P. A. (2010). Evaluation of fuel properties of butanol-biodiesel-diesel blends and their impact on engine performance and emissions. *Industrial & Engineering Chemistry Research*, *49*(16), 7660–7665. doi:10.1021/ie1006257

Morais, S., Mata, T. M., Martins, A. A., Pinto, G. A., & Costa, C. A. V. (2010). Simulation and life cycle assessment of process design alternatives for biodiesel production from waste vegetable oils. *Journal of Cleaner Production*, *18*(13), 1251–1259. doi:10.1016/j.jclepro.2010.04.014

Nabi, M. N., Rahman, M. M., & Akhter, M. S. (2009). Biodiesel from cotton seed oil and its effects on engine performance and exhaust emissions. *Applied Thermal Engineering*, *29*(11-12), 2265–2270. doi:10.1016/j.applthermaleng.2008.11.009

Nair, P., Singh, B., Upadhyay, S. N., & Sharma, Y. C. (2012). Synthesis of biodiesel from low FFA waste frying oil using calcium oxide derived from Mereterix mereterix as a heterogeneous catalyst. *Journal of Cleaner Production*, *29-30*, 82–90. doi:10.1016/j.jclepro.2012.01.039

Nirmala, N., Dawn, S. S., & Harindra, C. (2019). Analysis of performance and emission characteristics of Waste cooking oil and Chlorella variabilis MK039712.1 biodiesel blends in a single cylinder, four strokes diesel engine. Renewable Energy, doi: j.renene.2019.08.133.

Plamondon, E., & Seers, P. (2019). Parametric study of pilot–main injection strategies on the performance of a light-duty diesel engine fueled with diesel or a WCO biodiesel–diesel blend. *Fuel*, *236*, 1273–1281. doi:10.1016/j.fuel.2018.09.111

Praveen, A., Rao, G. L. N., & Balakrishna, B. (2018). Performance and emission characteristics of a diesel engine using Calophyllum Inophyllum biodiesel blends with TiO_2 nanoadditives and EGR. *Egyptian Journal of Petroleum*, *27*(4), 731–738. doi:10.1016/j.ejpe.2017.10.008

Rahman, M. M., Hassan, M. H., Kalam, M. A., Atabani, A. E., Memon, L. A., & Rahman, S. M. A. (2014). Performance and emission analysis of jatropha curcas and Motinga oleifers methyl ester fuel blends in a multi-cylinder diesel engine. *Journal of Cleaner Production*, *65*, 304–310. doi:10.1016/j.jclepro.2013.08.034

Rahman, S. M. A., Masjuki, H. H., Kalam, M. A., Abedin, M. J., Sanjid, A., & Imtenan, S. (2014). Effect of idling on fuel consumption and emissions of a diesel engine fueled by jatropha biodiesel blends. *Journal of Cleaner Production*, *69*, 208–215. doi:10.1016/j.jclepro.2014.01.048

Rakopoulos, D. C., Rakopoulos, C. D., Giakoumis, E. G., Dimaratos, A. M., & Kyritsis, D. C. (2010). Effects of butanol-diesel fuel blends on the performance and emissions of a high-speed DI diesel engine. *Energy Conversion and Management*, *51*(10), 1989–1997. doi:10.1016/j.enconman.2010.02.032

Rakopoulos, D. C., Rakopoulos, C. D., Kakaras, E. C., & Giakoumis, E. G. (2008). Effects of ethanol-diesel fuel blends on the performance and exhaust emissions of heavy duty DI diesel engine. *Energy Conversion and Management*, *49*(11), 3155–3162. doi:10.1016/j.enconman.2008.05.023

Ramadhas, A. S., Jayaraj, S., & Muraleedharan, C. (2004). Use of vegetable oils as I.C engine fuels- A review. *Renewable Energy*, *29*(5), 727–742. doi:10.1016/j.renene.2003.09.008

Ramadhas, A. S., Muraleedharan, C., & Jayaraj, S. (2005). Performance and emission evaluation of a diesel engine fueled with methyl esters of rubber seed oil. *Renewable Energy*, *30*(12), 1789–1800. doi:10.1016/j.renene.2005.01.009

Sanjid, A., Masjuki, H. H., Kalam, M. A., Rahman, A. S. M., Abedin, M. J., & Palash, S. M. (2014). Production of palm and jatropha based biodiesel and investigation of palm-jatropha combined blend properties, performance, exhaust emission and noise in an unmodified diesel engine. *Journal of Cleaner Production*, *65*, 295–303. doi:10.1016/j.jclepro.2013.09.026

Shahid, E. M., & Jamal, Y. (2008). A review of biodiesel as vehicular fuel. *Renewable & Sustainable Energy Reviews*, *12*(9), 2484–2494. doi:10.1016/j.rser.2007.06.001

Sharon, H., Jai Shiva Ram, P., Jenis Fernando, K., Murali, S., & Muthusamy, R. (2013). Fuelling a direct injection diesel engine with diesel-used palm oil-butanol blends – an experimental study. *Energy Conversion and Management*, *73*, 95–105. doi:10.1016/j.enconman.2013.04.027

Shivakumar., Pai, P.S., Rao, B.R.S., (2011). Artificial Neural Network based prediction of performance and emission characteristics of a variable compression ratio CI engine using WCO as a biodiesel at different injection timings. *Applied Energy 88, 2344–2354.*

Singh, P. J., Khurma, J., & Singh, A. (2010). Preparation, characterization, engine performance and emission characteristics of coconut oil-based hybrid fuels. *Renewable Energy*, *35*(9), 2065–2070. doi:10.1016/j.renene.2010.02.007

Solubility of alcohols. (2019). Retrieved from https://www.solubilityofthings.com/water/alcohols

Sukjit, E., Herreros, J. M., Dearn, K. D., Garcia-Contreras, R., & Tsolakis, A. (2012). The effect of the addition of individual methyl esters on the combustion and emissions of ethanol and butanol-diesel blends. *Energy, 42*(1), 364–374. doi:10.1016/j.energy.2012.03.041

Supple, B., Howard-Hildige, R., Gonzalez-Gomez, E., & Leahy, J. J. (2002). The effect of steam treating waste cooking oil on the yield of methyl ester. *Journal of the American Oil Chemists' Society, 79*(2), 175–178. doi:10.100711746-002-0454-1

Tabatabaei, M., Aghbashlo, M., Dehhaghi, M., Panahi, H. K. S., Mollahosseini, A., Hosseini, M., & Soufiyan, M. M. (2019). Reactor technologies for biodiesel production and processing: A review. *Progress in Energy and Combustion Science, 74*, 239–303. doi:10.1016/j.pecs.2019.06.001

Yadav, A. K., Khan, M. E., Pal, A., & Ghosh, U. (2018). Performance and emission characteristics of a stationary diesel engine fuelled by Schleichera Oleosa Methyl Ester (SOME) produced through hydrodynamic cavitation process. *Egyptian Journal of Petroleum, 27*(1), 89–93. doi:10.1016/j.ejpe.2017.01.007

Yassin, F. A., El Kady, F. Y., Ahmed, H. S., Mohamed, L. K., Shaban, S. A., & Elfadaly, A. K. (2015). Highly effective ionic liquids for biodiesel production from waste vegetable oils. *Egyptian Journal of Petroleum, 24*(1), 103–111. doi:10.1016/j.ejpe.2015.02.011

ADDITIONAL READING

Bittle, J. A., Knight, B. M., & Jacobs, T. J. (2010). Interesting behavior of biodiesel ignition delay and combustion duration. *Energy & Fuels, 24*(8), 4166–4177. doi:10.1021/ef1004539

Chhabra, M., Sharma, A., & Dwivedi, G. (2017). Performance evaluation of a diesel engine using rice bran biodiesel. *Egyptian Journal of Petroleum, 26*(2), 511–518. doi:10.1016/j.ejpe.2016.07.002

Gumus, M. (2010). A comprehensive experimental investigation of combustion and heat release characteristics of a biodiesel (hazelnut kernel oil methyl ester) fueled direct injection compression ignition engine. *Fuel, 89*(10), 2802–2814. doi:10.1016/j.fuel.2010.01.035

Heywood, J. (2017). *Internal combustion engines fundamentals* (Indian Edition). Tata Mc-graw Hill.

Sahoo, P. K., & Das, L. M. (2009). Combustion analysis of jatropha, karanja and polanga based biodiesel as fuel in a diesel engine. *Fuel, 88*(6), 994–999. doi:10.1016/j.fuel.2008.11.012

Sharon, H., Jayaprakash, R., Karthigai Selvan, M., Soban Kumar, D. R., Sundaresan, A., & Karuppasamy, K. (2012). Biodiesel production and prediction of engine performance using SIMULINK model of trained neural network. *Fuel, 99*, 197–203. doi:10.1016/j.fuel.2012.04.019

Sharon, H., Karuppasamy, K., Soban Kumar, D. R., & Sundaresan, A. (2012). A test on direct injection diesel engine fuelled by methyl esters of used palm oil. *Renewable Energy, 47*, 160–166. doi:10.1016/j.renene.2012.04.032

Yu, C. W., Bari, S., & Ameen, A. (2002). A comparison of combustion characteristics of waste cooking oil with diesel as fuel in a direct injection diesel engine. *Proceedings of IMechE Part D: Journal of Automobile Engineering*, *216*(3), 237–243. doi:10.1243/0954407021529066

KEY TERMS AND DEFINITIONS

Calorific value: Amount of energy released when 1 kg of fuel is burned.

Cetane number: Cetane number of the test fuel is the volume percent of n-hexadecane (cetane number 100) in a mixture of n-hexadecane and 1-methylnaphthalene (cetane number 0) that gives similar ignition delay as test fuel.

Cloud Point: Cloud point is the temperature at which saturated fatty acids in biodiesel begins to separate when biodiesel is cooled and the fuel looks cloudy. Lower cloud point is always preferred.

Engine Emissions: Exhaust gases released from engine after combustion of fuel inside engine chamber.

Flash Point: Flash point is the minimum temperature at which a liquid fuel forms ignitable vapor mixture with the air near its surface. It gives an indication on the ease of flammability of fuel.

Pour Point: Pour point is the temperature at which biodiesel ceases to flow or loses its flowing properties. Cloud and pour point are essential for identifying suitable locations for applicability of particular biodiesel.

Trans-esterification: Trans-esterifcation is the process in which triglyceride reacts with alcohol in the presence/absence of catalyst to produce esters (biodiesel) and glycerol. Non-catalyst trans-esterifcation requires high temperature and pressure.

Viscosity: Viscosity refers to resistance offered by adjacent layers of fluid during its flow. Viscosity is an important fuel property which gives an idea of spray characteristics of fuel and overview of ease of combustion.

Waste Cooking Oil: Oil discarded after frying edible items.

Chapter 10
Behaviour of Oxygenated Biofuels in Engines:
Engine Features of Oxygenate Mixtures

A. Prabu
https://orcid.org/0000-0002-2284-6707
SRM Institute of Science and Technology, India

ABSTRACT

An experimental investigation was conducted to disclose the outcomes of oxygenate mixture as additives in Jatropha biodiesel on the performance, combustion, and emission characteristics of a direct injection compression ignition engine. The experiments were conducted in an instrumented single-cylinder, air-cooled, four-stroke, direct-injection diesel engine, equipped with data acquisition system, AC alternator, and an electric loading device. Four oxygenate additives, namely, Ethylene Glycol (C2H6O2), Di methyl Carbonate (C3H6O3), 2-Butoxyethanol (C6H14O2), & Propylene Glycol (C3H8O2) were selected and nine different combinational oxygenate test fuels were prepared attaining ratios of 1, 2, and 4% volume of oxygenates with biodiesel. A significant reduction of emissions such as CO by 60%, Unburned HC by 11%, and smoke emissions by 27% were observed. Substantial improvement in brake thermal efficiency by 6% was observed, while NO emission increased marginally by 4%.

HIGHLIGHTS

- Four oxygenates namely, Dimethyl Carbonate ($C_3H_6O_3$), Ethylene Glycol ($C_2H_6O_2$), Propylene Glycol ($C_3H_8O_2$) and 2-Butoxyethanol ($C_6H_{14}O_2$) were selected as fuel additives.
- Ratios of 1, 2 and 4% volume of oxygenates were blended with Jatropha biodiesel.
- The brake thermal efficiency for oxygenates dispersed test fuels was substantially improved compared to that of neat biodiesel operation.
- The oxygenate-dispersed test fuels reduced CO emission by 60%, Unburned HC emission by 11% and smoke emission by 27%.

DOI: 10.4018/978-1-7998-2539-5.ch010

Table 1. The commonly used additives in fuel

S.No	Types of Additives	Function
1	Anti-Foam Agents	Reduce foaming during tank filling
2	Antioxidants	Inhibit oxidation, reduce precipitate and gum formation, extend storage life,
3	Biocides	Inhibit fungi and bacterial growth, prevent fuel filter plugging
4	Cetane Number Improver	Improves ignition quality by increasing cetane number, easy start of engine, reduces white smoke
5	Cloud Point Depressants (Suppressants)	Reduces temperature at which paraffin solubilize
6	Demulsifiers / Dehazers	Increase the rate of water separation from the fuel
7	Detergents / Dispersants	Improves spray patterns and clean injectors
8	Lubricity Improvers	Improve lubricity, better injector and pump lubrication
9	Metal Deactivators	Deactivate copper compounds in fuel and promotes longer storage life
10	Nanoparticles	Enhanced surface area to volume ratio improves the evaporation characteristics.
11	Oxygenates	Improves the oxygen content in the fuel
12	Pour Point Depressants	Improve cold-flow properties and low temperature operability
13	Rust Preventors	Reduce formation of rust in fuel systems and storage tanks
14	Smoke Suppressants	Reduce exhaust smoke and promote more complete combustion,
15	Stabilizers	Inhibits oxidation and extends storage life

- The addition of oxygenates as additives in biodiesel resulted in enhanced performance and emission characteristics of a diesel engine.

INTRODUCTION

Biodiesels are conceived as substitute fuels for petroleum products. Various efforts are established to examine the potency and suitability of the fuel. Biofuels have received more attention than other renewable fuels, as it has better fuel properties than those of petroleum fuel and suppress the environmental pollution. Lower emissions (aldehydes, polycyclic aromatic hydrocarbons and carbon monoxide), lower toxicity and sulphur dioxide emissions are the advantage of biodiesel fuel, having some of the disadvantages such as higher fuel consumption, lower calorific value, higher nitrous oxide (NOx) emissions, low stability and higher freezing point than diesel fuel. Blending biodiesel with additives significantly eliminates the disadvantages of it (Prabu and Anand, 2018). An additive for the fuel has to be selected carefully based on its functions. The commonly used additives in fuel are listed in Table 1.

Need for Performance and Emission

The engine performance parameter; brake thermal efficiency of biodiesel is lower than the diesel fuel due to its high density, viscosity and poor atomization of fuel. In general, biodiesel supplies enough oxygen for the combustion of fuel. Despite the fact, due to its lower calorific value, poor atomization and vaporization of fuel, an increment in brake thermal efficiency of the engine cannot be achieved when compared to the diesel fuel. So as to improve the brake thermal efficiency of the engine, fuel formulations is needed.

Biodiesel emits higher emission of NO when compared to diesel fuel. When sufficient oxygen is present complete combustion takes place, which in turn raises the levels of adiabatic flame temperature, resulting in Thermal NOx formation. The free radicals that are formed during the reaction of fuel reacts with N_2 eventually form prompt NO (Prabu and Anand, 2016). So to suppress the NOx formation, fuel formulations is needed.

Oxygenate Additives for Engine Improvement

Biodiesel containing fuel-borne oxygen content promotes the complete combustion of fuel with the diminution of brake thermal efficiency due to its high density and viscosity. It causes poor atomization and vaporization of fuel when compared to diesel fuel. Oxygenates contain oxygen as a part of their chemical structure; tend to vaporize more abruptly than aromatic fuel and increase clean burning of fuel (CFDC, 2003). So, in order to improve the fuel vaporization of fuel, oxygenate additives are used.

Advantages of Oxygenates

- Oxygenates are useful in controlling greenhouse gas emissions
- Oxygenates provides better combustion of gasoline, reducing exhaust emissions.
- Substantially reduces primary particulate matter by 25 to 30%.
- Reduces the long-term reactivity of CO and ozone formation in the atmosphere
- Blending properties of oxygenates dilute harmful aromatic and improve the gasoline blend

Oxygenates can be used as the main fuel (Semelsberger *et al.,* 2006; Park and Lee, 2013) and blending fuel (Cheng and Dibble, 1999; Yanfeng *et al.,* 2007) in diesel engines. The addition of oxygenates as additive in diesel fuel leads to reduction in exhaust emissions (CO emission (Dayang and Yun 2012; Tsolakis *et al.* 2010), HC (Chen *et al.,* 2008; Dayang and Yun 2012), NO (Ren *et al.,* 2008; Cinar *et al.,* 2010; Dayang and Yun 2012) and Smoke (Kim and Cho, 2013)) due to the presence of more oxygen atom causing easier combustion effectively.

Qi *et al.,* (2011) conducted an experimental investigation to assess the effects of Diethyl ether and ethanol as additives in biodiesel-diesel blends and observed a slight increase in cylinder pressure due to the low viscosity and higher volatility of diethyl ether and ethanol. Hulwan and Joshi (2011) investigated the performance, emission and combustion characteristic of a multi cylinder DI diesel engine running on diesel–ethanol–biodiesel blends of high ethanol content and observed an appreciable improvement in brake thermal efficiency and smoke for the blends especially at medium and high loads. They also ascertained rapid premixed combustion and improved diffusive combustion for high ethanol content blends that helped to form better air–fuel mixture, resulting in a larger percentage of fuel burned in the premixed burning phase causing peak cylinder pressure and heat release rate closer to the TDC. Sivalakshmi and Balusamy (2013) together conducted experiments using neem biodiesel and diethyl ether fuel blends and observed a significant improvement in the brake thermal efficiency and concluded that addition of oxygenates up to 5% (by vol.) could be a promising technique for using biodiesel efficiently in diesel engines without any modifications in the engine.

In recent years, progress in oxygenate additives in fuels was observed for the betterment of engine efficiency. Basha *et al.,* 2019 studied the effects of Hexanol and Methyl Acetate as oxygenated additives in diesel Fuel and observed better performance characteristic than that of neat diesel. Basha., 2018 experimented with oxygenate diethyl ether along with biodiesel emulsion fuels in a diesel engine and resulted with higher brake thermal efficiency and reduced NO and smoke emissions than that of pure diesel and biodiesel. Improvement in the engine performance characteristics was observed in various studies with usage of oxygenates such as diethyl ether in diesel (Ergen *et al.,*2013) and diethyl ether blended with Jatropha biodiesel (Sivalakshmi and Balusamy, 2013), ethanol blended with biodiesel / diesel (Qi *et al.,* 2011) and Dimethoxymethane ($C_3H_8O_2$) and Diethyl ether $(C_2H_5)_2O$ blended with diesel (Adelbert and Robert, 1999). Effective smoke and CO emission reduction are witnessed in various studies with oxygenates such as ethylene glycol mono-n-butyl ether ($C_6H_{14}O_2$) in biodiesel (Cho and Kim, 2013), 2-methoxyethyl acetate ($C_5H_{10}O_3$) with diesel (Yanfeng *et al.,* 2013), di-n-pentyl ether ($C_{10}H_{22}O$) with hydrotreated vegetable oils (Happonen et al. 2013) and Dimethoxymethane ($C_3H_8O_2$), Diglyme ($C_6H_{14}O_3$), Dimethyl carbonate ($C_3H_6O_3$), Diethyl Carbonate ($C_5H_{10}O_3$), and Diethyl adipate ($C_{10}H_{18}O_4$) with diesel (Ren *et al.,* 2008). Dhamodaran and Esakkimuthu, 2018 used diisopropyl ether and n-butanol with gasoline at 5, 15, and 25% by volume in a four-stroke four-cylinder multipoint fuel injection spark-ignition engine and detected higher nitrogen oxide, lower hydrocarbon and carbon monoxide emissions for oxygenate blend fuels. Jehad et al., 2019 blended Dimethoxy ethane, Tri-propylene glycol methyl ether and Diethylene glycol monoethyl ether of 20% volume with diesel fuel and resulted with 28% reduction of particulate matter emission in Toyota Diesel Engine. Similarly oxygenates such as Ethyl Aceto Acetate, Diethyl Carbonate and Diethylene Glycol were blended at 2.5%, 5% and 7.5% by volume with diesel fuel in a single cylinder naturally aspirated direct injection diesel engine and noticed lower brake specific fuel consumption with lower smoke emission compared with diesel fuel.

From the previous studies, it was observed that the addition of oxygenates in fuel had shown improvement in the brake thermal efficiency, and cuts down the emissions significantly. The effect of combined addition of oxygenates with biodiesel were prepared and the performance, combustion and emission characteristics of the engine were tested in a single cylinder air-cooled direct injection diesel engine, with an AC alternator for loading, data acquisition system for measuring the in-cylinder pressure, heat release rate and a crank-angle indicator to find the crank angle. The observed test results are compared with those of neat diesel and neat biodiesel as base fuels. Oxygenates, such as Dimethyl Carbonate (DMC), Ethylene Glycol (EG), Propylene Glycol (PG) and 2-Butoxyethanol (BE) are selected as oxygenate additives for Jatropha biodiesel. The detailed specifications of oxygenates are listed in Table 2.

MATERIALS AND METHODS

The preparation of oxygenate test fuels, stability of test fuels; engine experimental set-up, test procedures and uncertainty analysis are discussed under this section.

Preparation and Stability of Test Fuel

The crude jatropha oil was purchased from the market and the preparation of jatropha biodiesel was carried out by both acid esterification and base transesterification process, since the free fatty acid value (8.1%) was found to be greater than 1%. At the start of the fuel preparation process, the acid esterification

Table 2. Detailed specifications of oxygenates

	Oxygenates			
Physical and Chemical Properties	**Dimethyl Carbonate**	**Ethylene Glycol**	**Propylene Glycol**	**2-Butoxyethanol**
CAS. NO	616-38-6	110-74-4	57-55-56	111-76-2
Molecular formula	$C_3H_6O_3$	$C_2H_6O_2$	$C_3H_8O_2$	$C_6H_{14}O_2$
Molecular Weight, g/mol	90.08	62.07	76.09	118.17
Boiling Point °C	90	152	188	199
Melting Point °C	2 to 4 °C	-12.9 °C	-59 °C	-77 °C
Density, g/cm³	1.07	1.11	1.04	0.9
Oxygen Content wt %	53.3	26.9	24.2	18.6
Calorific Value MJ/kg	20.2	32.4	33.5	35.2

Table 3. Compositions of test fuels

	Test fuels								
Blends	**DMCBE1**	**DMCBE2**	**DMCBE4**	**DMCPG1**	**DMCPG2**	**DMCPG4**	**EGBE1**	**EGBE2**	**EGBE4**
Biodiesel (ml)	1000	1000	1000	1000	1000	1000	1000	1000	1000
DMC (ml)	5	10	20	5	10	20	-	-	-
BE (ml)	5	10	20	-	-	-	5	10	20
PG (ml)	-	-	-	5	10	20	-	-	-
EG (ml)	-	-	-	-	-	-	5	10	20

process was carried out (molar ratio of methanol/oil 6:1) by heating the jatropha oil to 65 °C in a borocil flask with the presence of sulphuric acid, methanol and the mixture was stirred continuously at a speed of 700 rpm for a reaction period of 1 h using the mechanical stirrer. Then the mixture was allowed for settlement in a separation funnel and the collected mixture FFA value was measured to be 0.90%. Then the base transesterification process was carried out with a specified quantity of KOH and methanol with the treated oil at 65 °C and molar ratio of 6:1. The mixture was stirred continuously stirred for 1 h using the mechanical stirrer and the mixture was transferred to the separation funnel. The jatropha biodiesel containing low FFA value (0.5%) was carefully collected and the fuel properties were tested as per the ASTM fuel standards.

The maximum yield achieved for the molar ratio as 6:1 was found to be 96%. Oxygenates, such as Dimethyl carbonate (DMC), Ethylene glycol (EG), Propylene glycol (PG) and 2-Butoxyethanol (BE) were selected as oxygenate additives for Jatropha biodiesel, and the test fuels were prepared by varying the oxygenates proportion attaining 1, 2 and 4% volume for the nine test fuels. The detailed composition of the nine different test fuels are listed in Table 3 and their fuel properties are tabulated in Table 4. A phase separation test was conducted for all the test fuels to ensure the stability characteristics and the test fuels were found to be stable for 4 hours.

Table 4. Properties of test fuels

Properties	ASTM Std.	Neat Diesel	Neat Biodiesel	DMCBE1	DMCBE2	DMCBE4	DMCPG1	DMCPG2	DMCPG4	EGBE1	EGBE2	EGBE4
Density @ 15 °C (kg/m^3)	D1298	835	873	873	872	870	872	871	870	871	870	868
Kinematic Viscosity @ 40 °C (cSt)	D445	2.20	4.10	3.90	3.80	3.80	3.90	3.80	3.70	3.90	3.70	3.50
Flash point (°C)	D93	48	85	82	78	75	83	80	78	82	78	77
Calorific value (MJ/kg)	D240	44.3	39.5	40	40.6	41.7	40	40.6	41.6	40.2	40.9	42.2

Engine Experimental Set-Up

The performance, combustion and emission characteristics for the nine test fuels were studied in an instrumented engine setup that was used by Basha and Anand, 2013; Basha and Anand, 2014. The schematic layout of engine experimental set-up is shown in Figure 1. The whole experimental investigation was carried out in a single cylinder air-cooled direct injection diesel engine, with an AC alternator for loading the engine. The detailed engine specification is listed in Table 5. All the experimental trails were performed by applying different loads at fuel injection timing of 26° bTDC (before top dead centre) and a constant speed of 1500 rpm. A Kistler piezoelectric pressure transducer, which was flush mounted at the engine cylinder head for measuring the in-cylinder pressure and a crank-angle indicator to find the crank angle. The levels of pollutants in the engine exhaust (NO, CO and Unburned HC) were measured by using an AVL 444 Di-gas analyzer and the smoke intensity was measured by using an AVL 437 smoke meter. The specification of AVL 444 Di-gas analyzer and AVL 437 smoke meter was listed in Table 6.

Figure 1. Schematic layout of experimental set-up

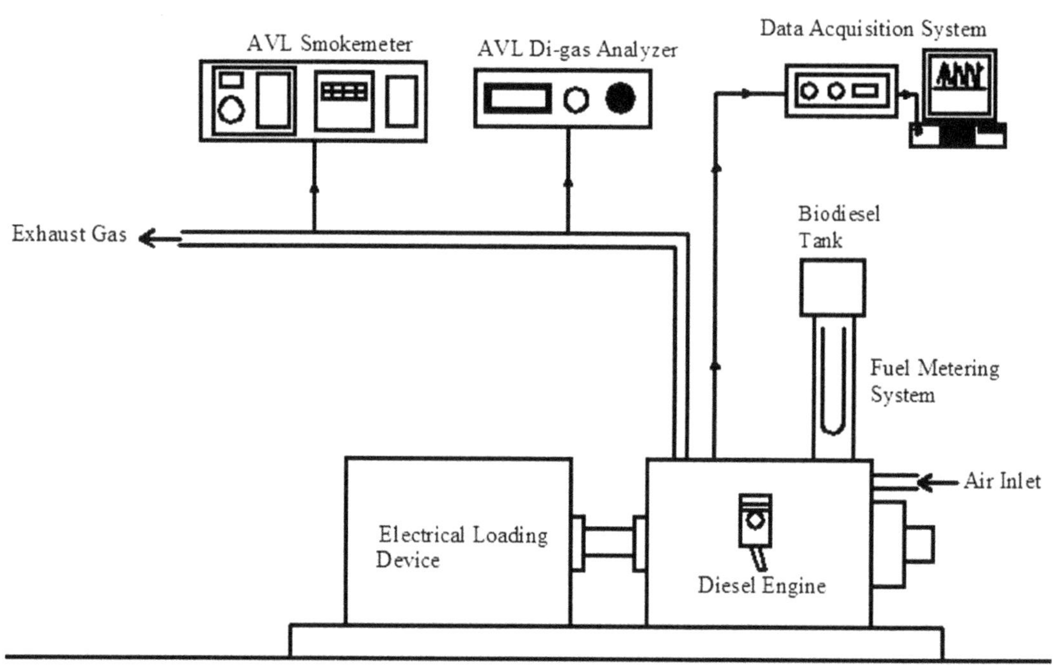

Table 5. Specification of diesel engine

Make	Kirloskar, India
Type	Single cylinder, four stroke, air cooled, Direct injection engine
Bore × Stroke	87.5 × 110 mm
Compression ratio	17.5:1
Swept volume	661 cm^3
Combustion chamber	Open hemispherical
Nozzle holes	3
Spray hole diameter	0.25 mm
Cone angle	110°
Rated output	4.4 kW at 1500 rpm,
Injection timing	26° btdc

Test Procedures and Uncertainty Analysis

The experiments were conducted in two phases: In the first phase, neat diesel and neat biodiesel were used as the base fuels, followed by the second phase with nine test fuels. The experimental trials were carried out from no-load to rated-load, for analyzing the performance, combustion and emission characteristics of the engine for nine different test fuels. The observed test results were compared with those of neat diesel and neat biodiesel as base fuels. The brake thermal efficiency (BTE), brake specific fuel consumption of the engine (BSFC), and emission characteristics such as Unburned Hydro Carbon (HC),

Table 6. Specification of gas analyzer and smoke meter

AVL 444 DIGAS Analyser						
Measured Parameter	**Measurement Principle**	**Measuring Range**	**Resolution**	**Accuracy**		**% uncertainty**
CO	NDIR	0 – 10% vol.	0.01% vol.	<0.6% vol: ≥0.6% vol:	±0.03% vol ±5% of value	±0.2 ±0.3
HC	NDIR	0 – 20,000 ppm vol.	≤2000: 1 ppm vol.	<200 ppm vol: ≥200 ppm vol:	±10 ppm. ±5% of value	±0.1 ±0.2
NOx	Electro chemical sensor	0 – 5000 ppm vol.	1 ppm vol.	<500 ppm vol: ≥500 ppm vol:	±50 ppm vol ±10% of value.	±0.2 ±0.9
AVL 437 Smoke meter						
Make			AVL			
Principle			Partial flow opacity measurement method			
Measurement			Smoke absorption coefficient			
Measuring range			0 – 100			
Accuracy			± 2% relative			
Resolution			0.1%			
Ambient temperature range			-5 to +45 °C			

Nitrogen Oxide (NO), Carbon monoxide (CO) & Smoke opacity were plotted against the brake mean effective pressure (bmep). The combustion characteristics such as Heat release rate (HRR), Cylinder pressure were plotted against the crank angle.

An uncertainty analysis as suggested by Holman (2011) was adopted for finding the uncertainty of this experimental research work. The uncertainty of the various measured and calculated quantities for Unburned HC, CO, NO, Smoke opacity and brake thermal efficiency are ±1 ppm, ±0.01% vol, ±1 ppm, ± 1% and ± 1.5% respectively.

RESULTS AND DISCUSSION

The results on the performance, combustion and emission characteristics of the CI engine fuelled with Jatropha biodiesel and oxygenates dispersed test fuels were presented and compared with neat biodiesel and neat diesel.

Performance Characteristics

Figure 2 depicts about the variation of brake specific fuel consumption (BSFC) and brake thermal efficiency (BTE) under bmep. At full load, due to its lower calorific value and high viscosity of the biodiesel fuel, lower BTE (%) and higher BSFC (kg/kWh) were observed for neat biodiesel fuel, when compared with the BSFC and BTE of neat diesel (Neat Diesel = 0.262 kg/kWh, Neat Biodiesel = 0.317 kg/kWh, Neat Diesel = 32.2% and Neat Biodiesel = 28.5%). Nevertheless, improvement in BSFC were observed for the oxygenate test fuels due to the supply of more oxygen molecule during combustion. The values observed for DMCBE4, DMCPG4 and EGBE4 test fuels as 0.297, 0.298 and 0.301 kg/kWh respectively. Improvement in BSFC reflects increase of BTE for DMCBE4, DMCPG4 and EGBE4 test fuels as 30.8, 30.6 and 30.4% respectively.

Combustion Characteristics

Heat Release Rate and The Cylinder Pressure

Heat release rate and the cylinder pressure characteristics assist in understanding the combustion processes for CI engine. The variations of cylinder pressure and heat release rate with respect to crank angle for oxygenates dispersed test fuels at the full load were illustrated in Figure 3-5. The rich oxygen content of neat biodiesel increased the cylinder pressure than neat diesel. Qi *et al.,* (2011) described that the combustion starts earlier for biodiesel fuel, burning large amount of fuel during the premixed combustion causes a higher cylinder pressure and heat release rate (Tesfa *et al.,* 2013; Tse *et al.*, 2015; Rajasekar and Selvi, 2014). However, for oxygenate dispersed test fuels; lesser premixed part of the combustion was observed for the test fuels than neat biodiesel, which reflected in lowered cylinder peak pressure (Imtenan *et al.,* 2015). At the full load, the cylinder pressure observed for neat biodiesel was 73.2 bar, whereas it was 70.4, 70.9, 71.6, 71.0, 70.8, 71.2, 71.2, 71.2 and 69.8 bar for EGBE1, EGBE2, EGBE4, DMCBE1, DMCBE2, DMCBE4, DMCPG1, DMCPG2 and DMCPG4 test fuels, respectively. The heat release rate is defined as the rate of release of chemical energy of the fuel during the CI combustion process (Gopal and Karupparaj, 2015). Peak pressure intends the large quantity of fuel burned

Figure 2. Variations of test fuels on BSFC and BTE

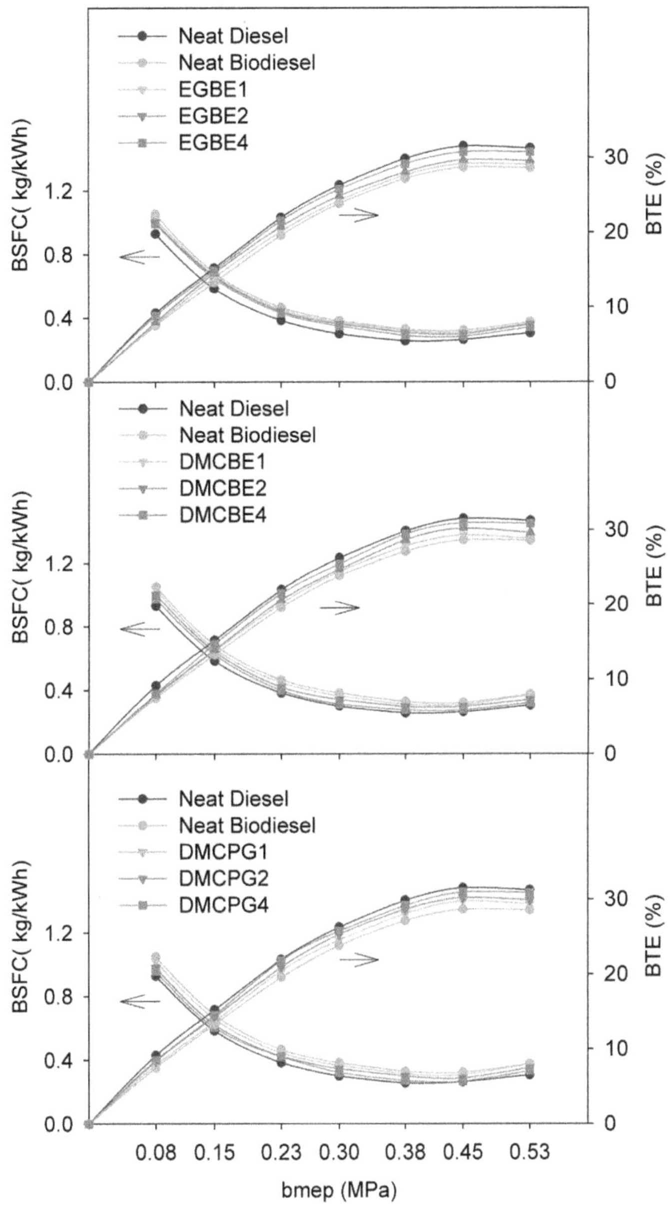

in the premixed combustion stage (Rajasekar and Selvi, 2014; Tse *et al.*, 2015). Whereas for oxygenates dispersed test fuel, slender decrement in heat release rate and cylinder pressure characteristics were observed. The incorporation of oxygenate additives in neat biodiesel ensued an considerable decrement of heat release rate and cylinder pressure. At the full load, the heat release rate observed for the neat biodiesel was 20.8 J/deg.CA, whereas it was 20.3, 20.7, 21.3, 20.1, 20.4, 20.5, 19.5, 20.3 and 20.8 J/deg.CA for EGBE1, EGBE2, EGBE4, DMCBE1, DMCBE2, DMCBE4, DMCPG1, DMCPG2 and DMCPG4 test fuels respectively.

Figure 3. Variations of cylinder pressure and heat release rate for EGBE test fuels

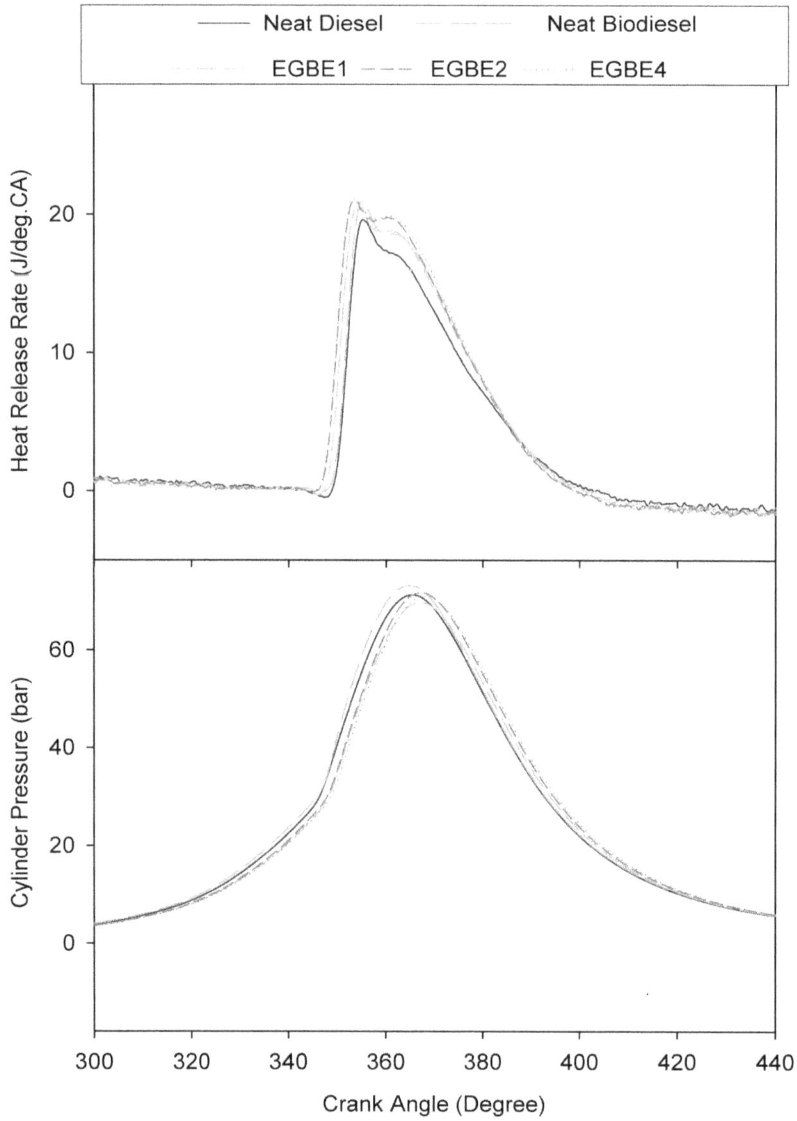

Emission Characteristics

NO and Exhaust Gas Temperature (EGT)

The variations of test fuels on nitric oxide (NO) and Exhaust gas temperature (EGT) under bmep were shown in Figure 6. The NO emission recorded for DMCBE4, DMCPG4, DMCBE2, EGBE4, EGBE2, DMCBE1, DMCPG2, EGBE1 and DMCPG1 are 9.67, 9.65, 9.54, 9.53, 9.41, 9.34, 9.32, 9.27 and 9.26 g/kWh respectively. Whereas the NO emission recorded for neat biodiesel is 9.21 g/kWh. Increase in NO emission was observed for the oxygenate test fuels due to the chemical reaction between oxygen molecule and nitrogen, at higher temperature (Prabu and Anand, 2015). Similar trend is witnessed for

Figure 4. Variations of cylinder pressure and heat release rate for DMCBE test fuels

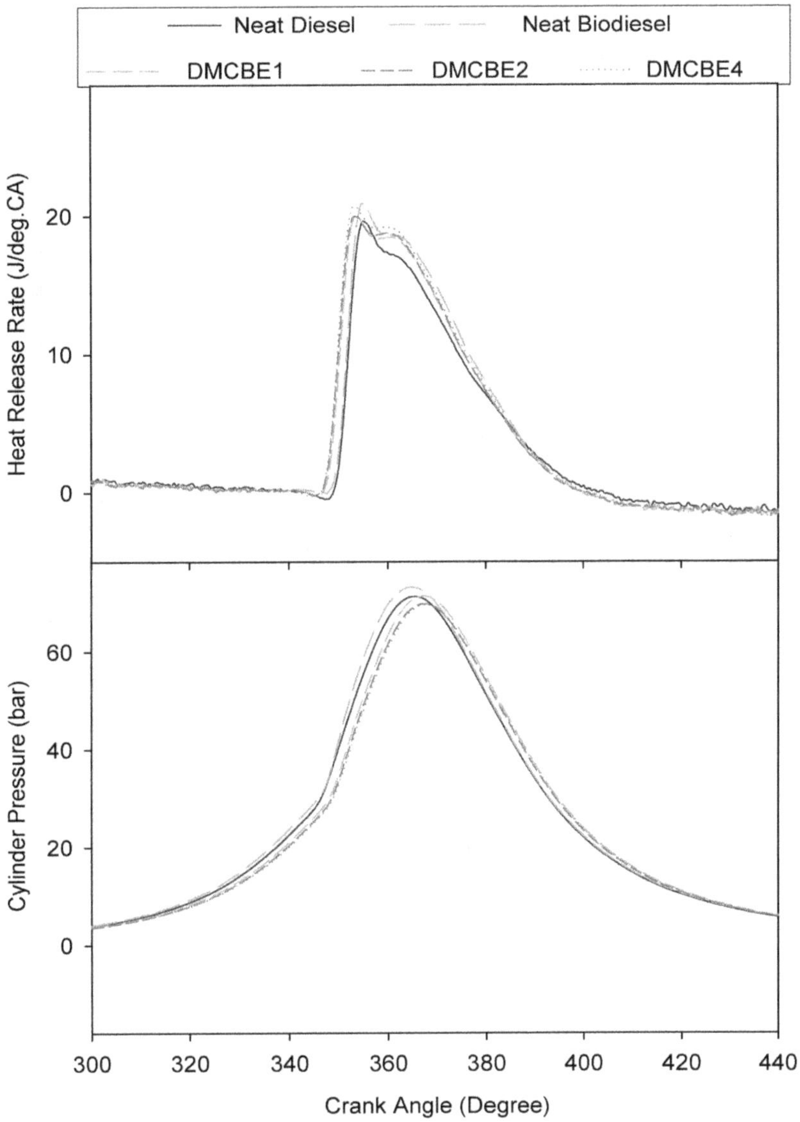

EGT of oxygenate test fuels due to the presence of rich oxygen molecule in the oxygenate test fuels, causing an increased oxidation reaction. The values observed for DMCBE4, DMCPG4, DMCBE2, EGBE4, EGBE2, DMCBE1, DMCPG2, EGBE1 and DMCPG1 of 352, 351, 348, 348, 344, 342, 341, 340 and 339 °C respectively, which was higher than neat biodiesel of 329 °C.

CO, Unburned Hydrocarbon and Smoke Opacity

The CO, Unburned HC and smoke opacity for the oxygenate test fuels under bmep were shown in Figure 7. It was observed from the figure that CO (g/kWh) emission decreases for EGBE1, DMCBE1, DMCPG1, EGBE2, DMCBE2, DMCPG2, DMCBE4, DMCPG4 and EGBE4 test fuels as 0.000143,

Figure 5. Variations of cylinder pressure and heat release rate for DMCPG test fuels

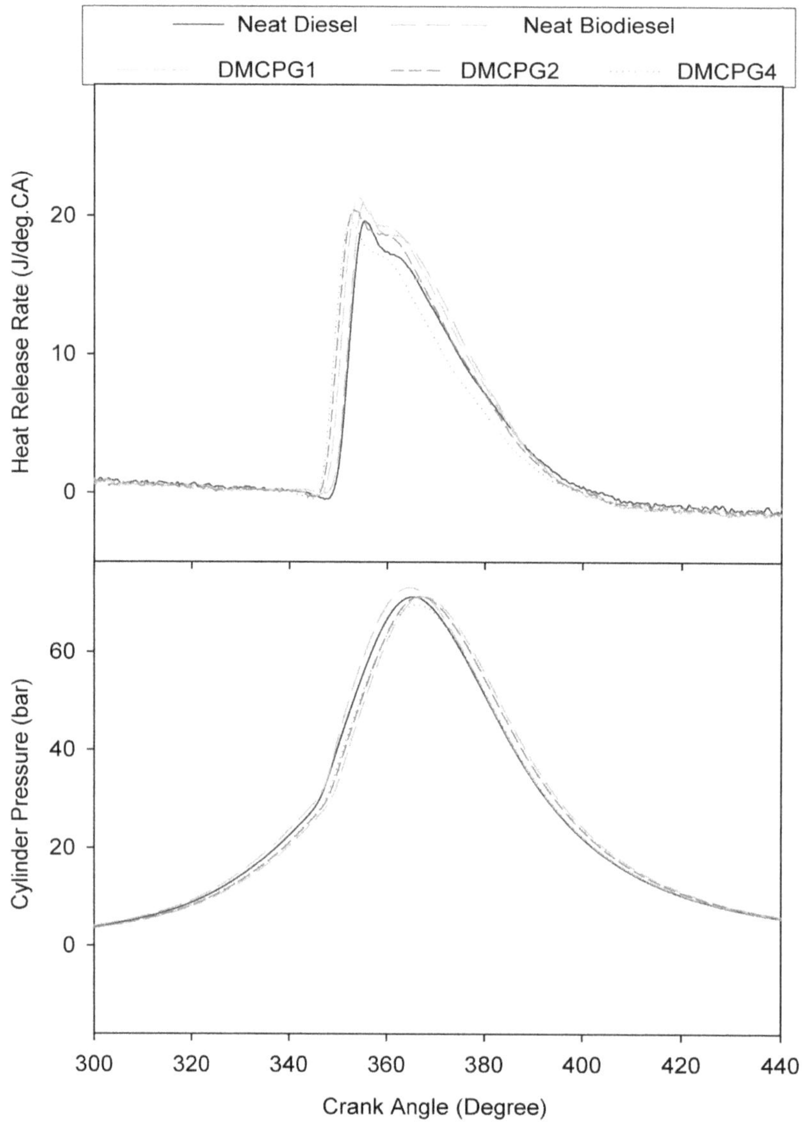

0.000143, 0.000107, 0.000107, 0.000107, 0.000107, 0.000107, 0.000071 and 0.000071 g/kWh respectively, compared with the CO emission for neat biodiesel as 0.00018 g/kWh and neat diesel as 0.00032 g/kWh. Oxygenates ameliorated the oxidation process with enough oxygen molecules ensuing lower CO emission (Hulwan and Joshi, 2011).

Decrement in Unburned HC (ppm) emission was observed for the oxygenate test fuels, when compared with neat biodiesel of 0.03 g/kWh and neat diesel as 0.04 g/kWh. A substantial reduction in Unburned HC emissions were observed for DMCBE4, DMCBE2, EGBE1, DMCBE1 and DMCPG1 test fuels are 0.031, 0.033, 0.033, 0.033 and 0.033 g/kWh respectively. Merely for the other oxygenate test fuels, gain in Unburned HC emissions were observed for DMCPG4 of 0.041 g/kWh, DMCPG2 of 0.039 g/kWh

Figure 6. Variations of test fuels on NO and EGT

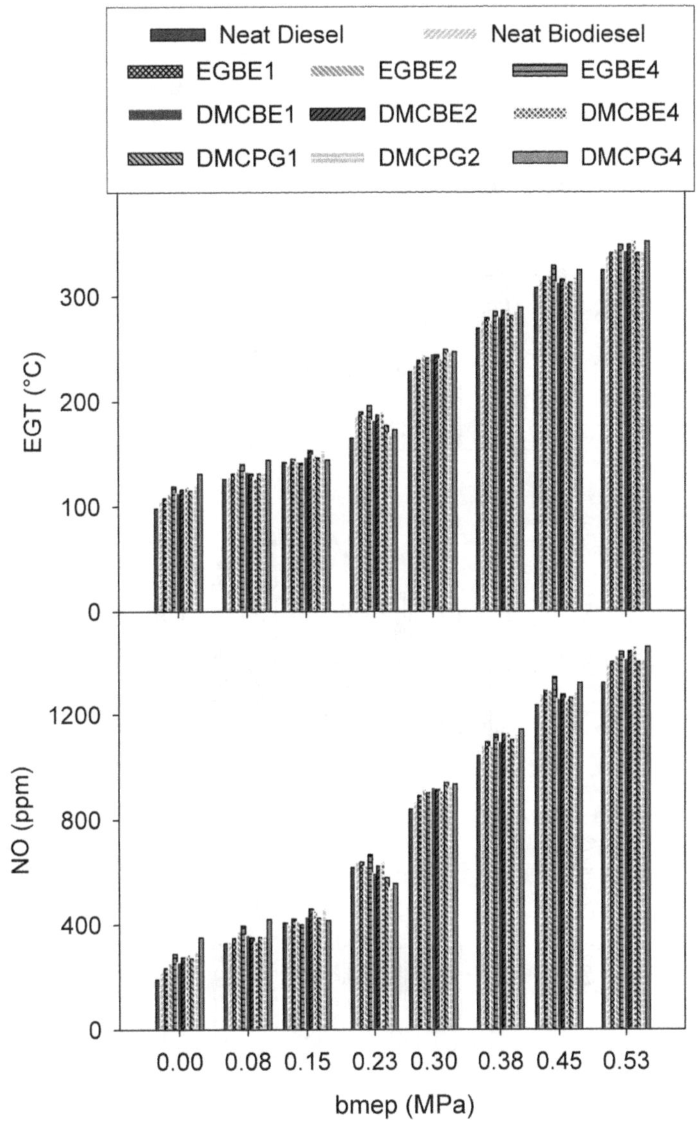

and EGBE4 of 0.037 g/kWh, ascribable to its lower viscosity of the oxygenate test fuel, inducing fuel slip from the injector at the final position of the expansion stroke (Yanfeng *et al.,* 2013).

It was observed from the figure 7, that the smoke opacity for neat biodiesel (37.4%) was lower when compared with neat diesel (43.4%). For the oxygenate test fuels, reduction in smoke opacity values were observed for EGBE1, EGBE2, EGBE4, DMCBE1, DMCBE2, DMCBE4, DMCPG1, DMCPG2 and DMCPG4 test fuels are 37.1, 36.7, 34.9, 36.0, 34.2, 31.1, 29.3, 28.8 and 27.3% respectively, imputable to the ample oxygen content in the oxygenate test fuels (Bhale *et al.*, 2009).

Figure 7. Variations of test fuels on CO, unburned HC and smoke opacity

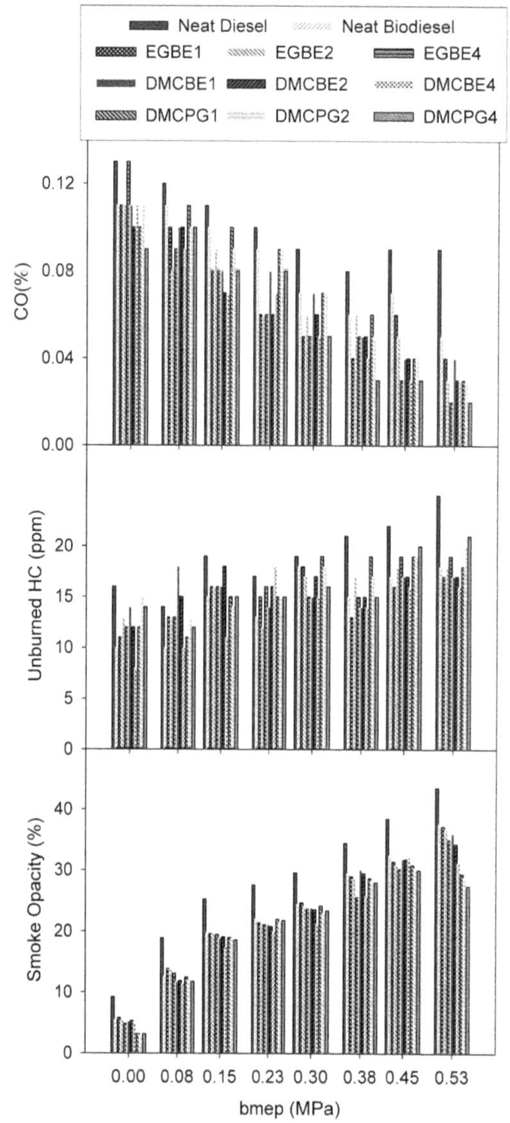

CONCLUSION

The performance parameters namely brake thermal efficiency (BTE) and brake specific fuel consumption (BSFC), Combustion parameter namely heat release rate (HRR) and cylinder pressure and emission parameters; CO, HC, NO and smoke opacity were obtained by conducting an experimental investigation on a diesel engine fuelled with nine oxygenate dispersed test fuels at various proportions, and the following conclusions were drawn

- The brake thermal efficiency for oxygenates dispersed test fuels improved marginally when compared to that of neat biodiesel operation

- The oxygenates dispersed test fuels has reduced Unburned HC emission by 11% CO emission by 60% and Smoke emission by 27%, with increased emission of NO by 4%.
- On the whole, the addition of oxygenates as additives in biodiesel resulted in enhanced performance, combustion and emission characteristics of a diesel engine.

Scope for Future Work

- The similar investigations may be extended to study the effect of mixtures of oxygenate additives on the performance, combustion and emission characteristics of a CI by using diesel-biodiesel blends.
- The experimental investigation can be carried out by varying the injection pressure; injection timing and Exhaust Gas Recirculation (EGR) engines with a mixture of oxygenate test fuels.

REFERENCES

Basha, J. S. (2018). Impact of Carbon Nanotubes and Di-Ethyl Ether as additives with biodiesel emulsion fuels in a diesel engine – An Experimental Investigation. *Journal of the Energy Institute*, *91*(2), 289–303. doi:10.1016/j.joei.2016.11.006

Basha, J. S., & Anand, R. B. (2013). The influence of nano additive blended biodiesel fuels on the working characteristics of a diesel engine. *Journal of the Brazilian Society of Mechanical Sciences and Engineering*, *35*(3), 257–264. doi:10.100740430-013-0023-0

Basha, J. S., & Anand, R. B. (2014). Performance, emission and combustion characteristics of a diesel engine using Carbon Nanotubes blended Jatropha methyl ester emulsions. *Alexandria Engineering Journal*, *53*(2), 259–273. doi:10.1016/j.aej.2014.04.001

Basha, J. S., Montaha, A. S. A., Khadija, M. M. A., Manal, Y. M. A., Sulaiman, A. S. A., Shahad, M. K. A., & Rashid, N. S. A. (2019). The Influence of Hexanol and Methyl Acetate as Oxygenated Additives with Diesel Fuel in a Diesel Engine Generator. *International Journal of I.C. Engines and Gas Turbines*, *5*(1), 1–9.

Bhale, P. V., Deshpande, N. V., & Thombre, B. S. (2009). Improving the low temperature properties of biodiesel fuel. *Renewable Energy*, *34*(3), 794–800. doi:10.1016/j.renene.2008.04.037

Chen, H., Wang, J., Shuai, S., & Chen, W. (2008). Study of oxygenated biomass fuel blends on a diesel engine. *Fuel*, *87*(15-16), 3462–3468. doi:10.1016/j.fuel.2008.04.034

Cinar, C., Ozer, C., Fatih, S., & Serdar, Y. H. (2010). Effects of premixed diethyl ether (DEE) on combustion and exhaust emissions in a HCCI-DI diesel engine. *Applied Thermal Engineering*, *30*(4), 360–365. doi:10.1016/j.applthermaleng.2009.09.016

Dayong, J., & Yun, B. B. (2012). Performances on engine-out emissions and combustion of ethylene glycol monobutyl ether palm oil monoester as a new type biodiesel. *Journal of Renewable and Sustainable Energy*, *4*, 1–10.

Dhamodaran, G., & Esakkimuthu, G. (2018). A Comparison Study on Emission Characteristics of Using Higher Alcohol Oxygenates with Gasoline in a Multipoint Fuel Injection Spark-Ignition Engine. *Journal of Testing and Evaluation*, *48*(1), 1–13.

Gopal, K. N., & Karupparaj, R. T. (2015). Effect of pongamia biodiesel on emission and combustion characteristics of DI compression ignition engine. *Ain Shams Engineering Journal*, *6*(1), 297–305. doi:10.1016/j.asej.2014.10.001

Happonen, M., Heikkila, J., Saksa, P. A., Murtonen, T., Lehto, K., Rostedt, A., ... Virtanen, A. (2013). Diesel exhaust emissions and particle hygroscopicity with HVO fuel-oxygenate blend. *Fuel*, *103*, 380–386. doi:10.1016/j.fuel.2012.09.006

Holman, J. P. (2011). Experimental methods for engineers, Eighth edition, McGraw-Hill: Company, New York.

Hulwan, D. B., & Joshi, S. V. (2011). Performance, emission and combustion characteristic of a multi-cylinder DI diesel engine running on diesel–ethanol–biodiesel blends of high ethanol content. *Applied Energy*, *88*(12), 5042–5055. doi:10.1016/j.apenergy.2011.07.008

Imtenan, S., Masjuki, H. H., Varman, M., Rizwanul Fattah, I. M., Sajjad, H., & Arbab, M. (2015). Effect of n-butanol and diethyl ether as oxygenated additives on combustion–emission-performance characteristics of a multiple cylinder diesel engine fuelled with diesel–jatropha biodiesel blend. *Energy Conversion and Management*, *94*, 84–94. doi:10.1016/j.enconman.2015.01.047

Jehad, A. A. (2019). Effect of Some Oxygenates on the Opacity Level of a DI Diesel Engine with and without DPF. *Modern Applied Science*, *13*(3), 35–41. doi:10.5539/mas.v13n3p35

Kim, H. G., & Cho, S. H. (2013). The characteristics of biodiesel and oxygenated additives in a compression ignition engine. *Journal of Mechanical Science and Technology*, *27*(5), 1539–1543. doi:10.100712206-013-0334-1

Oxygenates Fact Book. (2003). *A compilation of information on the benefits of oxygenates in gasoline, clean fuels development coalition*. CFDC.

Park, S. H., & Lee, C. S. (2013). Combustion performance and emission reduction characteristics of automotive DME engine system. *Progress in Energy and Combustion Science*, *39*(1), 147–168. doi:10.1016/j.pecs.2012.10.002

Prabu, A., & Anand, R. B. (2015). Influence of oxygenate additives on the performance and emission characteristics of Jatropha fuelled direct injection diesel engine. *Biofuels*, 667–672.

Prabu, A., & Anand, R. B. (2016). Emission control strategy by adding alumina and cerium oxide nano particle in biodiesel. *Journal of the Energy Institute*, *89*(3), 366–372. doi:10.1016/j.joei.2015.03.003

Prabu, A., & Anand, R. B. (2018). Influence of Antioxidant Addition in Jatropha Biodiesel on the Performance, Combustion and Emission Characteristics of a DI Diesel Engine. *J. Inst. Eng. India Ser. C*, *99*(2), 207–216. doi:10.100740032-017-0350-5

Qi, D. H., Chen, H., Geng, L. M., & Bian, Y. Z. (2011). Effect of diethyl ether and ethanol additives on the combustion and emission characteristics of biodiesel-diesel blended fuel engine. *Renewable Energy*, *36*(4), 1252–1258. doi:10.1016/j.renene.2010.09.021

Rajasekar, E., & Selvi, S. (2014). Review of combustion characteristics of CI engines fuelled with biodiesel. *Renewable & Sustainable Energy Reviews*, *35*, 390–399. doi:10.1016/j.rser.2014.04.006

Rao, P. V., Ramesh, S., & Kumar, S. A. (2019). Effects of Oxygenated Additives with Diesel on the Performance of D I Diesel Engine. *Journal of Energy Research and Reviews*, *2*(3), 1–9. doi:10.9734/jenrr/2019/v2i330076

Ren, Y. Z. H., Huang, D., Jiang, M., Li, W., Liu, B., & Wang, X. B. (2008). Effects of the addition of ethanol and cetane number improver on the combustion and emission characteristics of a compression ignition engine. *Proceedings of the Institution of Mechanical Engineers. Part D, Journal of Automobile Engineering*, *222*(6), 1077–1087. doi:10.1243/09544070JAUTO516

Semelsberger, T. A., Borup, R. L., & Howard, L. (2006). Dimethyl ether (DME) as an alternative fuel. *Journal of Power Sources*, *156*(2), 497–511. doi:10.1016/j.jpowsour.2005.05.082

Sivalakshmi, S., & Balusamy, T. (2013). Effect of biodiesel and its blends with diethyl ether on the combustion, performance and emissions of a diesel engine. *Fuel*, *106*, 106–110. doi:10.1016/j.fuel.2012.12.033

Tesfa, B., Mishra, R., Gu, F., & Ball, A. (2013). Combustion and performance characteristics of CI engine running with biodiesel. *Energy*, *51*(1), 101–115. doi:10.1016/j.energy.2013.01.010

Tse, H., Leung, C. W., & Cheung, C. S. (2015). Investigation on the combustion characteristics and particulate emissions from a diesel engine fueled with diesel biodiesel- ethanol blends. *Energy*, *83*, 343–350. doi:10.1016/j.energy.2015.02.030

Tsolakis, A., Rounce, P., Leung, P., & York, A. P. E. (2010). A comparison of diesel and biodiesel emissions using dimethyl carbonate as an oxygenated additive. *Energy & Fuels*, *24*(9), 4812–4819. doi:10.1021/ef100103z

Yanfeng, G., Liu, S., Guo, H., Tiegang, H., & Longbao, Z. (2013). A new diesel oxygenate additive and its effects on engine combustion and emissions. *Applied Thermal Engineering*, *27*(1), 202–207. doi:10.1016/j.applthermaleng.2006.04.021

Yanfeng, G., Shenghua, L., Hejun, G., Tiegang, H., & Longbao, Z. (2007). A new diesel oxygenate additive and its effects on engine combustion and emissions. *Applied Thermal Engineering*, *27*(1), 202–207. doi:10.1016/j.applthermaleng.2006.04.021

Yongcheng, H., Longbao, Z., Shangxue, W., & Shenghua, L. (2006). Study on the performance and emissions of a compression ignition engine fuelled with Fischer– Tropsch diesel fuel. *Proc. Institution of Mechanical Engineers Part D: Journal of Automobile Engineering*, *220*, 827–835.

KEY TERMS AND DEFINITIONS

ASTM: American Society for Testing and Materials
BE2: Butoxy ethanol
Bmep: Brake Mean Effective Pressure
BSFC: Brake Specific Fuel Consumption
Btd: Before Top Dead Centre
BTE: Brake Thermal Efficiency
CO: Carbon Monoxide
DMC: Dimethyl Chloride
DMCBE1: Biodiesel + 0.5% 2-Butoxy ethanol + 0.5% Dimethyl Chloride
DMCBE2: Biodiesel + 1% 2-Butoxy ethanol + 1% Dimethyl Chloride
DMCBE4: Biodiesel + 2% 2-Butoxy ethanol + 2% Dimethyl Chloride
DMCPG1: Biodiesel + 0.5% Propylene glycol + 0.5% Dimethyl Chloride
DMCPG2: Biodiesel + 1% Propylene glycol + 1% Dimethyl Chloride
DMCPG4: Biodiesel + 2% Propylene glycol + 2% Dimethyl Chloride
EGBE1: Biodiesel + 0.5% Ethylene Glycol + 0.5% 2-Butoxy Ethanol
EGBE2: Biodiesel + 1% Ethylene Glycol + 1% 2-Butoxy Ethanol
EGBE4: Biodiesel + 2% Ethylene Glycol + 2% 2-Butoxy Ethanol
EGT: Exhaust Gas Temperature
FFA: Free Fatty Acid
HC: Hydrocarbon
NO: Nitric Oxide
PG: Propylene Glycol

Chapter 11
Investigation of Alternative Fuels as Low Reactivity Fuel in Port-Charged Compression Ignition (PCCI) Engine

Karthickeyan V.
Sri Krishna College of Engineering and Technology, India

Thiyagarajan S.
SRM Institute of Science and Technology, India

Ashok B.
Vellore Institute of Technology, India

ABSTRACT

In this chapter, four alternative fuels were obtained from non-edible oils, namely Moringa oleifera seed oil, pumpkin seed oil, waste cooking palm oil, and lemon oil. The existing diesel engine intake manifold was converted into port charged compression ignition engine by adopting necessary supporting components and control mechanics. In this study, two modes of injection were carried out, namely main injection with conventional fuel and pilot injection with the prepared alternative fuel samples. Due to characteristic fuel properties, lemon oil biofuel in pilot fuel injection experienced high thermal efficiency and low fuel consumption. At all loads, lemon oil biofuel in pilot fuel injection exhibited lower emission than other alternative fuel samples. Lemon oil biofuel in pilot fuel injection and conventional fuel in main injection showed superior combustion characteristics. On the whole, this work recommends the application of the alternative fuel admission in pilot injection mode by adopting PCCI technique to achieve improved engine characteristics.

DOI: 10.4018/978-1-7998-2539-5.ch011

HIGHLIGHTS

- Compression ignition engine was converted into port charged compression ignition engine.
- *Moringa oleifera* seed oil biodiesel, pumpkin seed oil biodiesel, waste cooking palm oil biodiesel and straight lemon oil were considered as an alternative fuel resource.
- Intake manifold was equipped with air preheater and PE3 ECU for pilot fuel injection.
- Straight lemon oil exhibited superior engine characteristics than other fuel samples.

INTRODUCTION

In past few decades, the usage of fossil based resources was increasing drastically due to growing energy requirement. Diesel powered engines were widely used in energy production sectors, transportation, industry, agriculture and the like owing to high brake thermal efficiency and reduced fuel consumption. Conversely, the application of diesel engine causes emission of harmful greenhouse gases which in-turn leads to varying climatic conditions and global warming. About 25% of greenhouse gases were contributed by transportation sectors (Diórdinis et al., 2019). Many countries were imposing stringent emission norms for engine to defend the environment. Thus, environmental pollution and increasing price of resource tends to induce the application of renewable based fuel to satisfy the demand. One such acceptable renewable alternative fuel considered was biodiesel for engine applications. It was important to mention that countries like Europe and Brazil shifted to renewable based fuel usage for energy, transportation and many sectors (Mattson, Burnete, Depcik, Moldovanu, & Burnete, 2019). In 2017, the worldwide production of biodiesel was predicated as 36 billion litre for various applications whereas in 2027, 39 billion was predicted for energy production due to extensive need and increasing environmental concerns (Karthickeyan, 2019c). Many research works were carried out in renewable based alternative fuel for operation in diesel engine application.

Biodiesel sample was developed from *Moringa oleifera* oil, a non-edible source of feedstock (Karthickeyan, 2019b, 2019a). Conversion of biodiesel was performed by transesterification reaction under two stages owing to high acid value of oil. Engine study revealed that the performance characteristics were improved with low concentration of emissions due to cetane improver and ceramic coating (Material: Partially stabilized zirconia, thickness: 500 µm and method: Plasma spray technique) inside combustion chamber. Similarly, *Moringa oleifera* alternative sample was blended with diesel with various proportions and investigated in common rail direct injection diesel engine (Teoh, How, Masjuki, Kalam, & Alabdulkarem, 2019). It was important to mention that the biodiesel experienced high viscosity, density and cetane number and low calorific value compared to diesel. Declined engine efficiency with increased fuel consumption was noticed with biodiesel sample. Conversely, biodiesel sample exhibited lower emission characteristics (except NOx) than other fuel samples. Similar engine combustion characteristics were observed with diesel at all loads. Also, investigation of *Moringa oleifera* biodiesel blends in multi cylinder compression ignition engine (Mofijur et al., 2014) experienced low performance characteristics than other fuel samples owing to its inferior fuel properties. But, the availability of oxygen in biodiesel leads to reduced CO, HC and smoke emissions. Pumpkin seeds were chosen as one of the prime source for oil production and application in engines (Karthickeyan, 2019c). Low performance and combustion characteristics were noticed with neat pumpkin seed oil biodiesel owing to its inferior physical and chemical properties. The presence of oxygen in biodiesel sample leads to reduced emission characteristics. Pumpkin seed oil with different bowl geometries showed better engine characteristics than conventional engine operation (Karthickeyan & Balamurugan, 2017). It was identified that the low concentration of pumpkin seed oil biodiesel with diesel (B25) experienced higher engine performance than other blending conditions. B25 showed lower engine exhaust emissions like CO, HC and smoke than diesel. Conversely, rise in carbon dioxide and NOx emission was noticed with B25 sample. Further, the pumpkin seed oil biodiesel was investigated in ceramic coated engine (Material: Partially stabilized zirconia, thickness: 500 μm and method: Plasma spray technique). Improved engine characteristics with excess NOx emission was noticed with B25 in coated engine owing to rapid increase in combustion chamber temperature. Also, it was important to mention that minor cracks were observed in ceramic components due to exposure in high temperature ambience. Further, neat pumpkin seed oil biodiesel was explored in diesel engine at different temperatures of fuel inlet by heating with available heat at engine exhaust (Ramakrishnan, Rathinam, & Viswanathan, 2018). Biodiesel sample with fuel preheating technique showed high performance characteristics than diesel due to availability of high input energy in the form of heat. Also, gradual reduction in emissions was noticed with preheated biodiesel sample due to high combustion chamber temperature. Waste cooking oil was considered as one of the prominent sources for fuel production owing to its low cost, availability, not having any impact on normal food chain and the like. This oil was collected from restaurants after processing of food. The high acid value was noticed with waste cooking palm oil (42 mg NaOH/gm). Thus, conversion of waste cooking oil into biodiesel was achieved using two stage transesterification processes. AVL5402 model single cylinder diesel engine was experimented with waste cooking oil biodiesel, ultra-low sulphur diesel and winter grade diesel (Mattson et al., 2019). Lower in-cylinder pressure and heat release rate was noticed with waste cooking palm oil biodiesel than conventional owing to its diminished properties. Reduced engine emissions were noticed with biodiesel sample. Similar trend of results were noticed with waste cooking palm oil biodiesel with different compositions of magnesium oxide (Ranjan et al., 2018) and waste cooking palm oil methyl ester in Partially stabilized zirconia coated engine (Karthickeyan, Balamurugan, & Senthil, 2017).

The feedstocks discussed earlier were high viscous oils and two stage transesterification was employed for the production of biodiesel. Low viscous oil namely straight lemon oil was investigated. Lemon oil was obtained from the steam distillation of waste lemon peels under controlled conditions (Karthickeyan, Thiyagarajan, et al., 2019). The presence of oxygen in biodiesel leads to production of NOx emission and low heating value causes reduced engine output parameters. Therefore, straight lemon oil was attempted with thermal barrier coating and selective catalytic reduction technology in engine exhaust which helps to achieve controlled emission level with engine output (Ashok, Thundil Karuppa Raj, Nanthagopal, Krishnan, & Subbarao, 2017). Enhanced engine performance and combustion characteristics were noticed with straight lemon oil owing to its promising physical and chemical properties. Declined engine emissions were noticed with straight lemon oil with increment level of NOx at all loads. Table 1 shows the different feedstock investigation in compression ignition engine and its outcomes.

In recent years, progress in engine investigation was observed either as port charged or reactivity controlled or homogeneous charged compression ignition engine. Specifically, port charged compression ignition engine models with biodiesel were considered as the advanced and easily feasible approach compared to other approaches. Recently, polyoxymethylene dimethyl ether was explored in RCCI engine with diesel and methanol in common rail direct injection three-cylinder turbocharged compression ignition engine (Duraisamy, Rangasamy, & Govindan, 2020). In the intake manifold, separate arrangements were carried for injection of methanol using Bosch make fuel injector, pump, filter and distributor

Table 1. Different feedstock investigation in compression ignition engine and its outcomes

Researchers	Production technique	Engine selected	Engine characteristics	Outcomes	Ref.
Feedstock: Moringa Oleferia					
Karthickeyan (2019)	Transesterification 69.86% composed of Octadecadienoic acid ($C_{20}H_{36}O_2$)	Kirloskar TV1, single cylinder, direct injection, diesel engine	BTE (↓), BSFC (↑), NOx (↑), CO (↓), HC (↓), ICP (↑), HRR (↑)	Engine was converted into low heat rejection model and used pyrogallol as cetane improved. Engine study found to produce improved characteristics.	(Karthickeyan, 2019b, 2019a)
Teoh et al. (2019)	Transesterification	Turbocharged direct injection, CRDi with 140 MPa injection pressure	BTE (↑), BSFC (↓), ICP (↓), HRR (↑), NO (↑), other emission (↓)	Without modifying the engine, biodiesel with limited can be used for improved engine characteristics.	(Teoh et al., 2019)
Mofijur et al. (2014)	Transesterification 74.1% composed of Cis-9-Octadecenoic ($C_{18}H_{34}O_2$)	Mistsubishi Pajero 4D56T model 4 cylinder indirect injection	BSFC (↑), BTE (↓), CO (↓), HC (↓), NOx (↑)	It was recommended that B10 and B20 sample can be admitted as promising fuel combination for engine applications.	(Mofijur et al., 2014)
Feedstock: Pumpkin seed oil					
Karthickeyan (2019)	Transesterification 69.05% composed of Octadecadienoic acid ($C_{19}H_{38}O_2$)	Kirloskar TV1, single cylinder, direct injection, diesel engine	BTE (↓), BSFC (↑), ICP (↓), HRR (↓), NO (↑), other emission (↓)	Modification of combustion bowl geometry improved the engine characteristics with pumpkin seed oil biodiesel.	(Karthickeyan, 2019c)
Karthickeyan (2019)	Transesterification 69.05% composed of Octadecadienoic acid, methyl ester (CAS) ($C_{19}H_{38}O_2$)	Kirloskar TV1, single cylinder, direct injection, diesel engine	BSFC (↑), BTE (↓), CO_2 (↑), HC (↓), NOx (↑), smoke (↓)	B25 fuel investigation found superior that other samples and also investigation in low heat rejection found in good argument than conventional engine.	(Karthickeyan & Balamurugan, 2017)
Karthickeyan (2017)	Transesterification	Kirloskar TV1, single cylinder, direct injection, diesel engine	BTE (↓), BSFC (↑), ICP (↓), HRR (↓), NO (↑), other emission (↓)	Fuel preheating adaptation helps to improve the engine characteristics owing to increase in combustion rate.	(Ramakrishnan et al., 2018)
Feedstock: Waste cooking oil					
Mattson et al. (2019)	Transesterification	AVL 5402, single cylinder, water cooled, compression ignition engine	ICP (↑), HRR (↓), CO (↓), HC (↓), NO (↑)	Converted alternative fuel sample was investigated in diesel engine with ultra low sulphur diesel and found with improve engine characteristics.	(Mattson et al., 2019)
Ranjan et al. (2018)	Transesterification	TV1, water cooled, compression ignition engine	BTE (↓), BSFC (↑), ICP (↑), HRR (↑), NOx (↑), CO_2 (↓)	Magnesium oxide was blended with waste cooking oil biodiesel increase the engine characteristics in all aspect.	(Ranjan et al., 2018)
Karthickeyan (2017)	Transesterification 73.58% composed of methyl 11-docosenoate ($C_{23}H_{44}O_2$)	Kirloskar TV1, single cylinder, direct injection, diesel engine	BTE (↓), BSFC (↑), NOx (↑), CO (↓), HC (↓), smoke (↓)	Thermal barrier coating in combustion chamber helps to improve the engine performance and emission characteristics.	(Karthickeyan et al., 2017)
Feedstock: Lemon oil					
Karthickeyan et al. (2019)	Steam distillation 84.66% composed of Trans-Isolimonene ($C_{10}H_{16}$)	Kirloskar TV1, single cylinder, direct injection, diesel engine	BTE (↑), BSFC (↓), ICP (↑), HRR (↑), NO (↑), other emission (↓)	Addition of additive and investigation in low heat rejection engine further improves the engine characteristics.	(Karthickeyan, Thiyagarajan, et al., 2019)
Ashok et al. (2017)	Steam distillation 61.76% composed of M-Mentha-4,8-Diene ($C_{10}H_{16}$)	Kirloskar TAF1, single cylinder, air cooled, compression ignition engine	BTE (↑), BSFC (↓), ICP (↑), HRR (↑), NO (↑), other emission (↓)	Neat lemon oil can be admitted in diesel engine but control of NOx can be achieved with selective catalytic reduction technique.	(Ashok et al., 2017)

BTE- Brake thermal efficiency, BSFC- Brake specific fuel consumption, ICP- In-cylinder pressure, HRR- Heat release rate, NO- Nitric oxide, NOx- Oxides of nitrogen, CO- Carbon monoxide, HC- Hydrocarbon, CO_2- Carbon dioxide, ↑- Increase, ↓- Decrease.

block. High combustion chamber pressure and heat release rate was noticed with methanol injection than normal operation. Improved engine brake thermal efficiency was observed with RCCI working condition. In addition, declined engine exhaust emissions were noticed with RCCI than normal operation owing to improved in-cylinder pressure parameters of injected fuel. Similarly, light duty compression ignition engine manifold was modified to achieve reactivity controlled compression ignition engine with the help of vaporizer and pilot fuel injector unit (Charitha, Thirumalini, Prasad, & Srihari, 2018). The injection of fuel in manifold and heater temperature was monitored and controlled with the help of TWR-K40D100M microcontroller unit as per the engine specification. Main fuel injection was performed with diesel and pilot fuel injection was performed with cotton seed oil biodiesel under various volume proportions. Better engine performance and combustion characteristics were noticed with RCCI than conventional operation. Simultaneous reduction of smoke and oxides of nitrogen emissions were noticed owing to reduced combustion gas temperature and presence oxygen in sample. Engine operation was performed with premixed charged condition using various proportions of ethanol injection in intake manifold with help of fuel vaporizer unit (Srihari & Thirumalini, 2017). Testing was performed in GL400 model compression ignition engine at constant speed of 3600 rpm. On comparing with diesel, biodiesel blends showed low NOx emission due to absorption of available heat in the combustion chamber and usage of absorbed heat for vaporization of ethanol. Except CO and HC emissions, other emissions were declined with gradual increase in fuel consumption at all loads. Similarly, the same engine was operated with diesel and cotton seed oil biodiesel as main fuel injection sample and diethyl ether as pilot fuel injection (Srihari, Thirumalini, & Prashanth, 2017). The control of fuel injection in intake manifold was monitored and controlled using ATmega328P model engine microcontroller unit. Reduced oxides of nitrogen were recorded with biodiesel sample owing to high cetane number which in-turn leads to increased combustion quality and high latent heat of vaporization of diethyl ether injection in intake manifold. The presence of oxygen in biodiesel sample leads to reduced engine exhaust emissions. Enhanced thermal efficiency and in-cylinder pressure was noticed at maximum operating condition. Table 2 shows the different feedstock investigation in reactivity controlled and port charged compression ignition engine and its outcomes

From the literature review, very few works were identified in the context of PCCI and RCCI engine with biodiesel. High viscous oils like *Moringa oleifera* seed oil, pumpkin seed oil and waste cooking palm oil were converted into its respective biodiesel samples by means of transesterification process. Low viscous lemon oil was obtained based on steam distillation approach. Thus, four different alternative fuels namely *Moringa oleifera* seed oil biodiesel, pumpkin seed oil biodiesel, waste cooking palm oil biodiesel and straight lemon oil were investigated in PCCI engine. The physical and chemical properties were determined and examined based on ASTM biodiesel standards and compared with diesel. The conventional engine was converted into PCCI with necessary equipment's. Two modes of fuel injection was carried out namely main injection by diesel and pilot injection by alternative fuel samples. The aim of the present work is to investigate the impact of biodiesels as low reactivity fuel on engine performance, combustion and emission characteristics in PCCI.

Table 2. Different feedstock investigation in reactivity controlled and port charged compression ignition engine and its outcomes

Researchers	Working technique	Engine selected	Engine characteristics	Outcomes	Ref.
Ganesh et al (2019)	RCCI MI: Diesel PI: Methanol / Ploxymethylene dimethyl ethers	Three cylinder, direct injection, diesel engine	ICP (↑), HRR (↑), BTE (↑), CO (↓), HC (↓), Soot (↓), NO (↑)	Ployxymethylene dimethyl ethers investigation in RCCI mode supports for improvement in engine characteristics.	(Duraisamy et al., 2020)
Charitha et al. (2018)	RCCI MI: Diesel PI: Cotton seed biodiesel	Greaves GL 400 model, single cylinder, direct injection, compression ignition engine	BTE (↑), BSFC (↓), ICP (↓), NO (↓), smoke (↓), HC (↑), CO_2 (↑)	Significant reduction in emission was noticed and thus helpful in simplifying the exhaust after treatment systems.	(Charitha et al., 2018)
Srihari and Thirumalini (2017)	PCCI MI: Diesel PI: E15D (15% Ethanol with diesel)	GL 400, compression ignition, direct injection diesel engine	NOx (↓), smoke (↓), HC (↑), CO (↑), BSFC (↑)	In low temperature combustion technique helpful to reduce engine exhaust emissions.	(Srihari & Thirumalini, 2017)
Srihari et al. (2017)	PCCI MI: Diesel PI: Cottonseed oil biodiesel and diethyl ether	Greaves GL 400 model, direct injection, diesel engine	NOx (↓), smoke (↓), HC (↑), CO (↑), BSFC (↑), BTE (↑), ICP (↓)	Engine operating in PCCI model found to exhibit reduced engine emission with improved combustion and performance characteristics.	(Srihari et al., 2017)

RCCI- Reactivity controlled compression ignition, PCCI- Port charged compression ignition, MI- Mail injection, PI- Pilot injection, ICP- In-cylinder pressure, HRR- Heat release rate, BTE- Brake thermal efficiency, CO- Carbon monoxide, HC- Hydrocarbon, NO- Nitric oxide, CO_2- Carbon dioxide, BSFC- Brake specific fuel consumption, NOx- Oxides of nitrogen.

MATERIALS AND METHODS

This section provides the information related to selection of oil, production of a biodiesel. Development and adopting of PCCI (Port Controlled Compression Ignition) engine technique in existing DI diesel engine. Further, the engine setup, experimentation procedure and uncertainty analyses were discussed.

Production of Biodiesel

Due to negative impact on food chain, the application of edible oil for the production of biodiesel was not recommended. Four non-edible oils namely *Moringa oleifera* seed oil, pumpkin seed oil, waste cooking palm oil and lemon oil were chosen as the prime sources for biodiesel production. The above said oils were selected based on the factors like range of availability, cost of production and usage. *Moringa oleifera* seed oil and pumpkin seed oil were produced from its respective seeds using mechanical expeller mechanism. The waste cooking palm oil was chosen for the fuel production. This concept paves the way for the conversion of waste resource into useful energy resource. To end with, lemon oil obtained from waste lemon peels and seeds using steam distillation process were used for the fuel production. It was important to mention that the viscosity of the extracted lemon oil was lower than the other oil samples. The high viscosity oils namely, *Moringa oleifera* seed oil, pumpkin seed oil and waste cooking palm oil were converted into biodiesel by using transesterification process. Figure 1 represents the production of biodiesel. The acid value of all the oils was determined based on the titration process. One gram of sodium hydroxide was dissolved in deionized water and filled in graduated burette. One gram of non-edible oil was mixed with 10 ml of isopropyl alcohol and taken in Erlenmeyer flask and certain quantity of phenolphthalein (indicator) was added to the solution. The titration process was commenced by allowing drop by drop of burette solution into the Erlenmeyer flask. When the colour change was observed, the titration process was stopped and the corresponding reading was recorded. The aforesaid process was repeated thrice to find the precise amount of required reagent. Similarly, the titration process was repeated three times to obtain the required amount of reagent. The same procedure was performed for all three non-edible oils to calculate the amount catalyst requirement. The acid value of the three oils was higher than the prescribed bounds. Therefore, two stage transesterification processes was selected for fuel production. One litre of non-edible was taken and filled in the round bottom flask which was attached to the condenser to inhibit the escape of alcohol. The heater was switched on and temperature was fixed as 60 to 65 °C at constant stirring speed of 650 rpm continuously. After attainment of particular temperature with a period of time, one by four times of methanol (alcohol) was added and the solution was allowed to react for a period of 30 minutes. 2% of sulphuric acid with 0.2 normality was added to the solution. The process was allowed for about one hour continuously and the solution was transferred to glass made gravity separation funnel. The acquired solution was undisturbed for one day and two layers were identified. The top layer contains useful crude was taken separately and filled in round bottom flask to perform base esterification process. To begin with, 250 ml of methanol and 2% of sodium hydroxide was taken in the flask and dissolved to achieve homogeneity of mixture. The solution was transferred to heated crude ester (at 55°C) and stirred continuously at stirring speed of 650 rpm. After a period of one hour, the solution was transferred to gravity separation funnel. Owing to difference in density of solution, glycerol rises to top of the funnel and ester was settled at bottom portion. The separated ester was allowed to react with de-ionized water for more than three times to obtain neat biodiesel sample.

Figure 1.

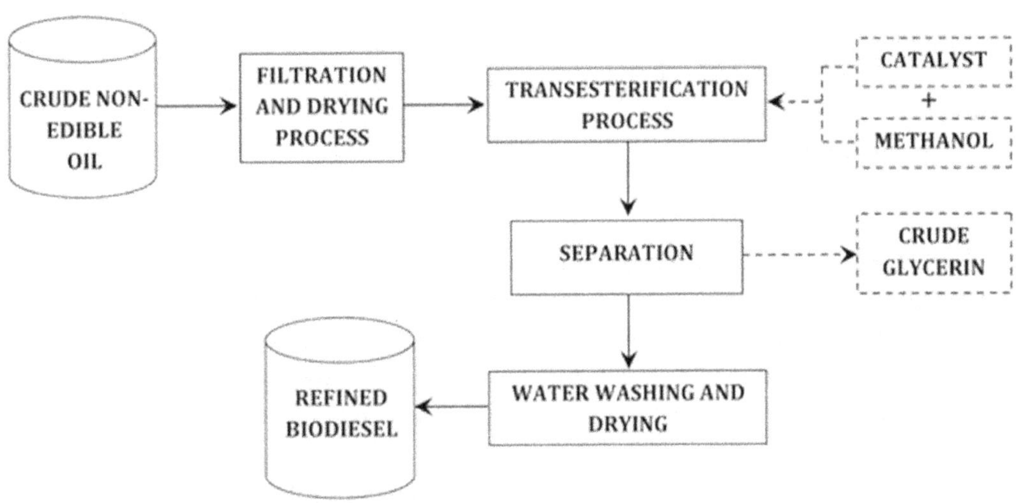

Table 3. Physical and chemical properties of alternative fuels

Properties	Instrument used	ASTM standards	Diesel	Moringa oleferia biodiesel	Pumpkin seed biodiesel	Waste cooking palm biodiesel	Lemon oil
Molecular formula			$C_{12}H_{24}$	$C_{18}H_{34}O_2$	$C_{19}H_{38}O_2$	$C_{18.5}H_{34.8}O_2$	$C_{10}H_{16}$
Density (kg/m^3)	Density meter	D1298	835	859	897	892	848
Kinematic viscosity at 40°C (cSt)	Redwood viscometer	D445	2.5	5.04	5.9	5.32	1.02
Flash point (°C)	Pensky-Martens open cup apparatus	D93	56	150	176	172	48
Fire point (°C)		D93	62	162	192	190	54
Gross calorific value (MJ/kg)	Bomb calorimeter	D240	43.26	40.5	40.06	39.78	42.76
Calculated cetane index		D613	48	56	54	52	12
Proximity analysis (2400 Series II Perkin Elmer)							
Carbon (mass %)			-	76.32	75.69	75.41	88.62
Hydrogen (mass %)			-	12.21	13.98	13.06	9.87
Oxygen (mass %)			-	11.46	10.33	11.53	1.51
C/H			-	6.25	5.41	5.77	8.97

Further, the ester solution was heated at about 50- 60 °C for a period time of 5 minutes to eradicate the presence of water molecules. Same conversion procedure was repeated for all three non-edible oils namely *Moringa oleifera* seed oil, pumpkin seed oil and waste cooking palm oil. As lemon oil was the low viscous oil, no transesterification process was performed. The raw oil after filtration process was used for the present study. Table 3 depicts the physical and chemical properties of biodiesels.

Pilot Fuel Injection Unit

PCCI (Port Controlled Compression Ignition) technique was considered as one the important method to achieve low temperature combustion. Without modifying the engine components, regulated and unregulated engine emission were controlled by adjusting the reactivity of the admitted fuel. In PCCI approach, three methods namely fuel injection using port admission technique, fumigation technique and late DI were carried out. Based on literature works (Gowtham, Mohan, & Prakash, 2019), the fumigation technique was recommended to obtain reduction in emission level. Generally, the alcohols were allowed in the intake manifold along with air for combustion process. On the other hand, biodiesel samples were allowed as the fumigation material in the present study. Two injection techniques were performed such as engine main and primary injection of diesel as per manufacture catalogue directly into the combustion chamber. Secondly, biodiesel sample was injected into the intake manifold as secondary fuel injection system based on PCCI technique. Secondary injection system consists of storage tank, fuel pump and fuel injection unit and overflow return to tank. The fuel flow was fixed as 0.06 grams per second for all loads and engine operating speed conditions. Based on the operating condition of fuel pump and injector unit, 3.5 bar was fixed as fuel injection pressure. Air preheater unit was purchased from Dewalt and allowed to operate at temperature limit between 50°C to 600°C. Also, the maximum air flow was 650 liters per

Figure 2.

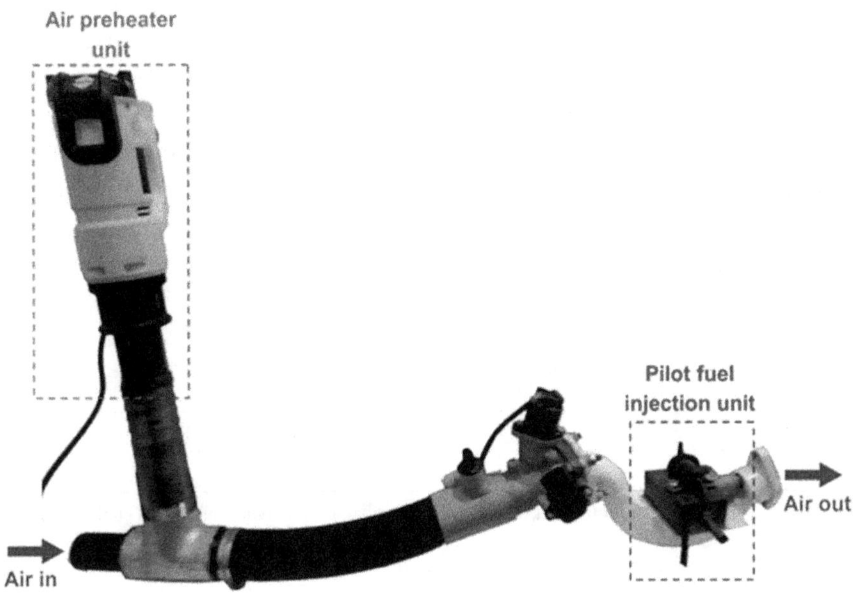

minute and can be modified. The equipment was provided with LED display to view the temperature of air and quantity of air flow. The air was preheated using pre-heater unit at the temperature range of 200°C ± 10°C and measured with the help of laser IR infrared thermometer in-built. The equipment was built up with temperature indicator and air speed controller. The air pipe connection was considered as the mixing chamber where preheated air mixes with the injected fuel in the intake manifold of the engine. The secondary fuel injection was controlled using PE3 ECU series for programmable injection for fuel control system. Figure 2 shows the photographic view of pilot fuel injection unit.

Experimental Setup and Procedure

In the present investigation, water cooled computerized single cylinder direct injection diesel engine was used. Brake power was produced using water cooled eddy current dynamometer. Fuel tank was connected to three-way flow control value junction with electronically controlled fuel flowmeter. Also, the fuel flow was manually calculated using graduated burette for estimating fuel consumed by the engine. Initially, the fuel was allowed to flow to filter and taken by the pump to the fuel injector unit. Air flow to engine was determined by using electronic air flowmeter and manual reading was obtained by using U-tube manometer difference. The engine was established with water cooling technology where the flow was measured by using rotometer device and controlled using valve. The water-cooled dynamometer was controlled using separate controller unit and junction to engine management system. K-type thermocouples were used at different positions of engine to measure and record the temperature readings. In-cylinder pressure was measured using piezo-electric based pressure transducers. The measurable voltage readings obtained from pressure transducer were converted into digital signals by using analog-digital converter. Table 4 shows the technical specification of test engine. The engine speed was determined by using speed sensor and can be viewed using digital indicator mounted on the control panel. In the crank of engine, crank angle sensor was installed to determine the position of the piston during combustion process. National Instrument based engine management system was used for collecting information about various sensors for processing and storing. Then, Matlab based engine performance and combustion software was used for data viewing and recording purpose. Tail pipe regulated emissions were explored by means of AVL model 444 di-gas analyzer was employed. Carbon monoxide, oxides of nitrogen and hydrocarbon were evaluated by introducing the probe of the analyser in the tail pipe. Before observation, leak proof test and zero calibration tests were performed in the analyser. In experimental engine, the estimation of smoke emission was carried out by using AVL model 437C smokemeter. Figure 3 depicts the schematic diagram of experimental setup.

Table 4. Technical specification of test engine

Make and model	**Kirloskar / TV1**
Details of engine	
Type	Direct injection, water cooled, vertical, diesel, naturally aspired engine
No. of cylinder / stroke	01 / 04
Rated power	5.2 kW
Bore (mm) / stroke (mm)	87.5 / 110
Ignition type	Compression Ignition (CI)
Compression ratio	17.5:1
Injection pressure	210 bar
Injection timing	21° before Top Dead Centre (bTDC)
Speed	1500 Rev/min
Diameter/ no. of nozzle hole	0.3 mm/ 3
Details of dynamometer	
Make and model	Technomech Pvt. Ltd. / TMEC10
Type	Eddy current dynamometer
Cooling	Water cooled
Speed	1500- 6000 Rev/min
Details of combustion analyser	
Cylinder pressure sensor	Kistler (M111A22), Piezo electric sensor
Crank angle encoder	Kistler (2614C11)
Details of data acquisition system	
Make and model	National instruments / USB-6210
Type	16AI; 4DI; 4DO USB- multifunction I/O devic
Details of other instruments	
Air flow measurement	Make: Wika / Model: SL1
Fuel flow measurement	Make: Broiltech / Model: FCM and Differential pressure transmitter
Load cell	Make: Sensortronics / Model: 60001

Figure 3.

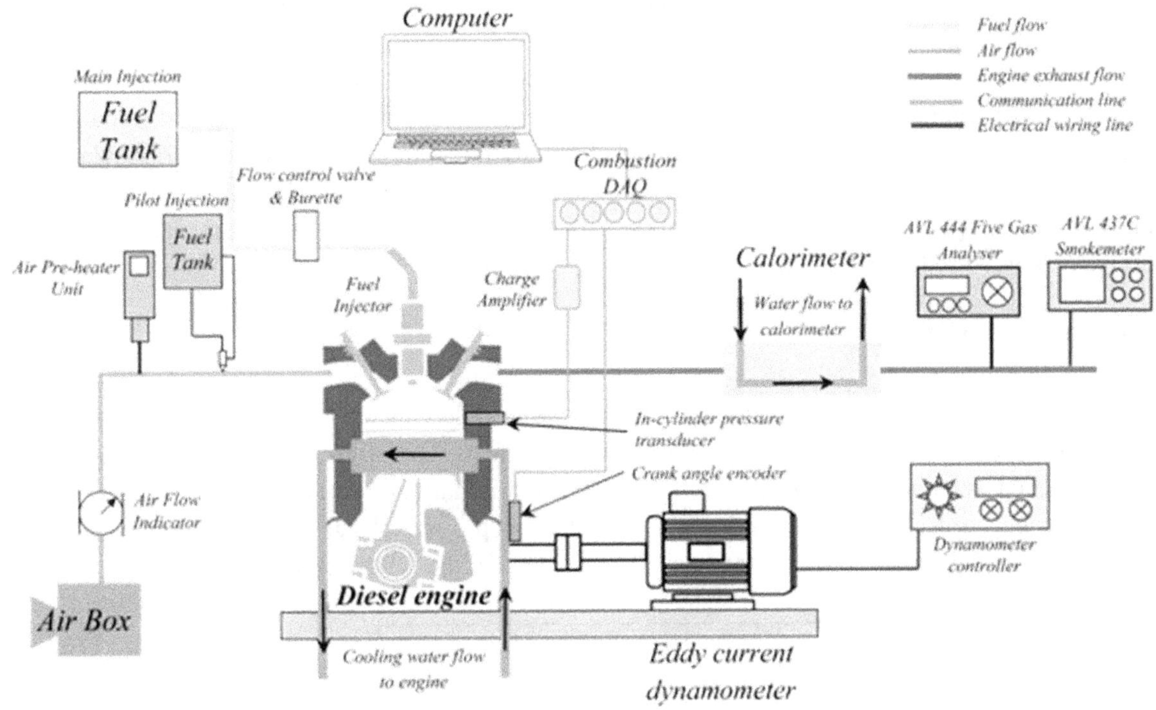

Initially, the diesel namely was filled in the storage tank and the engine was allowed to run using electrical starting technique. Leakage of water, lubrication oil and fuel were determined for safety operation. Initially, no load was applied on the engine and allowed to run for few minutes. The outlet temperature of cooling water was used to determine the steady state condition. The engine was loaded from no load to maximum load condition with an interval of $1/4^{th}$ load condition. The engine operating parameters like measuring of air flow rate, fuel flow rate, load, engine exhaust emissions were recorded using online mode. Using the measured value, engine thermal efficiency, fuel consumption, carbon monoxide, smoke, oxides of nitrogen, hydrocarbon and in-cylinder pressure data were calculated. The readings measured and plotted were considered as baseline observation of the compression ignition engine. Then, the engine intake manifold was disassembled and converted into reactivity controlled compression ignition engine by incorporating the fuel admission unit and pre-heater arrangements.

Fuel Sample Admission in Pilot Injection

PCCI mode contains major components like storage tank, fuel pre-heater unit, fuel injection unit and PE3 electronic control unit. The prepared alternative fuel sample namely *Moringa oleifera* seed oil biodiesel was filled in the storage tank and diesel was filled in the primary fuel tank. Initially, the engine was allowed to run with diesel based on the aforesaid experimental procedure. Using PE3 software, the fuel pump was powered and biodiesel sample was injected into the air intake. On the other hand, air preheater was switched on and temperature of air intake was rapidly increased. At no load condition, main injection was performed with diesel as per manufacture specification and pilot injection was performed using fuel injector attached in the intake manifold. Irrespective of engine operating speed and load applied to engine, fuel injection in intake manifold was fixed as 3.5 bar and flow rate as 0.06 grams per second (Srihari & Thirumalini, 2017). The load applied in baseline condition was repeated for PCCI mode operation and various operating parameter data were recorded. After completing the observation with *Moringa oleifera biodiesel* sample, remaining sample in tank was drained out and refuelled the tank with biodiesel sample. After completing the observation of varying the pilot injection fuel sample and main injection with diesel, the measured data was stored in the computer for further processing. To improve the quality of result, engine observation with all operating conditions were repeated thrice and average data was used for the calculation.

Uncertainty Analysis

Uncertainty or error during experimentation occurs due to following various factors like observation, equipment selection, service, calibration and unidentified effects. Therefore, uncertainty or error analysis was considered as an important parameter to support the originality of obtained value. For uncertainty analysis, propagation of error technique was used and it was proposed by J. P. Holman (Holman, 2007). The uncertainty values of measured data was calculated using the following equation,

$$UAV = \sqrt{\left[\left(\frac{\partial R}{\partial A_1}\right)W_1\right]^2 + \left[\left(\frac{\partial R}{\partial A_2}\right)W_2\right]^2 + \ldots\ldots + \left[\left(\frac{\partial R}{\partial A_n}\right)W_n\right]^2}$$

where, UAV is the total uncertainty value, and R is the Main function parameter which depends on the independent variables like A_1, A_2, A_n.

The other parameter like W_1, W_2, W_n denotes the uncertainty of each independent variable. The different experimental uncertainties were tabulated in Table 5 with accuracy and error values. Based on the analysis of propagation error techniques, the total uncertainty was determined as ±2.27%.

RESULTS AND DISCUSSION

In the present experimentation, biodiesel were admitted at pilot injection and diesel was admitted at main injection. The engine characteristics like brake thermal efficiency (in %), brake specific fuel consumption (kg/kWh), carbon monoxide (in % volume), hydrocarbon (in ppm), oxides of nitrogen (in ppm), smoke (in HSU) and in-cylinder pressure (in bar) were presented.

Performance Characteristics

Figure 4 depicts about the variation of engine thermal efficiency against different loads. At full load, brake thermal efficiency for operating conditions like MI(D)-PI(D), MI(D)-PI(M), MI(D)-PI(P), MI(D)-PI(W) and MI(D)-PI(L) were observed as 31.42, 31.16, 30.95, 30.84 and 31.47% respectively. With all operating condition, diesel was fixed as main injection with varying biodiesel as pilot injection fuel. In general, the conversion of chemical energy in fuel into useful mechanical energy was depicted

Table 5. Experimental uncertainty

Parameters	Measuring technique	Accuracy	Errors (±)
Load	Strain gauge type load cell	10 N	± 0.2
Speed	Magnetic pickup principle	10 rpm	± 0.1
Fuel flow measurement	Volumetric measurement	0.1 cc	± 1
Temperature	Thermocouple	1 °C	± 0.15
Crank angle encoder	Magnetic pickup principle	1 deg	± 0.2
Pressure	Magnetic pickup principle	0.1 kg	± 0.1
Time	Stop watch (Manual)	0.1 s	± 0.2
Manometer deflection	Balancing of column of liquid	1 mm	± 1
BTE	-	0.4	± 0.04
BSFC	-	0.02 kg/kWh	± 1.1
CO	NDIR technique	0.02% volume	± 0.2
HC	NDIR technique	10 ppm	± 0.1
NOx	NDIR technique	12 ppm	± 0.2
Smoke	Opacimeter	1 HSU	± 1

as thermal efficiency. During engine operation, diesel at main injection and pilot injection exhibited high brake thermal efficiency till 75% of engine load. But, the admission of biodiesel experienced high engine thermal efficiency. The admission of biodiesel as pilot injection showed low engine thermal efficiency. At full load, maximum BTE was observed due to high combustion chamber temperature. Amongst biodiesel, low viscous lemon oil in pilot injection experienced complete combustion which in-turn leads to maximum thermal efficiency at all loads. *Moringa oleifera* seed oil, pumpkin seed oil and waste cooking oil showed low thermal efficiency due to its inferior fuel properties like high viscosity and low calorific value. Also, it was important to mention that the preheating of atmospheric air with the admission of biodiesel leads to attain the auto-ignition property of fuel. At full load, high combustion chamber temperature and preheated fuel admission leads to achieve enhanced combustion phase. In addition, the presence of oxygen content in biodiesel supports the combustion process. Similarly, diesel as main injection and cotton seed oil as pilot injection was investigated in reactivity-controlled compression ignition engine and observed an improved thermal efficiency owing to improved atomization and combustion (Charitha et al., 2018).

Figure 5 depicts about the variation of engine brake specific fuel consumption against different loads. At full load, brake specific fuel consumption for operating conditions like MI(D)-PI(D), MI(D)-PI(M), MI(D)-PI(P), MI(D)-PI(W) and MI(D)-PI(L) were noticed as 0.362, 0.375, 0.382, 0.397 and 0.351 kg/kWh respectively. In general, fuel consumption helps to estimate the amount of fuel consumed to achieve the rated power output of the engine. At increasing loads, declined fuel consumption was noticed with all the fuel samples. Diesel showed lower fuel consumption than biodiesel samples. This declined fuel consumption was due to influencing physical and chemical properties of diesel. Amongst alternative fuel samples, lemon oil experienced low fuel consumption due to its characteristic fuel properties. The admission of biodiesel sample in intake manifold at increased temperature of air exhibited rise in fuel consumption. Admission of waste cooking oil biodiesel in pilot injection mode leads to increased

Figure 4.

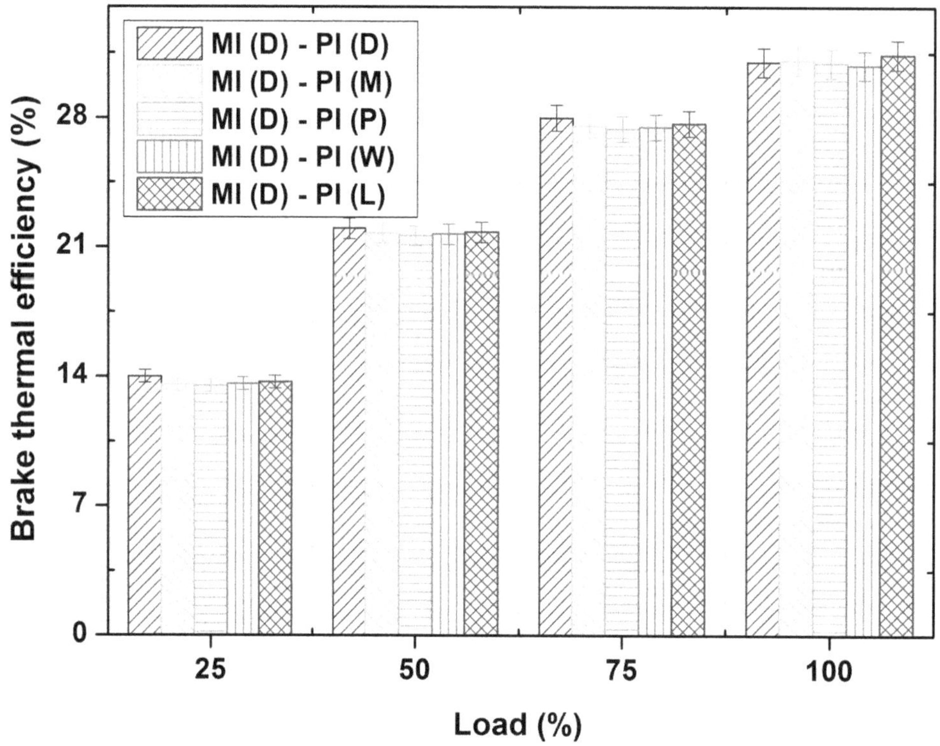

fuel consumption due to high kinematic viscosity and low calorific value. In addition, high flash point denotes the occurrence of longer ignition delay period during fuel combustion. The pilot injected fuel leads to attain auto-ignition temperature rapidly due to high air temperature. The low viscosity lemon oil admission as pilot fuel experienced superior fuel consumption than other fuel samples. Similar trend of results were observed with canola-safflower in common rail direct injection diesel engine with oxygenated additives like solketal and ethanol under main and pilot fuel injection (Alptekin, 2017).

Emission Characteristics

Figure 6 depicts about the variation of carbon monoxide against different loads. At maximum load, CO emission for engine operating conditions like MI(D)-PI(D), MI(D)-PI(M), MI(D)-PI(P), MI(D)-PI(W) and MI(D)-PI(L) were noticed as 0.05, 0.06, 0.08, 0.11 and 0.04% volume respectively. CO emission was considered as the important emission parameter which denotes the loss of available chemical energy in the admitted fuel. Air and fuel equivalent was considered as the parameter for the development of CO emission. During engine observation, diesel experienced lower CO emission than high viscous fuel samples. Thus, the diesel at main injection attains rated power output of engine. Among the high viscous biodiesels, lower CO emission was identified with *Moringa oleifera* admission as pilot fuel. Low viscous lemon oil namely lemon oil exhibited lower CO emission than other fuel samples owing to low viscosity, improved air-fuel mixing, high combustion chamber temperature and the like. The admission of lemon oil with preheated air at intake manifold leads to enhanced combustion process. The admis-

Figure 5.

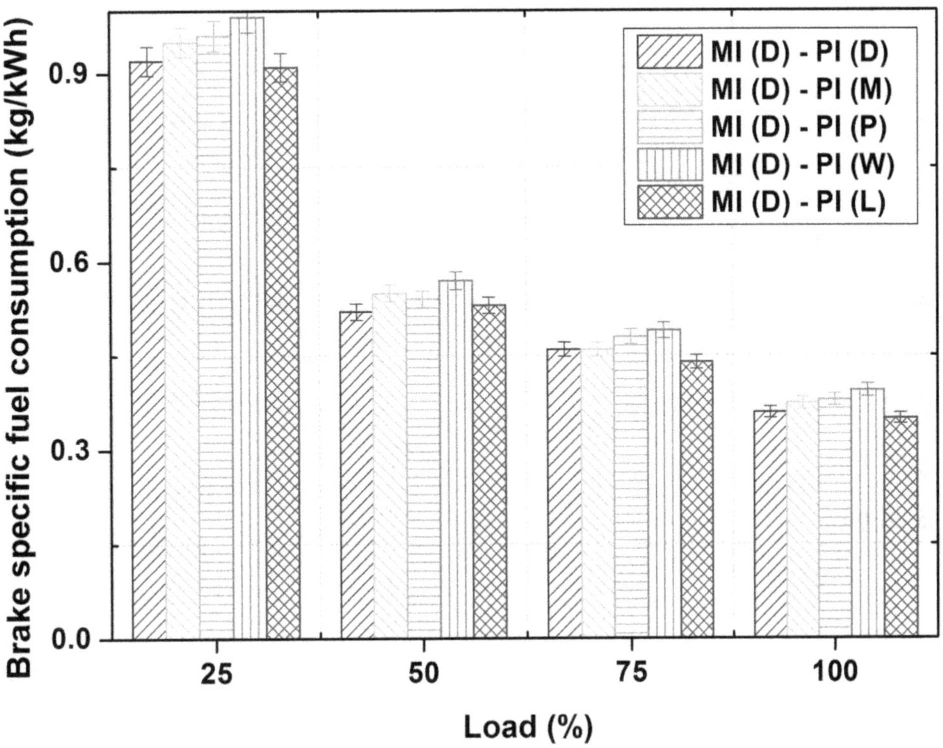

sion of high viscous fuel as pilot low reactivity fuel experienced partial oxidation during combustion. Consequently, high viscous fuel experienced high CO emission. Similarly, premixed charge compression ignition condition was experimented with diesel and di-methyl ether under dual fuel combustion found to increase the carbon monoxide emission owing to reduction in fuel vaporization condition (Wang, Liu, Huang, & Ke, 2016).

Figure 7 depicts about variation of hydrocarbon against different loads. At full load, HC emission for engine operating conditions like MI(D)-PI(D), MI(D)-PI(M), MI(D)-PI(P), MI(D)-PI(W) and MI(D)-PI(L) were noticed as 25, 22, 18, 14 and 12 ppm respectively. In general, HC emission was due to incomplete combustion of the fuel. Biodiesel samples experienced lower HC emission than diesel. The presence of oxygen in the fuel supports the oxidation of hydrocarbon molecules and thus leads to the improved emission of carbon dioxide and water vapour in the exhaust tail pipe with low HC emission. Among alternative fuel samples, lemon oil showed low HC emission. This decreased HC emission was due to its distinguishing fuel properties, improved air-fuel mixing. Pre-heating of air leads to enhanced fuel temperature during engine intake stroke and short ignition delay was achieved at all engine loads. Thus, presence of oxygen, air-fuel mixing, air preheating and the like causes complete combustion of fuel which in-turn leads to low HC emission. Similarly, diesel-cotton seed oil blend in premixed charge compression ignition engine with additive diethyl ether experienced low HC emission owing to presence of oxygen and improved oxidation process (Srihari et al., 2017).

Figure 6.

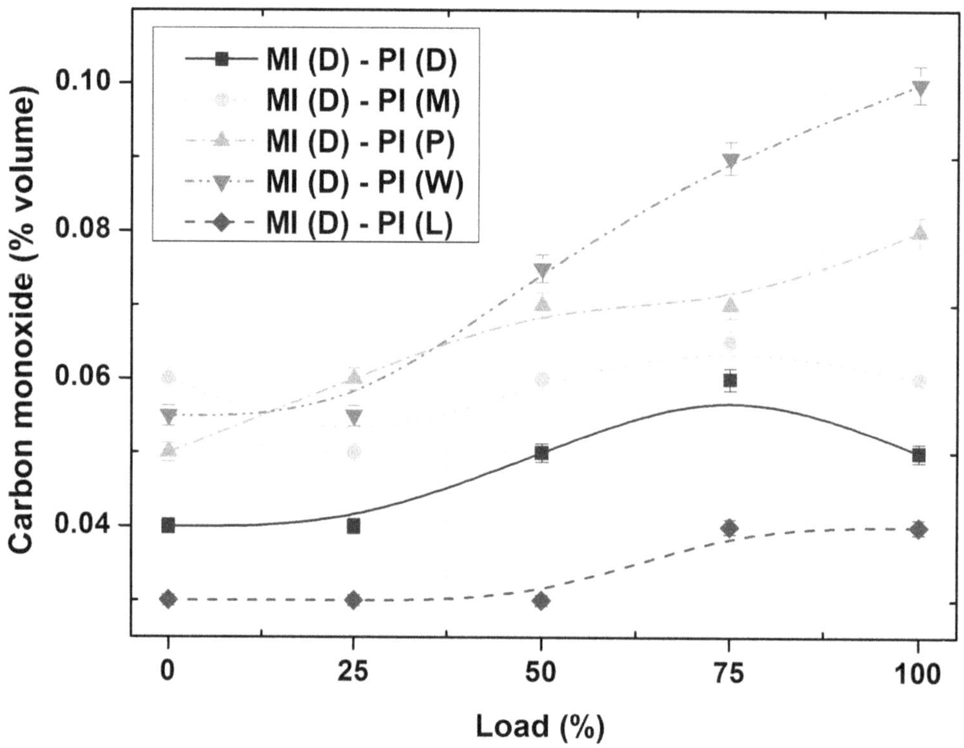

Figure 7.

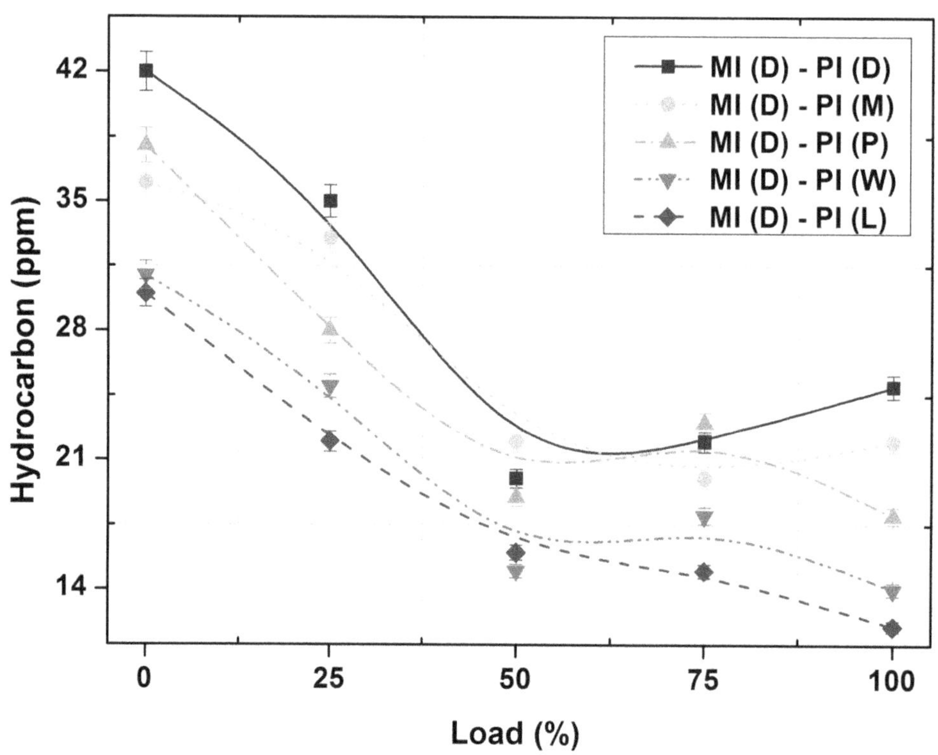

Figure 8 depicts about variation of oxides of nitrogen against different loads. At maximum load, NOx emission for engine operating conditions like MI(D)-PI(D), MI(D)-PI(M), MI(D)-PI(P), MI(D)-PI(W) and MI(D)-PI(L) were noticed as 623, 601, 557, 493 and 463 ppm respectively. In internal combustion engines, the occurrence of NOx emission was due to reaction between oxygen and nitrogen during combustion process. The other parameters influencing NOx emission were air/fuel ratio, combustion chamber design, combustion chamber temperature, and combustion duration and in-cylinder pressure (Karthickeyan, 2019c). It was important to mention that the humidity level also plays a vital role in the formation of NOx emission. At increasing loads, diesel experienced higher NOx emission than other fuel samples. The high heating value of fuel was considered as the reason for high oxides of nitrogen emission. With pilot fuel injection, the prepared biodiesel samples showed low NOx emission. The admittance of low viscous fuel with preheated air in pilot fuel injection leads to short ignition delay period. Also, the atomization rate of low viscous alternative fuel sample was faster than other renewable fuel investigation. In low load condition, production of NOx was observed lower and increase in load tends to develop NOx emission owing to improvement in temperature of combustion chamber and pressure conditions. The oxygen available in high viscosity biodiesel resource was considered as one of the reason for NOx emission increment. The presence of oxygen, improved in-cylinder temperature and pressure leads to more formation of NOx emission with high viscous fuel. The presence of low oxygen content in Lemon oil and preheated air causes enriched the fuel atomization which in-turn leads to reduced NOx emission. Similarly, diesel and cotton seed oil biodiesel in reactivity-controlled compression ignition engine showed low oxides of nitrogen emission owing to reduced combustion gas temperature (Charitha et al., 2018).

Figure 8.

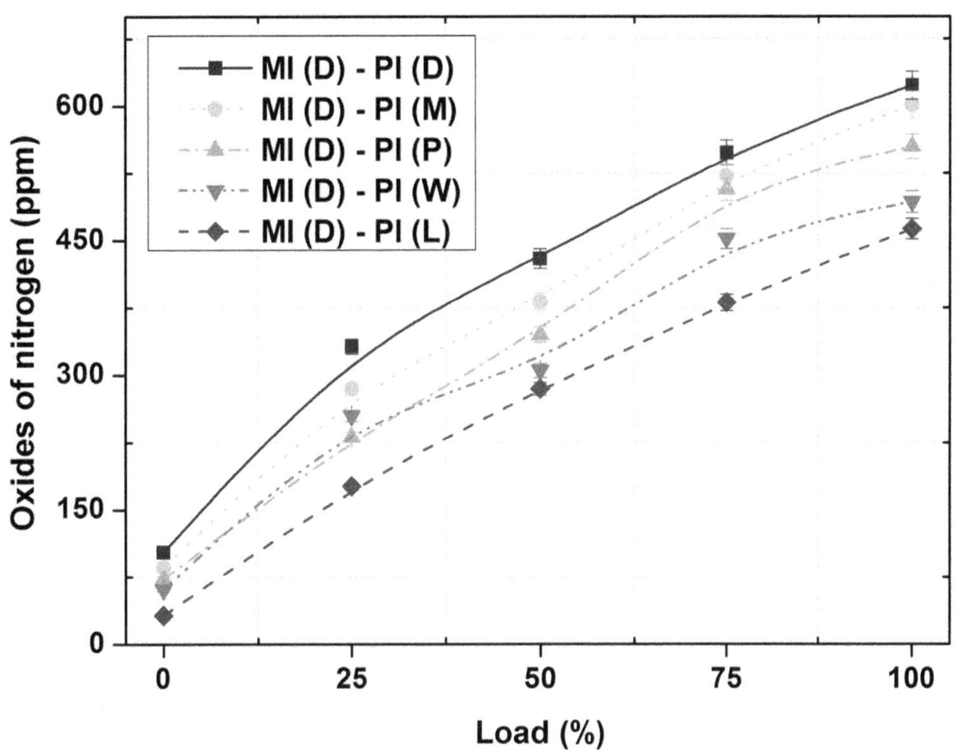

Figure 9 depicts about variation of smoke against different loads. At maximum load, smoke emission for engine operating conditions like MI(D)-PI(D), MI(D)-PI(M), MI(D)-PI(P), MI(D)-PI(W) and MI(D)-PI(L) were noticed as 54.25, 51.98, 48.05, 43.33 and 42.71 HSU respectively. The occurrence of smoke emission was due to the formation of many small solid particles from various sources associated during the combustion of fuel. Also, it was important to mention that the soot was considered as the major particulate present in the smoke emission. Incomplete or partial combustion was associated with the smoke emission. The presence of sufficient oxygen and time causes complete combustion of the admitted fuel sample which in-turn leads to declined formation of soot particles. It was noticed that the diesel exhibited higher smoke emission than other fuel samples. On the other hand, biodiesel samples exhibited lower smoke emission than diesel at all loads. This decreased smoke emission was due to presence of oxygen content, low kinematic viscosity and decreased atomization of fuel. In PCCI mode, the admission of biodiesel in intake manifold leads to enhanced combustion with diesel during compression and power stroke. Higher viscous fuel samples like *Moringa oleifera* seed oil, pumpkin seed oil and waste cooking oil exhibited higher smoke emission than lemon oil. The low viscous lemon oil tends to increase the spray inside the combustion chamber and leads to enhanced combustion efficiency of admitted fuel. Likewise, low cetane number of lemon oil tends to attain improved combustion characteristics. Similarly, *Pistacia khinjuk* biodiesel exhibited low trend of smoke emission level in all working loads owing to oxygen presence in sample (Karthickeyan, Ashok, Nanthagopal, Thiyagarajan, & Edwin Geo, 2019).

Figure 9.

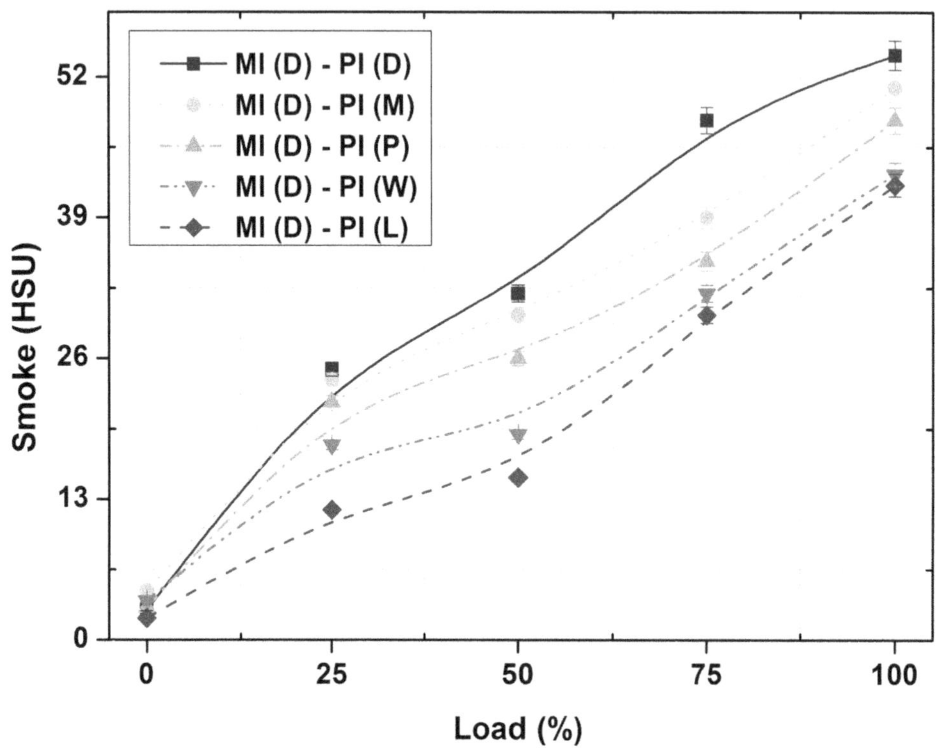

Combustion Characteristics

Figure 10 depicts about the variation of in-cylinder pressure against engine crank angle at full load. At maximum load, in-cylinder pressure for engine operating conditions like MI(D)-PI(D), MI(D)-PI(M), MI(D)-PI(P), MI(D)-PI(W) and MI(D)-PI(L) were noticed as 69.58, 68.88, 68.43, 66.96 and 69.11 bar respectively. The relationship between in-cylinder combustion chamber pressure with crank angle of engine provides information about combustion characteristics for diesel and biodiesels in PCCI mode of operation. In-cylinder pressure was measured using Kistler make M111A22 model piezo electric transducer. The analog signals of data were transferred into digital using charge amplifier and further processing was performed using Enginesoft software. Similarly, the engine crank angle was determined using Kistler make 2614C11 model crank angle encoder sensor. Diesel showed higher in-cylinder pressure at full load owing to high heating value of fuel and high temperature ambience inside combustion unit. On the other side, high viscous alternative fuel samples in pilot fuel injection exhibited low combustion pressure due to low energy content. With air preheating, reduced ignition delay period was identified with the attainment of auto-ignition temperature of fuel. High viscous biodiesel showed low in-cylinder pressure due to inferior fuel properties. Low viscous lemon oil with preheated air in pilot fuel injection exhibited higher in-cylinder than other fuel samples. The presence of oxygen leads to achieve the complete usage of available chemical energy in the straight lemon oil as useful work. It was noteworthy to mention that the application of straight lemon oil as pilot fuel injection as the promising method for

Figure 10.

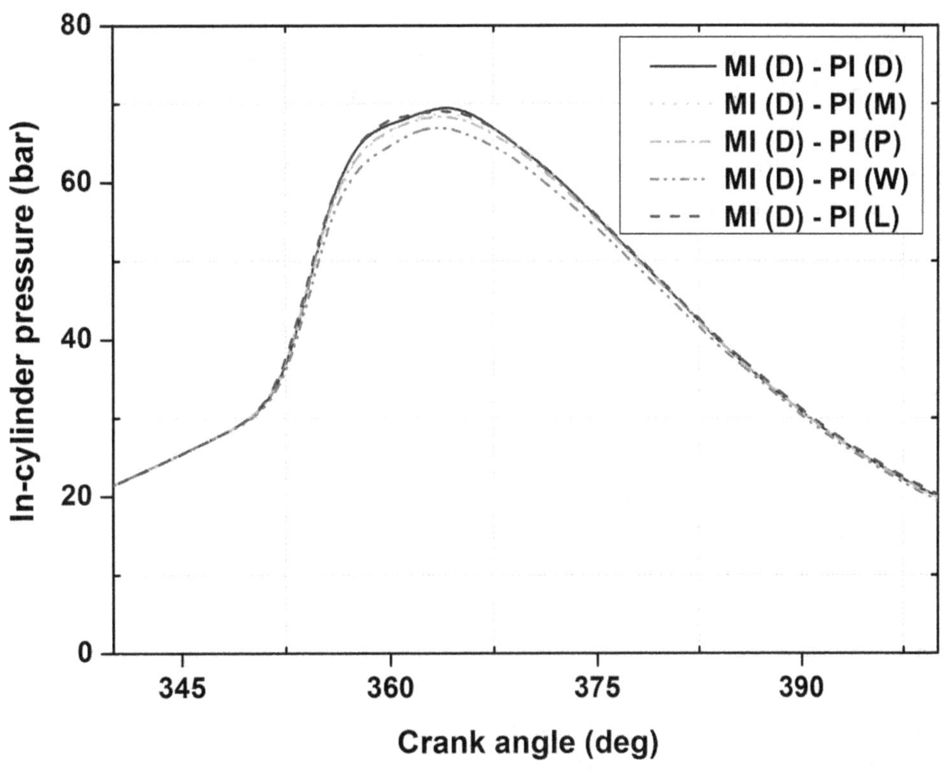

achieving improved combustion characteristics of engine. Similarly, PCCI investigation with cotton seed oil showed lower in-cylinder pressure than diesel owing to promising physical and chemical properties of fuel (Charitha et al., 2018).

Economic Analysis

Biodiesel was already considered as one of the main source for producing power and in India ethanol was blended with diesel for all modes of fuelling application. But more attention required for adopting biodiesel development from non-edible biomass for energy production. Many modes of engines were used like stationary or movable gensets, agriculture purpose engines and automobile engines. In the initial condition, fuel produced from non-edible biomass can be employed for power development form gensets and agriculture purpose engine and further can be taken into automotive engines for regular usage. In the present study, biodiesel feedstock's like *Moringa oleifera*, pumpkin, waste cooking oil and lemon oil were not used further for edible purpose and used as prime source for fuel production. The converted fuel direct or blended form investigation was performed earlier and the technology of reactivity control was not explored extensively for engine applications. Therefore, diesel injection as main admission inside the combustion chamber as per manufacture advice and modification was performed in intake manifold. The low reactivity fuel was considered as biodiesel resource produced through transesterification and high reactivity was diesel. In the point of fuel production, two approaches were performed for high viscosity fuel sample by using acid esterification and base esterification techniques. On the other side, lemon oil was produced using steam distillation approach under controlled production conditions. The quality of biodiesel was improved by adopting water washing technique using de-ionized water for biodiesels alone. Lemon oil was not water washed due to production by steam distillation approach. Usage of fresh water was not recommended due to water scarcity and may increase the production cost of biodiesel resource. The total cost for producing *Moringa oleifera* methyl ester, pumpkin seed oil methyl ester, waste cooking oil methyl ester and straight lemon oil were determined as INR 325, INR 296, INR 152 and INR 120 per kg respectively. The low production rate was observed with lemon oil and waste cooking oil owing to production through waste by products.

CONCLUSION

The present investigation focussed on the process of controlling the reactivity of fuel during combustion with conventional and biodiesels. Compression ignition engine was chosen and intake manifold was equipped with fuel preheater and injector arrangements to control the reactivity. Diesel was admitted as the main injection and also considered as high reactivity fuel. The prepared biodiesel samples were considered as low reactivity and admitted for combustion using pilot injection strategy. The following conclusions were obtained based on experimental investigation,

- Lemon oil in pilot fuel injection showed higher thermal efficiency than other alternative fuel samples owing to low viscosity and high energy content. In addition, lemon oil exhibited low fuel consumption owing to complete conversion of chemical energy in the fuel into useful work.
- Amongst biodiesels, lemon oil showed low CO and HC emission owing to presence of oxygen content in fuel and improved combustion chamber temperature with PCCI technology.

- PCCI technique experienced low concentration of NOx and smoke emission. Further, lemon oil exhibited lower NOx and smoke than other fuel samples owing to enhanced combustion characteristics.
- Higher in-cylinder pressure was noticed with diesel owing to significant fuel properties. Among biodiesel samples, lemon oil showed substantial combustion characteristics.

The present experimentation was proceeded to suggest an appropriate strategy to obtain the enhanced engine characteristics with utilization of available energy in biodiesel. Therefore, the practice of using biodiesels with Port Charged Compression Ignition technique may be considered as the prominent approach to attain superior engine characteristics. In addition, straight lemon oil as pilot fuel injection may be the prominent alternative fuel in accordance to its performance, combustion and emission characteristics. Further the study can be explored with bowl geometry modification and varying injection volume and injection pressure of pilot fuel to achieve improved engine characteristic

ACKNOWLEDGMENT

The authors would like to express their thanks to University Grants Commission- South Eastern Regional Office, Hyderabad, India for financial support through Minor research project for teachers with grant number 4-4/2013-14 (MRP- SEM/UGC- SERO).

REFERENCES

Alptekin, E. (2017). Emission, injection and combustion characteristics of biodiesel and oxygenated fuel blends in a common rail diesel engine. *Energy*, *119*, 44–52. doi:10.1016/j.energy.2016.12.069

Ashok, B., Thundil Karuppa Raj, R., Nanthagopal, K., Krishnan, R., & Subbarao, R. (2017). Lemon peel oil – A novel renewable alternative energy source for diesel engine. *Energy Conversion and Management*, *139*, 110–121. doi:10.1016/j.enconman.2017.02.049

Charitha, V., Thirumalini, S., Prasad, M., & Srihari, S. (2018). Investigation on performance and emissions of rcci dual fuel combustion on diesel - bio diesel in a light duty engine. *Renewable Energy*. doi:10.1016/j.renene.2018.09.048

Diórdinis, T., Lanzanova, M., Dalla, M., Eduardo, M., Martins, S., Romeu, P., ... Maria, S. (2019). The effects of residual gas trapping on part load performance and emissions of a spark ignition direct injection engine fuelled with wet ethanol. *Applied Energy*, *253*(January), 113508. doi:10.1016/j.apenergy.2019.113508

Duraisamy, G., Rangasamy, M., & Govindan, N. (2020). A comparative study on methanol/diesel and methanol/PODE dual fuel RCCI combustion in an automotive diesel engine. *Renewable Energy*, *145*, 542–556. doi:10.1016/j.renene.2019.06.044

Gowtham, M., Mohan, C. G., & Prakash, R. (2019). Effect of n-butanol fumigation on the regulated and unregulated emission characteristics of a diesel engine. *Fuel*, *242*(2), 84–95. doi:10.1016/j.fuel.2019.01.019

Holman, J. P. (2007). Experimental Techniques for Engineers (7th Ed). New Delhi, India: Tata McGraw Hill.

Karthickeyan, V. (2019a). Data set for effect of cetane enhancer on ceramic coated diesel engine fuelled with neat Moringa oleifera methyl ester. *Data in Brief, 24*. doi:10.1016/j.dib.2019.103932 PMID:31193703

Karthickeyan, V. (2019b). Effect of cetane enhancer on Moringa oleifera biodiesel in a thermal coated direct injection diesel engine. *Fuel, 235*(March 2018), 538–550. doi:10.1016/j.fuel.2018.08.030

Karthickeyan, V. (2019c). Effect of Combustion Chamber Bowl Geometry Modification on Engine Performance, Combustion and Emission characteristics of Biodiesel Fuelled Diesel Engine with its Energy and Exergy Analysis. *Energy, 176*, 830–852. doi:10.1016/j.energy.2019.04.012

Karthickeyan, V., Ashok, B., Nanthagopal, K., Thiyagarajan, S., & Edwin Geo, V. (2019). Investigation of Novel Pistacia khinjuk Biodiesel in DI Diesel Engine with Post Combustion Capture System. *Applied Thermal Engineering, 113969*. doi:10.1016/j.applthermaleng.2019.113969

Karthickeyan, V., & Balamurugan, P. (2017). Effect of thermal barrier coating with various blends of pumpkin seed oil methyl ester in DI diesel engine. *Heat and Mass Transfer/Waerme- Und Stoffuebertragung, 53*(10), 3141–3154. doi:10.100700231-017-2058-8

Karthickeyan, V., Balamurugan, P., & Senthil, R. (2017). Investigation on environmental effects of thermal barrier coating with waste cooking palm oil methyl ester blends in diesel engine. *Biofuels*, 1–14. doi:10.1080/17597269.2017.1316142

Karthickeyan, V., Thiyagarajan, S., Geo, V. E., Ashok, B., Nanthagopal, K., Hwai, O., & Vignesh, R. (2019). Simultaneous reduction of NOx and smoke emissions with low viscous biofuel in low heat rejection engine using selective catalytic reduction technique. *Fuel, 255*(January), 115854. doi:10.1016/j.fuel.2019.115854

Mattson, J., Burnete, N. V., Depcik, C., Moldovanu, D., & Burnete, N. (2019). Second law analysis of waste cooking oil biodiesel versus ULSD during operation of a CI engine. *Fuel, 255*(June), 115753. doi:10.1016/j.fuel.2019.115753

Mofijur, M., Masjuki, H. H., Kalam, M. A., Atabani, A. E., Arbab, M. I., Cheng, S. F., & Gouk, S. W. (2014). Properties and use of Moringa oleifera biodiesel and diesel fuel blends in a multi-cylinder diesel engine. *Energy Conversion and Management, 82*, 169–176. doi:10.1016/j.enconman.2014.02.073

Ramakrishnan, M., Rathinam, T. M., & Viswanathan, K. (2018). Comparative studies on the performance and emissions of a direct injection diesel engine fueled with neem oil and pumpkin seed oil biodiesel with and without fuel preheater. *Environmental Science and Pollution Research International, 25*(5), 4621–4631. doi:10.100711356-017-0838-9 PMID:29192402

Ranjan, A., Dawn, S. S., Jayaprabakar, J., Nirmala, N., Saikiran, K., & Sai Sriram, S. (2018). Experimental investigation on effect of MgO nanoparticles on cold flow properties, performance, emission, and combustion characteristics of waste cooking oil biodiesel. *Fuel, 220*(November 2017), 780–791. doi:10.1016/j.fuel.2018.02.057

Srihari, S., & Thirumalini, S. (2017). Investigation on reduction of emission in PCCI-DI engine with biofuel blends. *Renewable Energy*, *114*, 1232–1237. doi:10.1016/j.renene.2017.08.008

Srihari, S., Thirumalini, S., & Prashanth, K. (2017). An experimental study on the performance and emission characteristics of PCCI-DI engine fuelled with diethyl ether-biodiesel-diesel blends. *Renewable Energy*, *107*, 440–447. doi:10.1016/j.renene.2017.01.015

Teoh, Y. H., How, H. G., Masjuki, H. H., Kalam, M. A., & Alabdulkarem, A. (2019). Investigation on Particulate Emissions and Combustion Characteristics of a Common-Rail Diesel Engine Fueled with Moringa Oleifera Biodiesel-Diesel Blends. *Renewable Energy*, *136*, 521–534. doi:10.1016/j.renene.2018.12.110

Wang, Y., Liu, H., Huang, Z., & Ke, X. (2016). Particulate Size Distribution from Diesel–Dimethyl Ether Dual Fuel Premixed Compression Ignition Combustion Engine. *Journal of Energy Engineering*, *142*(4). doi:10.1061/(ASCE)EY.1943-7897.0000351

KEY TERMS AND DEFINITIONS

CI: Compression Ignition
CO: Carbon monoxide emission
DI: Direct Injection
ECU: Electronic Control Unit
HC: Hydrocarbon emission
MI: Main Injection
MI (D) – PI (D): Main Injection (Diesel) + Pilot Injection (Diesel)
MI (D) – PI (M): Main Injection (Diesel) + Pilot Injection (*Moringa oleifera* biodiesel)
MI (D) – PI (P): Main Injection (Diesel) + Pilot Injection (Pumpkin seed oil biodiesel)
MI (D) – PI (W): Main Injection (Diesel) + Pilot Injection (Waste cooking oil biodiesel)
MI (D) – PI (L): Main Injection (Diesel) + Pilot Injection (Straight lemon oil)
NOx: Oxides of nitrogen emission
PI: Pilot Injection
PCCI: Port Charged Compression Ignition
RCCI: Reactivity Controlled Compression Ignition

Chapter 12
Process Optimization Study of Alternative Fuel Production From Linseed Oil

Karthickeyan V.
Sri Krishna College of Engineering and Technology, India

Balamurugan S.
Sri Krishna College of Engineering and Technology, India

Ashok B.
Vellore Institute of Technology, India

Thiyagarajan S.
SRM Institute of Science and Technology, India

Mohamed Shameer P.
VV College of Engineering, India

Dhinesh Balasubramanian
https://orcid.org/0000-0003-2687-9095
Mepco Schlenk Engineering College, India

ABSTRACT

This chapter focuses on the selection of optimum parameters for transesterification of linseed oil biodiesel production in the presence of calcium oxide (CaO) obtained from the waste eggshells. The waste chicken eggshells were calcined at 900°C for 4 hours and it was characterized by X-ray diffractometer (XRD). The transesterification process was conducted according to L9 orthogonal array with selected input control parameters such as methanol to oil molar ratio, reaction temperature, and catalyst loading. The output parameters were biodiesel yield and viscosity. The multi-objective, decision-making technique called Technique for Order Preference by Similarity to Ideal Solution (TOPSIS) was used to identify the optimum transesterification process parameters to obtain maximum biodiesel yield with minimal viscosity. The optimized values for transesterification process parameters were depicted as methanol to oil ratio of 6:1, reaction temperature of 65°C, and catalyst loading of 5% w/w.

DOI: 10.4018/978-1-7998-2539-5.ch012

INTRODUCTION

Fossil fuel plays an important role in the transportation sector. Continuous usage of fossil fuel in the automobiles leads to emission of harmful gases like CO, NO_x, and HC to the environment (Karthickeyan, Thiyagarajan, et al., 2019; Viswanathan, Balasubramanian, Subramanian, & Varuvel, 2019). Emission of harmful green-house gases leads to global warming (Sandouqa & Al-Hamamre, 2019). On the other hand, continuous usage of fossil fuel leads to extinct of particular resource. Thus, demand for fossil fuel-based resources, emission of greenhouse gases induces the researches to focus on biodiesel (Varun, Singh, Tiwari, Singh, & Kumar, 2017). Biodiesel was eco-friendly, renewable and bio-degradable and non-toxic resource. Biodiesel has 12% of oxygen, this adds further advantage in the context of emissions in diesel engine (Karthickeyan, 2019a). Generally non-edible oil was used for biodiesel production. Trans-esterification is a chemical process used to reduce the viscosity of the non-edible oil (Pandit & Fulekar, 2017). In the transesterification process, non-edible oil along with alcohol and catalyst was properly mixed and heated at elevated temperature (Karthickeyan, 2019b; Karthickeyan, Ashok, Nanthagopal, Thiyagarajan, & Edwin Geo, 2019). As continuous transfer of heat, triglyceride of non-edible oil was converted into Fatty Acid Methyl Ester (FAME) and this FAME was termed as biodiesel (Sandouqa & Al-Hamamre, 2019). Two category of catalyst was widely used in the transesterification process, namely homogeneous and heterogeneous catalyst. Homogeneous catalyst has its own drawbacks as multi washing of transesterified oil was required to remove catalyst residues and also produces unwanted waste water which then creates excess soap by-product. Homogeneous catalyst was not reusable and corrosive to equipment. Heterogeneous catalyst overcomes homogeneous catalysts drawbacks. Homogeneous solid catalyst was easy to recover and reusable for several times and excess wastewater production was eliminated (Mansir, Hwa Teo, Lokman Ibrahim, & Yun Hin, 2017).

Generally, all catalyst produces its impact on environmental concerns. Therefore, development of solid catalyst has recently gained much attention among the researchers to overcome the aforementioned issues. In this regard a waste eggshell catalyst is a potential source for producing low cost biodiesel. Waste eggshell contains high amount of calcium components ($CaCO_3$) and this $CaCO_3$ could be easily converted to Calcium Oxide (CaO) by calcinations method. Calcined CaO is a potential catalyst as it is cheap, non corrosive, eco-friendly and reusable (Tan, Abdullah, & Nolasco-Hipolito, 2015).

Calcium oxide (CaO) catalyst was prepared from waste chicken eggshell by calcinations process. Transmission electron microscope (TEM) image showed spherical structure with average particle size as 46.1±2.1 mm. RSM was used to optimize the transesterification process parameters (Pandit & Fulekar, 2019). Calcination-hydration-dehydration technique was used to convert the waste eggshell into catalyst. TEM image showed that spherical shape with average particle size of 75 nm. The effect of prepared catalyst was analysed in the transesterification process of dry biomass into biodiesel (Pandit & Fulekar, 2017). Ostrich-eggshell and chicken-eggshell were calcined and used for transesterification process. Waste cooking oil was converted as bio-fuel. RSM and Taguchi method was applied to optimize the process parameters. The optimum biodiesel yield of approximately 98% (ostrich eggshell) and 96% (chicken eggshell) were achieved (Tan, Abdullah, Nolasco-Hipolito, & Ahmad Zauzi, 2017). CaO catalyst was derived from waste shells of egg, oyster and clam. Transesterification of soybean oil was performed. Sonication was done on CaO catalyst about five hours which resulted in reduction of particle size by 34% and this resulted in a 56% increase in the activity (Risso, Ferraz, Meireles, Fonseca, & Vital, 2018). A simple wet impregnation method was used to make solid bifunctional tungsten molybdenum supported calcium mixed oxide catalyst. The prepared catalyst was used in a transesterification of waste cooking oil. This catalyst was recorded the highest bio-fuel yield of 96.2% under optimized condition (Mansir et al., 2017). Simple calcinations process was used to convert eggshell into CaO catalyst at 900 °C for 4 hours. Chicken oil was used in the synthesis of bio-fuel. RSM was used to optimize the transesterification parameters. Maximum biodiesel yield of 90.41% was recorded in this study (Kirubakaran & Arul Mozhi Selvan, 2018). The waste eggshell was calcined at 800 °C and used in the transesterifiation of dairy industry waste scum oil. Maximum bio-fuel yield of 96% was obtained at the optimized parameters. An efficient catalyst was derived from the waste eggshell by calcination process. Date seed oil was used in transesterification process. 93.5% of bio-fuel at optimized condition was recorded in this study (Farooq et al., 2018). Table 1 shows the summary of previous works related to the biodiesel synthesis using waste eggshell as working catalysts.

The literature provides knowledge that still there is a space to develop an effective and efficient catalyst for transesterification process to produce economical and eco-friendly biodiesel in the future. Selection of feedstock for the biodiesel synthesis plays major role in the economical aspects. Continuous availability of feedstock and catalyst derived from waste eggshell is the key area to produce low cost bio-fuel. Considering continuous availability, linseed oil was considered in this work for the bio-fuel synthesis process. This oil is non-edible and used in the paint manufacturing industries. This work aimed to produce cheap and effective catalyst from the waste chicken eggshell by calcination process and also from the literature reviews; it can be observed that many researchers have focused on optimizing single response problem in the transesterification process. A scanty of work carried out in the multi-objective optimization of transesterification experiments. In order to fill this gap, the present study uses TOPSIS techniques to optimize the bio-fuel production from the linseed oil along with CaO catalyst prepared from the waste chicken eggshell.

MATERIALS AND METHODS

Materials

Linseed oil was extracted from the linseed by pre-expelling and hexane extraction of the press cake method. The obtained oil was yellowish in colour and the properties of linseed oil like viscosity, flash point, fire point, could point and pour point, density were shown in the Table. 2. The advantages of linseed oil were continuously availability in India, lower acid value and composed with different chemical components (like glycerides of oleic acid, linoleic acid, linolenic acid). Some of the disadvantages were also encountered as follows, requires high time period to remove moisture content and growth in mild dew weather condition.

Catalyst Preparation

Chicken eggshells were collected from nearby restaurants. The collected eggshells were washed several times to remove the impurities for the calcination process. Washed eggshells were dried in sunlight more than three days to remove the moisture content. It was then crushed in a grinder to obtain fine grained powder. This powder was calcined at 900 °C for 4 hours in the muffle furnace. Above 800 °C, calcium carbonate in the eggshell was decomposed and converted as Calcium Oxide. This calcium oxide was stored in an air-tight container for future use (Gupta & Rathod, 2018).

Table 1. Summary of previous works related to the biodiesel synthesis using waste eggshell as working catalysts

Type of oil	Type of catalyst	Optimization technique	Optimized Process Parameters	Biodiesel Yield (%)	Ref.
Chlorella vulgaris biomass	Calcium oxide (CaO) catalyst prepared using chicken egg shell waste	Response surface methodology (RSM) based on central composite design (CCD)	Reaction temperature 70 °C, methanol to dry biomass ratio 10:1, catalyst loading 1.39%, reaction time 3 h	92.03	(Pandit & Fulekar, 2019)
Dry biomass	Calcium oxide (CaO) catalyst prepared using chicken egg shell waste	Response surface methodology (RSM) based on central composite design (CCD)	Catalyst loading 1.7%, reaction time 3.6 h, stirring rate 140.6 rpm	86.41	(Pandit & Fulekar, 2017)
Waste cooking oil	Calcium oxide (CaO) catalyst prepared using Ostrich-eggshell and chicken-eggshell	RSM and Taguchi method	Molar to oil ratio 10:1, catalyst concentration of ~1.5%w/v, reaction temperature 65 °C, reaction time of ~2 hours	98% (ostrich eggshell) and (chicken eggshell)	(Tan et al., 2017)
Soybean oil	Calcium oxide (CaO) catalyst prepared using waste shells of egg, oyster and clam	Experimental trials	Eggshells were subjected to irradiation and mollusc shells were subjected to calcination-hydration-calcination cycle to increase the surface area of CaO and to improve catalytic activity and followed by 5 h of sonication reduced particle size.	Transesterification rate was 2.5 times higher than that obtained with the untreated samples	(Risso et al., 2018)
Chicken fat	Calcium oxide (CaO) catalyst prepared using chicken eggshells	Response surface methodology	Molar ratio 1:13, catalyst loading 8.5 wt % of oil, reaction time 57.5 °C, reaction time 5 h.	90.41	(Kirubakaran & Arul Mozhi Selvan, 2018)
Waste scum oil from dairy industry	Calcium oxide (CaO) catalyst prepared using waste eggshells	Experimental trials	Molar ratio to methanol to oil 6:1, catalyst amount 2.4 wt%, reaction temperature 65 °C, reaction time 3 h.	96.0	(Rahees & Meera, 2014)
Date seed oil	Calcium oxide (CaO) catalyst prepared using waste eggshells	Experimental trials	Methanol to oil molar ratio 12:1, catalyst amount 5 wt%, reaction time 1.5 h.	93.5	(Farooq et al., 2018)
Waste cooking oil	Calcium oxide (CaO) catalyst prepared using waste eggshells	Experimental trials	Methanol to oil molar ratio 10:1, catalyst loading 1.50 wt%, reaction temperature 60 °C, reaction time 50 min.	96.07	(Gupta & Rathod, 2018)
Aegel marmelos biomass	Pyrolysis	TOPSIS and Grey relational analysis	Heating temperature 600 °C, feedstock particle size 0.6 mm.	42.75 wt%	(Baranitharan et al., 2019)

Table 2. Properties of linseed oil

Properties	Linseed oil
Acid value	0.338 KOH/mg
Kinematic viscosity @ 40 °C	5.054 cSt
Density	0.885 g/cc
Moisture content	0.245%
Flash point by PMCC method	188 °C
Fire point by PMCC method	210 °C
Could point	2 °C
Pour point	-1 °C
% FFA	2.4

The chemical equation (1) represents the transformation of waste chicken eggshell to catalyst as follows:

$$CaCO_3(s) \xrightarrow{900^\circ C} CaO(s) + CO_2 \quad (1)$$

Catalyst Characterization

The elemental chemical contents of the synthesized catalysts before calcination and after calcination were analysed by the X-ray diffraction (XRD) at room temperature. Cu kα radiation (λ= 0.15406 nm) in a 2θ scan range of 10 - 90° was used for all samples.

Taguchi Experimental Design

Geneichi Taguchi designed and developed the best tool for an effective quality process or system. Experimental design and quality loss function was analyzed through the effective design of experiments with the Taguchi tool. This tool was used for optimizing the process parameters with less trial. Taguchi orthogonal array was the cost-effective and time saving technique for optimizing the experimental process parameters. In this study, three factors were selected namely methanol to oil molar ratio, reaction temperature and catalyst loading in the transesterification process. Input parameters and their levels were tabulated in Table 3. In the Taguchi method, the orthogonal array was selected based on degrees of freedom (DOF) of process parameters. Total input parameters of DOF should be less than the selected DOF of the orthogonal array. DOF of individual parameter was subtracting one from several levels of parameters that were showed in Table 3. Three DOF factors with three levels were 6; hence L9 orthogonal array was selected (Gupta & Rathod, 2018). DOF of L9 orthogonal array was 8; this was higher than the total DOF input parameters. Henceforth, this design accommodates all the factors for investigation of transesterification parameters of biodiesel yield and its viscosity.

Table 3. Input parameters and levels of transesterification process

Input parameters	Level 1	Level 2	Level 3	DOF
Molar ratio	6:1	7:1	8:1	2
Temp (°C)	55	60	65	2
Catalyst (%w/w)	5	6	7	2
Total DOF				6

Transesterification Process

Linseed oil was converted into linseed oil biodiesel by means of transesterification process with the help of 250 ml two-necked flask, a condenser and a magnetic stirrer in a water bath. Particular amount of linseed oil, methanol and CaO catalyst were introduced into the reaction flask. A magnetic stirrer was used to continuously stir the solution during entire reaction. Particularly, for optimum biodiesel yield, oil to methanol molar ratio of 1:8, reaction temperature 60 °C, reaction time 1.3 hours, catalyst loading 7%w/w were used. 600 rpm stirring speed was used for all the experiments. Figure 1 shows (a) linseed and (b) linseed oil. After the reaction, transesterified oil was transferred to separation funnel and left for 24 hours undisturbed for the separation glycerol and biodiesel. Glycerol was a by-product from this reaction and biodiesel was taken separately and washed with warm distilled water to remove un-reacted catalyst. Table 4 shows L9 orthogonal array used in the present work including input parameters of transesterification process with corresponding output response.

Technique for Order Preference by Similarity to Ideal Solution Method

TOPSIS technique was an effective and simple multi-objective decision-making approach followed by many industries like solar forms site selection, computer networks and to optimize the machining process parameters in the manufacturing field. TOPSIS method was used to evaluate the better results from the

Figure 1.

Table 4. L9 Orthogonal array with experimental data

Experiments	Molar ratio	Temperature (°C)	Catalyst (%w/w)	Viscosity (cSt)	Yield (%)
1	6:1	50	4	4.1	88
2	7:1	50	5	4.5	86
3	8:1	50	6	4.3	90
4	6:1	60	4	3.0	89
5	7:1	60	5	3.8	88
6	8:1	60	6	3.5	93
7	6:1	70	4	3.2	91
8	7:1	70	5	3.4	92
9	8:1	70	6	3.1	95

large size alternative solutions. TOPSIS idea was established to solve the multi-objective problem by choosing a minimum distance from the selected alternatives positive ideal consequences and maximum distance from the negative ideal outcomes. Positive ideal consequences were the group of all greatest values. In the TOPSIS technique, all experimental responses were separated into beneficial or non-beneficial. Further, the beneficial responses were categorized as higher the better and non-beneficial were categorized as lower the better. In the TOPSIS analysis, the following steps are used to rank the responses (Baranitharan, Ramesh, & Sakthivel, 2019).

Step I: TOPSIS Decision Making Matrix Formulation

The design matrix was created by using collected experimental data. This matrix contains p attributes and q alternative solutions. In this study, the output responses were attributes and experimental data were alternative solutions.

$$T_m = \begin{bmatrix} b_{11}b_{12}b_{13}\dots\dots b_{1n} \\ b_{21}b_{21}b_{21}\dots\dots b_{2n} \\ \dots\dots\dots\dots\dots\dots \\ \dots\dots\dots\dots\dots\dots \\ \dots\dots\dots\dots\dots\dots \\ b_{m1}b_{m2}b_{m3}\dots\dots b_{mn} \end{bmatrix} \quad (2)$$

where, b_{ij} is the measure of jth attribute to i[th] alternative.

Step II: Normalization of Decision Matrix

Normalization is the process of reorganizing the design matrix from larger data to smaller data. The normalization was done by using equation (3).

$$\mu_{ij} = \frac{b_{ij}}{\sqrt{\sum_{i=1}^{m} b_{ij}^2}} \tag{3}$$

where, μ_{ij} is the Normalized value for $i=1,2,3,\ldots,m$ and $j=1,2,3,\ldots,n$

Step III: Calculation of Weighted Normalized Matrix

Weighted normalized matrix was calculated by using equation (4) to compare the alternatives by considering multiple criteria of various levels of importance.

$$\tau_{ij} = w_j \mu_{ij} \tag{4}$$

where,

$$\sum_{j=1}^{n} w_j = 1$$

Step IV: Determination of Positive Ideal Solution and Negative Ideal Solution

The positive ideal solution was calculated by using equation (5) to maximize the benefit. Similarly, by using equation (6) negative ideal solution was calculated.

$$\tau^+ = \left(\tau_1^+, \tau_2^+, \tau_3^+, \ldots, \tau_n^+\right) = \left\{(\max \tau_{ij} | j \in J_1), \left(\min \tau_{ji} | j \in J_2, i = 1,2,..,n\right)\right\} \tag{5}$$

$$\tau^- = \left(\tau_1^-, \tau_2^-, \tau_3^-, \ldots, \tau_n^-\right) = \left\{(\min \tau_{ij} | j \in J_1), \left(\max \tau_{ji} | j \in J_2, i = 1,2,..,n\right)\right\} \tag{6}$$

where, J_1 is the set of beneficial attribute and J_2 is the set of non-beneficial attribute.

Step V: Determination of Separation Measures From Positive Ideal Solution and Negative Ideal Solution

Separation measures from positive ideal solution and negative ideal solution was calculated from equation (7) and (8).

$$S_i^+ = \sqrt{\sum_{j=1}^{n} \left(\tau_{ij} - \tau^+\right)^2} \tag{7}$$

$$S_i^- = \sqrt{\sum_{j=1}^{n}\left(\tau_{ij} - \tau^-\right)^2} \tag{8}$$

where, i= 1, 2,……, m

Step VI: Determination of the Relative Closeness Coefficient

Closeness coefficient was calculated from equation (9) to order the priority of the alternatives.

$$CC = \frac{S_i^-}{S_i^+ - S_i^-} \tag{9}$$

Step VII: Ranking the Best in the Descending Order

Coefficient of closeness arranged in the descending order to seeking for highest value. In the ranking, highest value ranked in the first position and least value ranked in the last position.

RESULT AND DISCUSSION

Characterization of Feedstock

The characterization of the linseed oil was shown in Table 2. Linseed oil contains 2.4% of FFA, which was suitable for the alkali based transesterification process. If the acid value or FFA of any feedstock was high, then certainly low TG was obtained in the transesterification reaction. Similar trend of low TG for high FFA was observed with Rice-bran oil. Acid-based transesterification or alkali based transesterification was chosen based on the acid value and FFA. Miss selection of catalysts in the transesterification process not only leads to cost expenditures but also time overheads.

Technique for Order Preference by Similarity to Ideal Solution Analysis

In the TOPSIS analysis, all experimental data was converted to decision making matrix. In this matrix, experimental rows were considered as alternatives and experimental output results considered as attributes. All experimental data was normalized by using equation (2). Similarly, biodiesel yield and viscosity was given equal significance in this study hence, relative weightage of ½ was allotted. Weighted normalized values were calculated by using equation (3). The positive ideal solution and negative ideal solution normalized matrix were calculated from the equation (4) and (5). The maximum value of positive ideal solution

$$\tau^+_{viscosity} = 0.2456, \tau^+_{yield} = 0.1946$$

was taken from the output response. Similarly, the minimum value of negative ideal solution

$$\tau^{-}_{viscosity} = 0.1637, \tau^{-}_{yield} = 0.1799.$$

Equation (6) and (7) used to calculate separation measures of positive ideal solutions and negative ideal solutions.

Closeness coefficients were calculated from every separation measures alternative by equation (8). Normalized values, weighted normalized values, separation measures values and closeness coefficient values of entire alternatives of output response was shown in Table5. Optimum solution was taken from the highest closeness coefficient (CC) value. In this study, experimental trial 4 has highest closeness coefficient value 0.48060794. On comparing with entire experimental trails, experimental trial 4 showed highest CC and next level optimum solutions were perceived with experimental trial 9 and 7. The mean closeness coefficient (MCC) of entire level of experimental process parameters were calculated by taking the average of closeness coefficient at the same factor level was shown in the Table 6. The effect of transesterification parameters on MCC was shown in Figure 2. By means of chart, the optimal process parameter for transesterification of linseed oil for high biodiesel yield and low viscosity was recognized as MR1RT3CL1, which are methanol to oil molar ratio of 6:1, a reaction temperature of 65 °C and catalyst loading of 5%w/w.

Figure 2.

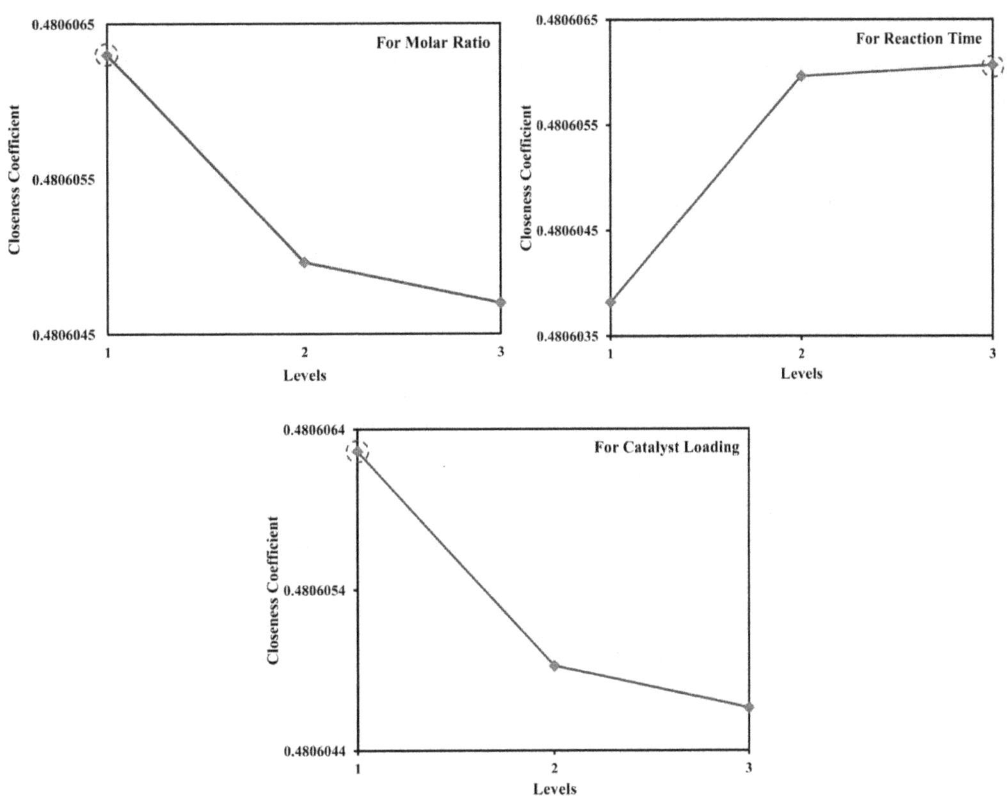

The confidence level of optimization for production of linseed oil biodiesel was performed as follows,

Sample size, $n = 9$

Sample proportions that supports biodiesel yield (Y), $\hat{p} = 0.88$

Standard deviation of sample proportion, $\tilde{A}_p = \sqrt{\frac{p(1-p)}{n}}$

$$\text{Standard error, } SE_{\hat{p}} = \sqrt{\frac{\hat{p}\left(1-\hat{p}\right)}{n}}$$

$$\text{Standard error, } SE_{\hat{p}} = 0.11$$

$$\text{Confidence level} = 1 - SE_{\hat{p}}$$

Confidence level = 89%

Figure 3 shows the percentage of error obtained from different experimental runs. Uncertainty of measured biodiesel yield was presented as error of measurement. ±2.12% was noticed as uncertainty value based on standard deviation approached with 89% of confidence level. Maximum yield of biodiesel was noticed at experimental run of 9 with catalyst loading of 6% (w/w), operating temperature of 70°C and catalyst to oil ratio of 8:1. It was important to mention that waste egg shell based catalyst found with higher level of biodiesel production. Also, least level of biodiesel yield was noticed with experimental run condition of 2.

Figure 4 shows the ranking of experiments based on TOPSIS technique. Experiment run performed in 4 was found to produce higher biodiesel yield of 89% with viscosity of 3.0 cSt. Also, experimental run of 9 found with higher yield of 95% with viscosity of 3.1 cSt and experiment run 3 found with 90% yield and viscosity of 4.3 cSt.

Characterization of Catalyst

XRD spectra of uncalcined and calcined eggshells sample was obtained with Cu radiation at 45 kV, 30 mA and a scan range of 10-80°. Figure 5 (a) and (b) shows the XRD profile of raw eggshell and calcined eggshell. For the raw eggshell, the most intense diffraction peak was seen at $2\theta = 29.36°$ with height of 2683.97 cts. The other typical peaks were noticed 31.38°, 35.56°, 39.35°, 43.07°. These peaks matched precisely with reported by (Gupta & Rathod, 2018). These peaks were characteristic of calcium carbonate. For the calcined catalyst as showed in Fig. 5 b, the peaks were appeared at $2\theta = 18.2681°$, $2\theta = 23.1808°$, $2\theta = 29.4704°$, $2\theta = 34.2237°$, $2\theta = 39.5250°$, $2\theta = 47.6788°$ and $2\theta = 57.5379°$, were matches with crystalline phase of CaO as a major phase and which were characteristics of calcium oxide. The minor peaks of CaO observed at $2\theta = 32.33°$, 37.64° showing partial change of eggshell $CaCO_3$ into CaO. However, it was further observed that the presence of calcium hydroxide at $2\theta = 18.2681$. The XRD pattern of the prepared waste eggshells sample calcined at 900 °C exhibited similar results to (Rahees & Meera, 2014)

Figure 3.

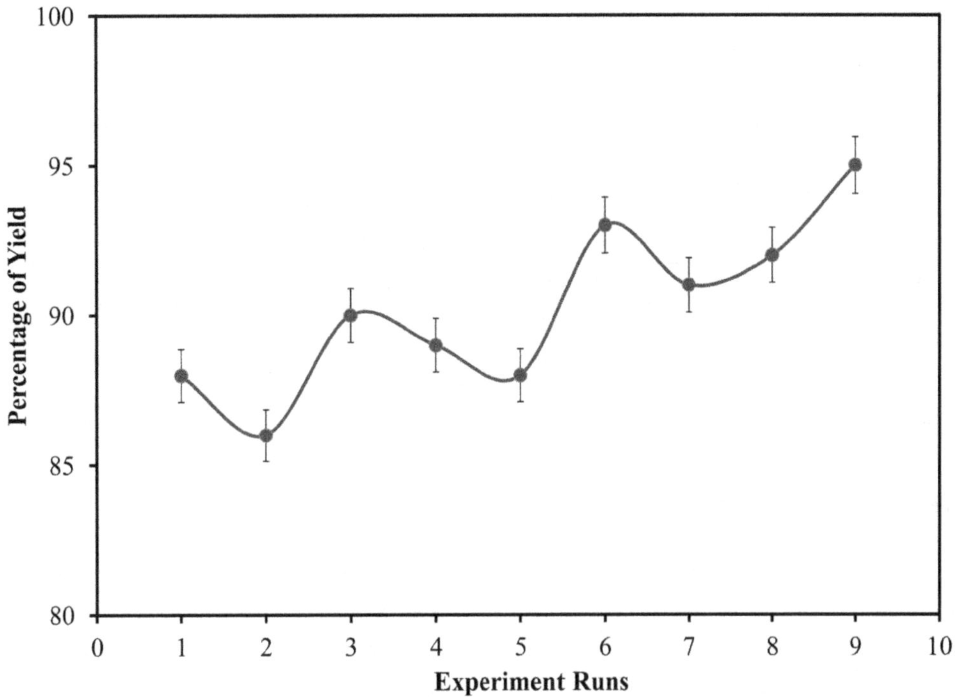

Figure 4.

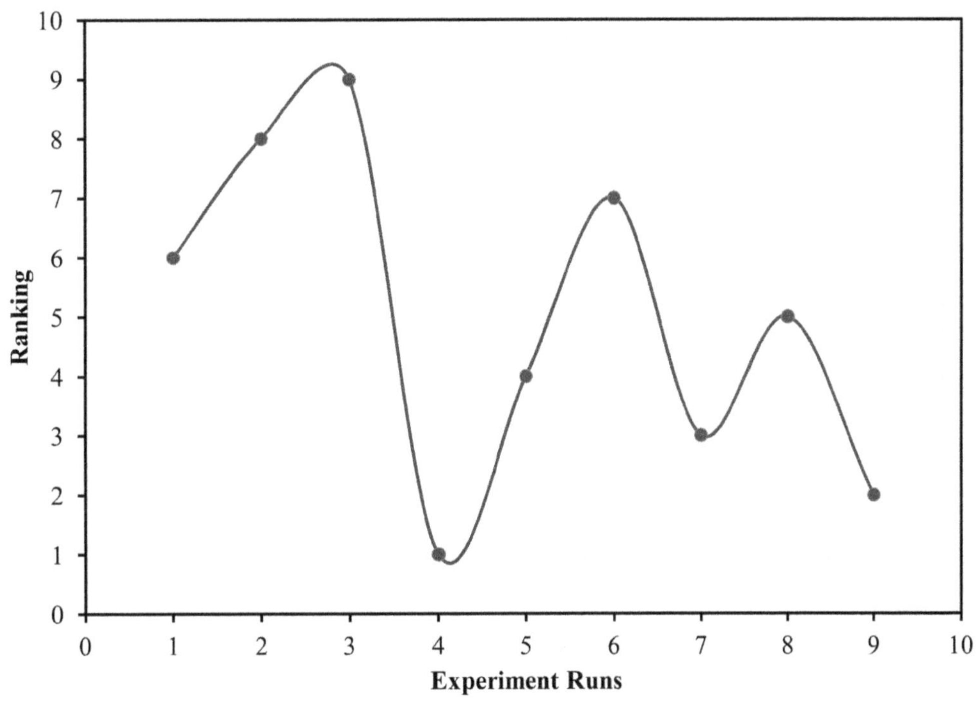

Table 5. TOPSIS analysis result

Trial No.	Normalization		Weighted Normalization		Separation		CC
	Viscosity (cSt)	Yield (%)	Viscosity (cSt)	Yield (%)	τ+	τ-	
1	0.447587	0.368320145	0.223793	0.18416	92.85735009	85.92232	0.480604521
2	0.491253	0.359949233	0.245627	0.179975	92.8608835	85.92541	0.480604018
3	0.46942	0.376691058	0.23471	0.188346	92.85284075	85.91759	0.480602922
4	0.327502	0.372505602	0.163751	0.186253	92.85707284	85.92323	0.480607937
5	0.414836	0.368320145	0.207418	0.18416	92.85784111	85.92313	0.480605573
6	0.382086	0.389247426	0.191043	0.194624	92.84787616	85.9135	0.480604382
7	0.349336	0.380876514	0.174668	0.190438	92.85255652	85.9185	0.480606335
8	0.371169	0.38506197	0.185585	0.192531	92.85013322	85.91587	0.480605182
9	0.338419	0.380876514	0.16921	0.190438	92.85272277	85.91878	0.480606687

Table 6- Mean Closeness Coefficient

	Molar Ratio	Reaction Temperature	Catalyst Loading
Level I	0.480606264*	0.480603821	0.480606264*
Level II	0.480604925	0.480605964	0.480604925
Level III	0.480604664	0.480606068*	0.480604664

*Optimal solution

Figure 5.

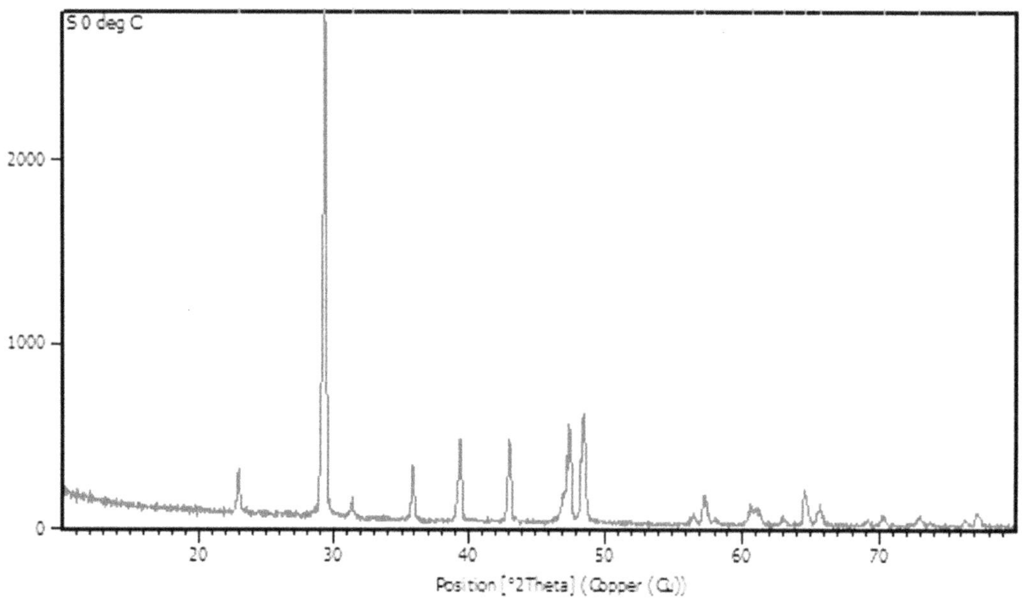

CONCLUSION

In the present work, waste eggshells were converted into catalyst for the biodiesel production. The waste eggshells were calcined at 900°C. The characteristics of uncalcined and calcined eggshells were studied with the help of XRD. Calcium carbonate ($CaCO_3$) in uncalcined eggshells was converted into calcium oxide (CaO). This CaO was effectively used in the biodiesel production process. Linseed oil was converted into linseed oil biodiesel using transesterification process. As the acid value of linseed oil was low, alkali catalyst transesterification was performed for biodiesel production. At this point, Taguchi orthogonal array was formed and L9 array was selected. Nine alkali catalyst based transesterification was conducted and its yield as well as viscosity was noted. TOPSIS multi-objective decision-making approach was used to optimize the conducted experiments. In this regard, experimental trial 4 has the highest closeness coefficient value 0.48060794 and it was ranked as the first position. The input transesterification parameters for experimental trial 4 were methanol to oil molar ratio of 6:1, reaction temperature of 60 °C and catalyst loading of 5% w/w. The mean closeness coefficient of entire level of experimental process parameters was calculated. Based on the closeness coefficient, transesterification of linseed oil with eggshells derived catalyst for producing high biodiesel yield and low viscosity was recognized as MR1RT3CL1, which are methanol to oil molar ratio of 6:1, a reaction temperature of 65 °C and catalyst loading of 5% w/w.

ACKNOWLEDGMENT

The authors would like to express their thanks to University Grants Commission- South Eastern Regional Office, Hyderabad, India for financial support through Minor research project for teachers with grant number 4-4/2013-14 (MRP- SEM/UGC- SERO).

REFERENCES

Baranitharan, P., Ramesh, K., & Sakthivel, R. (2019). Multi-attribute decision-making approach for Aegle marmelos pyrolysis process using TOPSIS and Grey Relational Analysis: Assessment of engine emissions through novel Infrared thermography. *Journal of Cleaner Production*, *234*, 315–328. doi:10.1016/j.jclepro.2019.06.188

Farooq, M., Ramli, A., Naeem, A., Mahmood, T., Ahmad, S., Humayun, M., & Islam, M. G. U. (2018). Biodiesel production from date seed oil (Phoenix dactylifera L.) via eggshell derived heterogeneous catalyst. *Chemical Engineering Research & Design*, *132*, 644–651. doi:10.1016/j.cherd.2018.02.002

Gupta, A. R., & Rathod, V. K. (2018). Waste cooking oil and waste chicken eggshells derived solid base catalyst for the biodiesel production: Optimization and kinetics. *Waste Management (New York, N.Y.)*, *79*, 169–178. doi:10.1016/j.wasman.2018.07.022 PMID:30343743

Karthickeyan, V. (2019a). Effect of cetane enhancer on Moringa oleifera biodiesel in a thermal coated direct injection diesel engine. *Fuel, 235*(March 2018), 538–550. doi:10.1016/j.fuel.2018.08.030

Karthickeyan, V. (2019b). Effect of Combustion Chamber Bowl Geometry Modification on Engine Performance, Combustion and Emission characteristics of Biodiesel Fuelled Diesel Engine with its Energy and Exergy Analysis. *Energy, 176*, 830–852. doi:10.1016/j.energy.2019.04.012

Karthickeyan, V., Ashok, B., Nanthagopal, K., Thiyagarajan, S., & Edwin Geo, V. (2019). Investigation of Novel Pistacia khinjuk Biodiesel in DI Diesel Engine with Post Combustion Capture System. *Applied Thermal Engineering, 113969*. doi:10.1016/j.applthermaleng.2019.113969

Karthickeyan, V., Thiyagarajan, S., Geo, V. E., Ashok, B., Nanthagopal, K., Hwai, O., & Vignesh, R. (2019). Simultaneous reduction of NOx and smoke emissions with low viscous biofuel in low heat rejection engine using selective catalytic reduction technique. *Fuel, 255*(January), 115854. doi:10.1016/j.fuel.2019.115854

Kirubakaran, K., & Arul Mozhi Selvan, V. (2018). Eggshell as heterogeneous catalyst for synthesis of biodiesel from high free fatty acid chicken fat and its working characteristics on a CI engine. *Journal of Environmental Chemical Engineering, 6*(4), 4490–4503. doi:10.1016/j.jece.2018.06.027

Mansir, N., Hwa Teo, S., Lokman Ibrahim, M., & Yun Hin, T. Y. (2017). Synthesis and application of waste eggshell derived CaO supported W-Mo mixed oxide catalysts for FAME production from waste cooking oil: Effect of stoichiometry. *Energy Conversion and Management, 151*(August), 216–226. doi:10.1016/j.enconman.2017.08.069

Pandit, P. R., & Fulekar, M. H. (2017). Eggshell waste as heterogeneous nanocatalyst for biodiesel production: Optimized by response surface methodology. *Journal of Environmental Management, 198*, 319–329. doi:10.1016/j.jenvman.2017.04.100 PMID:28494420

Pandit, P. R., & Fulekar, M. H. (2019). Biodiesel production from microalgal biomass using CaO catalyst synthesized from natural waste material. *Renewable Energy, 136*, 837–845. doi:10.1016/j.renene.2019.01.047

Rahees, K., & Meera, V. (2014). Production of Biodiesel from Dairy Waste Scum. *International Journal of Scientific & Engineering Research, 5*(7), 194–199. doi:10.1016/j.psep.2019.03.021

Risso, R., Ferraz, P., Meireles, S., Fonseca, I., & Vital, J. (2018). Highly active Cao catalysts from waste shells of egg, oyster and clam for biodiesel production. *Applied Catalysis A, General, 567*, 56–64. doi:10.1016/j.apcata.2018.09.003

Sandouqa, A., & Al-Hamamre, Z. (2019). Energy analysis of biodiesel production from jojoba seed oil. *Renewable Energy, 130*, 831–842. doi:10.1016/j.renene.2018.07.015

Tan, Y. H., Abdullah, M. O., & Nolasco-Hipolito, C. (2015). The potential of waste cooking oil-based biodiesel using heterogeneous catalyst derived from various calcined eggshells coupled with an emulsification technique: A review on the emission reduction and engine performance. *Renewable & Sustainable Energy Reviews, 47*, 589–603. doi:10.1016/j.rser.2015.03.048

Tan, Y. H., Abdullah, M. O., Nolasco-Hipolito, C., & Ahmad Zauzi, N. S. (2017). Application of RSM and Taguchi methods for optimizing the transesterification of waste cooking oil catalyzed by solid ostrich and chicken-eggshell derived CaO. *Renewable Energy, 114*(PB), 437–447. doi:10.1016/j.renene.2017.07.024

Varun, S., Singh, P., Tiwari, S. K., Singh, R., & Kumar, N. (2017). Modification in combustion chamber geometry of CI engines for suitability of biodiesel: A review. *Renewable & Sustainable Energy Reviews*, *79*, 1016–1033. doi:10.1016/j.rser.2017.05.116

Viswanathan, K., Balasubramanian, D., Subramanian, T., & Varuvel, E. G. (2019). Investigating the combined effect of thermal barrier coating and antioxidants on pine oil in DI diesel engine. *Environmental Science and Pollution Research International*, *26*(15), 15573–15599. doi:10.100711356-019-04649-6 PMID:30945081

KEY TERMS AND DEFINITIONS

τ_{ij} Weighted Normalized Matrix:
τ^- Negative Ideal Solution:
τ^+ Positive Ideal Solution:
μ_{ij} The Normalized Value Of The Design Matrix:
$CaCO_3$: Calcium Carbonate:
CaO: Calcium Oxide
CC: Closeness Coefficient
CL: Catalyst Loading
FAA: Free Fatty Acid
FAME: Fatty Acid Methyl Ester
MCC: Mean Closeness Coefficient
MR: Molar Ration
RT: Reaction Temperature
S_i^- Negative Separation Measures:
S_i^+ Positive Separation Measures:
T_m: TOPSIS Design matrix:
TOPSIS: Technique for Order Preference by Similarity to Ideal Solution
XRD: X ray Diffraction

Compilation of References

(2012). Ni, P., & Wang, X. Modelling the formation of NOx and Soot emissions in a diesel engine at different humidity. *International Journal of Green Energy*, *9*(8), 815–828. doi:10.1080/15435075.2011.641701

Aalam, C. S., Saravanan, C., & Kannan, M. (2015). Experimental investigations on a CRDI system assisted diesel engine fuelled with aluminium oxide nanoparticles blended biodiesel. *Alexandria engineering journal, 54*(3), 351-358.

Abbaszaadeh, A., Ghobadian, B., Omidkhah, M. R., & Najafi, G. (2012). Current biodiesel production technologies: A comparative review. *Energy Conversion and Management*, *63*, 138–148. doi:10.1016/j.enconman.2012.02.027

Abed, K. A., El Morsi, A. K., Sayed, M. M., El Shaib, A. A., & Gad, M. S. (2018). Effect of waste cooking oil biodiesel on performance and exhaust emissions of a diesel engine. *Egyptian Journal of Petroleum*, *27*(4), 985–989. doi:10.1016/j.ejpe.2018.02.008

Abu-Jrai, A., Tsolakis, A., & Megaritis, A. (2007). The influence of H2 and CO on diesel engine combustion characteristics, exhaust gas emissions, and after treatment selective catalytic NOx reduction. *International Journal of Hydrogen Energy*, *32*(15), 3565–3571. doi:10.1016/j.ijhydene.2007.02.014

Abu-Zaid, M., Badran, O., & Yamin, J. (2004). Effect of Methanol Addition on the Performance of Spark Ignition Engines. *Energy & Fuels*, *18*(2), 312–315. doi:10.1021/ef030103d

Agarwal, A. K. (2007). Biofuels (alcohols and biodiesel) applications as fuels for internal combustion engines. *Progress in Energy and Combustion Science*, *33*(3), 233–271. doi:10.1016/j.pecs.2006.08.003

Agarwal, D., Kumar, L., & Agarwal, A. K. (2008). Performance evaluation of a vegetable oil fuelled compression ignition engine. *Renewable Energy*, *33*(6), 1147–1156. doi:10.1016/j.renene.2007.06.017

Agarwal, D., Sinha, S., & Agarwal, A. K. (2006). Experimental investigation of controlNO$_x$ emissions in biodiesel-fuelled compression ignition engine. *Renewable Energy*, *31*(14), 2356–2369. doi:10.1016/j.renene.2005.12.003

Agbor, V. B., Cicek, N., Sparling, R., Berlin, A., & Levin, D. B. (2011). Biomass pretreatment: Fundamentals toward application. *Biotechnology Advances*, *29*(6), 675–685. doi:10.1016/j.biotechadv.2011.05.005 PMID:21624451

Ahmad, A. A., Zawawi, N. A., Kasim, F. H., Inayat, A., & Khasri, A. (2016). Assessing the gasification performance of biomass: A review on biomass gasification process conditions, optimization and economic evaluation. *Renewable & Sustainable Energy Reviews*, *53*, 1333–1347. doi:10.1016/j.rser.2015.09.030

Ahmed, S., Hassan, M. H., Kalam, M. A., Rahman, S. M. A., Abedin, M. J., & Shahir, A. (2014). An experimental investigation of biodiesel production, characterization, engine performance, emission and noise of Brassica juncea methyl ester and its blends. *Journal of Cleaner Production*, *79*, 74–81. doi:10.1016/j.jclepro.2014.05.019

Akarte, R. R. (2004). An overview of energy and issues of future', In *Proceedings First National Conference on Energy and Fuel Issues of Future, PIET, pp. 150-160,* Pune, India.

Alagu, R. M., & Ganapathy Sundaram, E. (2018). Preparation and characterization of pyrolytic oil through pyrolysis of neem seed and study of performance, combustion and emission characteristics in CI engine. *Journal of the Energy Institute*, *91*(1), 100–109. doi:10.1016/j.joei.2016.10.003

Alahmer, A., Yamin, J., Sakhrich, A., & Hamdan, M. A. (2010). Engine performance using emulsified diesel fuel. *Energy Conversion and Management*, *51*(8), 1708–1713. doi:10.1016/j.enconman.2009.11.044

Alenezi, R., Santos, R. C. D., Raymahasay, S., & Leeke, G. A. (2013). Improved biodiesel manufacture at low temperature and short reaction time. *Renewable Energy*, *53*, 242–248. doi:10.1016/j.renene.2012.11.019

Al-Farayedhi, A. A., Al-Dawood, A. M., & Gandhidasan, P. (2004). Experimental Investigation of SI Engine Performance Using Oxygenated Fuel. *Journal of Engineering for Gas Turbines and Power*, *126*(1), 178–191. doi:10.1115/1.1615254

Alonso, D. M., Bond, J. Q., & Dumesic, J. A. (2010). Catalytic conversion of biomass to biofuels. *Green Chemistry*, *12*(9), 1493–1513. doi:10.1039/c004654j

Alptekin, E. (2017). Emission, injection and combustion characteristics of biodiesel and oxygenated fuel blends in a common rail diesel engine. *Energy*, *119*, 44–52. doi:10.1016/j.energy.2016.12.069

Altun, S., Oner, C., Yasar, F., & Adin, H. (2011). Effect of n-Butanol blending with a blend of diesel and biodiesel on performance and exhaust emissions of a diesel engine. *Industrial & Engineering Chemistry Research*, *50*(15), 9425–9430. doi:10.1021/ie201023f

Annamalai, M., Dhinesh, B., Nanthagopal, K., SivaramaKrishnan, P., Lalvani, J. I. J., Parthasarathy, M., & Annamalai, K. (2016). An assessment on performance, combustion and emission behaviour of a diesel engine powered by ceria nanoparticle blended emulsified biofuel. *Energy Conversion and Management*, *123*, 372–380. doi:10.1016/j.enconman.2016.06.062

Armas, O., Yehliu, K., & Boehman, A. L. (2010). Effect of alternative fuels on exhaust emissions during diesel engine operation with matched combustion phasing. *Fuel*, *89*(2), 438–456. doi:10.1016/j.fuel.2009.09.022

Armenta, R. E., Vinatoru, M., Burja, A. M., Kralovec, J. A., & Barrow, C. J. (2007). Transesterification of fish oil to produce fatty acid ethyl esters using ultrasonic energy. JAOCS. *Journal of the American Oil Chemists' Society*, *84*(11), 1045–1052. doi:10.100711746-007-1129-2

Arockiasamy, P., & Anand, R. B. (2015). Performance, combustion and emission characteristics of a D.I. diesel engine fuelled with nanoparticle blended jatropha biodiesel. *Period. Polytech. Mech. Eng.*, *59*(2), 88–93. doi:10.3311/PPme.7766

Aro, E.-M. (2016). From first generation biofuels to advanced solar biofuels. *Ambio*, *45*(S1Suppl 1), S24–S31. doi:10.100713280-015-0730-0 PMID:26667057

Arroyo, J., Moreno, F., Muñoz, M., Monné, C., & Bernal, N. (2014). Combustion behavior of a spark ignition engine fueled with synthetic gases derived from biogas. *Fuel*, *117*, 50–58. doi:10.1016/j.fuel.2013.09.055

Ashok, B., Nanthagopal, K., Jeevanatham, A. K., Bhowmick, P., Malhotra, D., & Agarwal, P. (2017). An assessment of Calophyllum Inophyllum biodiesel fuelled diesel engine characteristics using novel antioxidant additives. *Energy Conversion and Management*, *148*, 953–943. doi:10.1016/j.enconman.2017.06.049

Ashok, B., Thundil Karuppa Raj, R., Nanthagopal, K., Krishnan, R., & Subbarao, R. (2017). Lemon peel oil – A novel renewable alternative energy source for diesel engine. *Energy Conversion and Management*, *139*, 110–121. doi:10.1016/j.enconman.2017.02.049

Atabani, A. E., Shobana, S., Mohammed, M. N., Uğuzd, G., Kumar, G., Arvindnarayan, S., ... Al-Muhtase, A. H. (2019). Integrated valorization of waste cooking oil and spent coffee grounds for biodiesel production: Blending with higher alcohols, FT–IR, TGA, DSC and NMR characterizations. *Fuel*, *244*, 419–430. doi:10.1016/j.fuel.2019.01.169

Atmanli, A., Ileri, E., Yuksel, B., & Yilmaz, N. (2015). Extensive analyses of diesel-vegetable oil-n-butanol ternary blends in a diesel engine. *Applied Energy*, *145*, 155–162. doi:10.1016/j.apenergy.2015.01.071

Attia, A. M. A., & Kulchitskiy, A. R. (2014). Influence of the structure of water-in-fuel emulsion on diesel engine performance. *Fuel*, *116*, 703–708. doi:10.1016/j.fuel.2013.08.057

Avinash, A., & Murugesan, A., (2018), Advantages & disadvantages of biodiesel, Indian copyright, ROC Number: L-76839/2018.

Avinash, A., & Murugesan, A. (2018). Prediction capabilities of mathematical models in producing a renewable fuel from waste cooking oil for sustainable energy and clean environment. *Fuel*, *216*, 322–329. doi:10.1016/j.fuel.2017.12.029

Avinash, A., Subramaniam, D., & Murugesan, A. (2014). Bio-diesel - A global scenario. *Renewable & Sustainable Energy Reviews*, *29*, 517–527. doi:10.1016/j.rser.2013.09.007

Aydin, H., & Bayindir, H. (2010). Performance and emission analysis of cottonseed oil methyl ester in a diesel engine. *Renewable Energy*, *35*(3), 588–592. doi:10.1016/j.renene.2009.08.009

Ayyasamy, T., Balamurugan, K., & Duraisamy, S. (2018). Production, performance and emission analysis of Tamanu oil-diesel blends along with biogas in a diesel engine in dual cycle mode. *International Journal of Energy Technology and Policy*, *14*(1), 4–19. doi:10.1504/IJETP.2018.088279

Azcan, N., & Danisman, A. (2007). Alkali catalyzed transesterification of cottonseed oil by microwave irradiation. *Fuel*, *86*(17–18), 2639–2644. doi:10.1016/j.fuel.2007.05.021

Azcan, N., & Danisman, A. (2008). Microwave assisted transesterification of rapeseed oil. *Fuel*, *87*(10–11), 1781–1788. doi:10.1016/j.fuel.2007.12.004

Azimov, U., Tomita, E., Kawahara, N., & Harada, Y. (2011). Effect of syngas composition on combustion and exhaust emission characteristics in a pilot-ignited dual-fuel engine operated in PREMIER combustion mode. *International Journal of Hydrogen Energy*, *36*(18). doi:10.1016/j.ijhydene.2011.04.192

Babu, R. V., & Sendilvelan, S. (2011). Reduction of pollutants in CI engine using emulsion fuels to reduce overall traffic-induced emissions. *Global Journal of Research Engineering, 11*(2).

Bailey, B., Eberhardt, J., Goguen, S., & Erwin, J. (1997). Diethyl Ether (DEE) as a Renewable Diesel Fuel. *SAE Transactions*. SAE Technical Paper 972978. doi:10.4271/972978

Bakar, S. A., Yusup, S., Ahmad, M. M., Quitain, A. T., Sasaki, M., Goto, M., ... Ahmad, J. (2014). Comparative Study on Two-Step Fatty Acid Methyl Ester (FAME) Production from High FFA Crude Palm Oil Using Microwave Technique and Conventional Technique. *Advanced Materials Research*, *917*, 87–95. doi:10.4028/www.scientific.net/AMR.917.87

Balaji, G., & Cheralathan, M. (2015). Effect of CNT as additive with biodiesel on the performance and emission characteristics of a DI diesel engine. *International Journal of Chemtech Research*, *7*(3), 1230–1236.

Balamurugan, K., Tamilvanan, A., Anbarasu, M., Akil, M., & Srihari, S. (2013). Nano-copper additive for reducing NOx emission in soya bean biodiesel-fuelled CI engine. *Journal of Biofuels*, *4*(1), 1–8. doi:10.5958/j.0976-4763.4.1.001

Balat, M. (2009). Gasification of biomass to produce gaseous products. *Energy Sources. Part A*, *31*(6), 516–526.

Balat, M. (2011). Production of bioethanol from lignocellulosic materials via the biochemical pathway: A review. *Energy Conversion and Management*, *52*(2), 858–875. doi:10.1016/j.enconman.2010.08.013

Balat, M., Balat, H., & Öz, C. (2008). Progress in bioethanol processing. *Progress in Energy and Combustion Science*, *34*(5), 551–573. doi:10.1016/j.pecs.2007.11.001

Banapurmath, N. R., Tewari, P. G., & Hosmath, R. S. (2008). Performance and emission characteristics of a DI compression ignition engine operated on Honge, Jatropha and sesame oil methyl esters. *Renewable Energy*, *33*(9), 1982–1988. doi:10.1016/j.renene.2007.11.012

Banga, S., & Varshney, P. K. (2010). Effect of impurities on performance of biodiesel: A review. *Journal of Scientific and Industrial Research*, *69*(8), 575–579.

Ban-Weiss, G. A., Chen, J. Y., Buchholz, B. A., & Dibble, R. W. (2007). A numerical investigation into the anomalous slight NO_x increase when burning biodiesel; A new (old) theory. *Fuel Processing Technology*, *88*(7), 659–667. doi:10.1016/j.fuproc.2007.01.007

Barabás, I., & Todoruţ, I.-A. (2011). Biodiesel Quality, Standards and Properties. In G. Montero (Ed.), *Biodiesel Quality, Emissions, and By-Products* (pp. 1–28). InTech. doi:10.5772/25370

Baranitharan, P., Ramesh, K., & Sakthivel, R. (2019a). Analytical characterization of the Aegle marmelos pyrolysis products and investigation on the suitability of bio-oil as a third generation bio-fuel for C.I engine. *Environmental Progress & Sustainable Energy*, *38*(4). doi:10.1002/ep.13116

Baranitharan, P., Ramesh, K., & Sakthivel, R. (2019b). Measurement of performance and emission distinctiveness of Aegle marmelos seed cake pyrolysis oil/diesel/TBHQ opus powered in a DI diesel engine using ANN and RSM. *Measurement*, *144*, 366–380. doi:10.1016/j.measurement.2019.05.037

Baranitharan, P., Ramesh, K., & Sakthivel, R. (2019c). Multi-attribute decision-making approach for Aegle marmelos pyrolysis process using TOPSIS and Grey Relational Analysis: Assessment of engine emissions through novel Infrared thermography. *Journal of Cleaner Production*, *234*, 315–328. doi:10.1016/j.jclepro.2019.06.188

Barik, D., & Sivalingam, M. (2013). Performance and emission characteristics of a biogas fueled DI diesel engine. Technical report, SAE Technical Paper 2013–01–2507. Retrieved from http://papers.sae.org/2013-01-2507/

Barik, D., & Sivalingam, M. (2014). Investigation on performance and exhaust emissions characteristics of a DI diesel engine fueled with Karanja methyl ester and biogas in dual fuel mode, Technical report, SAE Technical Paper 2014-01-1311 Retrieved from http://papers.sae.org/2014-01-1311/

Barik, D., & Murugan, S. (2014a). Investigation on combustion performance and emission characteristics of a DI (direct injection) diesel engine fueled with biogas-diesel in dual fuel mode. *Energy*, *72*, 760–771. doi:10.1016/j.energy.2014.05.106

Barik, D., & Murugan, S. (2014b). Simultaneous reduction of NOx and smoke in a dual fuel DI diesel engine. *Energy Conversion and Management*, *84*, 217–226. doi:10.1016/j.enconman.2014.04.042

Barik, D., & Murugan, S. (2016). Effects of diethyl ether (DEE) injection on combustion performance and emission characteristics of Karanja methyl ester (KME)-biogas fueled dual fuel diesel engine. *Fuel*, *164*, 286–296. doi:10.1016/j.fuel.2015.09.094

Bär, J., Phongpreecha, T., Singh, S. K., Kral Yilmaz, M., Foster, C. E., Crowe, J. D., & Hodge, D. B. (2018). Deconstruction of hybrid poplar to monomeric sugars and aromatics using ethanol organosolv fractionation. *Biomass Conversion and Biorefinery*, *8*(4), 813–824. doi:10.100713399-018-0330-x

Basha, J. S. (2018). Impact of Carbon Nanotubes and Di-Ethyl Ether as additives with biodiesel emulsion fuels in a diesel engine – An Experimental Investigation. *Journal of the Energy Institute*, *91*(2), 289–303. doi:10.1016/j.joei.2016.11.006

Basha, J. S., & Anand, R. (2011). An experimental investigation in a diesel engine using carbon nanotubes blended water–diesel emulsion fuel. *Proceedings of the Institution of Mechanical Engineers. Part A, Journal of Power and Energy*, *225*(3), 279–288. doi:10.1177/2041296710394247

Basha, J. S., & Anand, R. (2013). The influence of nano additive blended biodiesel fuels on the working characteristics of a diesel engine. *Journal of the Brazilian Society of Mechanical Sciences and Engineering*, *35*(3), 257–264. doi:10.100740430-013-0023-0

Basha, J. S., & Anand, R. B. (2012). Effects of nanoparticle additive in the water-diesel emulsion on the performance, emission and combustion characteristics of a diesel engine. *International Journal of Vehicle Design*, *59*(2/3), 164–181. doi:10.1504/IJVD.2012.048692

Basha, J. S., & Anand, R. B. (2014). An experimental study in a CI engine using nano additive blended water-diesel emulsion. *International Journal of Green Energy*, *8*(3), 332–348. doi:10.1080/15435075.2011.557844

Basha, J. S., & Anand, R. B. (2014). Performance, combustion, and emission characteristics of diesel engine using carbon nano tubes blended jatropha methyl ester emulsions. *Alexandria Engineering Journal*, *53*, 259–273.

Basha, J. S., & Anand, R. B. (2014). Performance, emission and combustion characteristics of a diesel engine using carbon nanotubes blended jatropha methyl ester emulsions. *Alexandria Engineering Journal*, *53*(2), 259–273. doi:10.1016/j.aej.2014.04.001

Basha, J. S., Montaha, A. S. A., Khadija, M. M. A., Manal, Y. M. A., Sulaiman, A. S. A., Shahad, M. K. A., & Rashid, N. S. A. (2019). The Influence of Hexanol and Methyl Acetate as Oxygenated Additives with Diesel Fuel in a Diesel Engine Generator. *International Journal of I.C. Engines and Gas Turbines*, *5*(1), 1–9.

Basinger, M., Reding, T., Rodriguez-Sanchez, F. S., Lackner, K. S., & Modi, V. (2010). Durability testing modified compression ignition engines fueled with straight plant oil. *Energy*, *35*(8), 3204–3220. doi:10.1016/j.energy.2010.04.004

Baskar, P., & Senthil Kumar, A. (2017). Experimental investigation on performance characteristics of a diesel engine using diesel-water emulsion with oxygen enriched air. *Alexandria Engineering Journal*, *56*(1), 137–146. doi:10.1016/j.aej.2016.09.014

Basu, S., Khan, A. L., Cano-Odena, A., Liu, C., & Vankelecom, I. F. J. (2010). Membrane-based technologies for biogas separations. *Chemical Society Reviews*, *39*(2), 750–768. doi:10.1039/B817050A PMID:20111791

Basu, S., & Miglani, A. (2016). Combustion and heat transfer characteristics of nanofluid fuel droplets: A short review. *International Journal of Heat and Mass Transfer*, *96*, 482–503. doi:10.1016/j.ijheatmasstransfer.2016.01.053

Bedoya, I. D., Saxena, S., Cadavid, F. J., & Dibble, R. W. (2012). Exploring Strategies for Reducing High Intake Temperature Requirements and Allowing Optimal Operational Conditions in a Biogas Fueled HCCI Engine for Power Generation. *Journal of Engineering for Gas Turbines and Power*, *134*(7). doi:10.1115/1.4006075

Bedoya, I. D., Saxena, S., Cadavid, F. J., Dibble, R. W., & Wissink, M. (2012). Experimental evaluation of strategies to increase the operating range of a biogas-fueled HCCI engine for power generation. *Applied Energy*, *97*, 618–629. doi:10.1016/j.apenergy.2012.01.008

Bedoya, I. D., Saxena, S., Cadavid, F. J., Dibble, R. W., & Wissink, M. (2012). Experimental study of biogas combustion in an HCCI engine for power generation with high indicated efficiency and ultra-low NOx emissions. *Energy Conversion and Management*, *53*(1), 154–162. doi:10.1016/j.enconman.2011.08.016

Bender, M. H. (2000). Potential conservation of biomass in the production of synthetic organics. *Resources, Conservation, and Recycling*, *30*(1), 49–58. doi:10.1016/S0921-3449(00)00045-8

Bhale, P. V., Deshpande, N. V., & Thombre, B. S. (2009). Improving the low temperature properties of biodiesel fuel. *Renewable Energy*, *34*(3), 794–800. doi:10.1016/j.renene.2008.04.037

Bhaskar, K., Nagarajan, G., & Sampath, S. (2014). The performance and emission characteristics of fish oil methyl esters and diesel blends in a partially premixed charge compression ignition engine. *International Journal of Green Energy*, *11*(4), 389–403. doi:10.1080/15435075.2013.773514

Bhattacharya, T., Mishra, T., & Singh, B. (1988). Techniques for removal of CO2 and H2S from biogas. *XXIV Annual Convention of ISAE, Held at PKV, Akola*. Retrieved from https://scholar.google.co.in/scholar?hl=en&q=Techniques+for+removal+of+CO2+and+H2S+from+biogas&btnG=#0

Bhowmick, P., Jeevanantham, A. K., Ashok, B., Nanthagopal, K., Perumal, D. A., Karthickeyan, V., ... Jain, A. (2019). Effect of fuel injection strategies and EGR on biodiesel blend in a CRDI engine. *Energy*, *181*, 1094–1113. doi:10.1016/j.energy.2019.06.014

Bhuiya, M. M. K., Rasul, M. G., Khan, M. M. K., Ashwath, N., Azad, A. K., & Hazrat, M. A. (2016). Prospects of 2nd generation biodiesel as a sustainable fuel – Part 2: Properties, performance and emission characteristics. *Renewable & Sustainable Energy Reviews*, *55*, 1129–1146. doi:10.1016/j.rser.2015.09.086

Binder, J. B., & Raines, R. T. (2010). Fermentable sugars by chemical hydrolysis of biomass. *Proceedings of the National Academy of Sciences of the United States of America*, *107*(10), 4516–4521. doi:10.1073/pnas.0912073107 PMID:20194793

Biofuel Systems Group Limited. Beech Croft, 51 Ruff Lane, Ormskirk, Lancashire, England, L39 4UL. Retrieved from https://www.biofuelsystems.com/ about.htm

Biswas, T., Kartha, A., & Pundarikakhadu, R. (1977). Removal of carbon dioxide from biogas. In *Proceedings of National Symposium on Biogas Technology and Uses*. New Delhi, India: IARI; Retrieved from https://scholar.google.co.in/scholar?q=Removal+of+carbon+dioxide+from+biogas+biswas&btnG=&hl=en&as_sdt=0%2C5#0

Blakey, S., Rye, L., & Wilson, C. W. (2011). Aviation gas turbine alternative fuels: A review. *Proceedings of the Combustion Institute*, *33*(2), 2863–2885. doi:10.1016/j.proci.2010.09.011

Boden, T., Marland, G., & Andres, R. (2017). *National CO2 emissions from fossil-fuel burning, cement manufacture, and gas flaring: 1751-2014*. Carbon Dioxide Information Analysis Center, Oak Ridge National Laboratory, US Department of Energy.

Bond, J. Q., Upadhye, A. A., Olcay, H., Tompsett, G. A., Jae, J., Xing, R., ... Huber, G. W. (2014). Production of renewable jet fuel range alkanes and commodity chemicals from integrated catalytic processing of biomass. *Energy & Environmental Science*, *7*(4), 1500–1523. doi:10.1039/C3EE43846E

Bora, B. J., & Saha, U. K. (2015). Improving the Performance of a Biogas Powered Dual Fuel Diesel Engine Using Emulsified Rice Bran Biodiesel as Pilot Fuel Through Adjustment of Compression Ratio and Injection Timing. *Journal of Engineering for Gas Turbines and Power*, *137*(9). doi:10.1115/1.4029708

Bora, B. J., Saha, U. K., Chatterjee, S., & Veer, V. (2014). Effect of compression ratio on performance, combustion and emission characteristics of a dual fuel diesel engine run on raw biogas. *Energy Conversion and Management*, *87*, 1000–1009. doi:10.1016/j.enconman.2014.07.080

Borris, C. C., Carlos, M., Aida, D.-S., & Paloma, A. (2014). Effects of the addition of oxygenated fuels as additives on combustion characteristics and particle number and size distribution emissions of a TDI diesel engine. *Fuel*, *132*, 93–100. doi:10.1016/j.fuel.2014.04.071

Bousbaa, H., Sary, A., Tazerout, M., & Liazid, A. (2012). Investigations on a compression ignition engine using animal fats and vegetable oil as fuels. *Journal of Energy Resources Technology*, *134*(2), 1–11. doi:10.1115/1.4005660

Brown, T. R., & Brown, R. C. (2013). A review of cellulosic biofuel commercial-scale projects in the United States. *Biofuels, Bioproducts, & Biorefining*, *7*(3), 235–245. doi:10.1002/bbb.1387

Brusca, S., Chiodo, V., Galvagno, A., Lanzafame, R., & Garrano, A. M. C. (2014). Analysis of reforming gas combustion in Internal Combustion Engine. *Energy Procedia*, *45*, 899–908. doi:10.1016/j.egypro.2014.01.095

Buyukkaya, E. (2010). Effects of biodiesel on a diesel engine performance, emission and combustion characteristics. *Fuel*, *89*(10), 3099–3105. doi:10.1016/j.fuel.2010.05.034

Cacua, K., Amell, A., & Cadavid, F. (2012). Effects of oxygen enriched air on the operation and performance of a diesel-biogas dual fuel engine. *Biomass and Bioenergy*, *45*, 159–167. doi:10.1016/j.biombioe.2012.06.003

Can, O., Celikten, I., & Usta, N. (2004). Effects of ethanol addition on performance and emissions of a turbocharged indirect injection diesel engine running at different injection pressures. *Energy Conversion and Management*, *45*(15-16), 2429–2440. doi:10.1016/j.enconman.2003.11.024

Cao, W., Han, H., & Zhang, J. (2005). Preparation of biodiesel from soybean oil using supercritical methanol and co-solvent. *Fuel*, *84*(4), 347–351. doi:10.1016/j.fuel.2004.10.001

Carlos, A., Guerrero, F., Guerrero-Romero, A., & Sierra, F. E. (2011). Biodiesel production from waste cooking oil, Biodiesel: Feedstocks and Processing Technologies InTech, pp. 35-36.

Cayli, G., & Kusefoglu, S. (2008). Increased yields in biodiesel production from used cooking oils by a two step process: Comparison with one step process by using TGA. *Fuel Processing Technology*, *89*(2), 118–122. doi:10.1016/j.fuproc.2007.06.020

Ceramicrotaryengines. (2019). Retrieved from www.ceramicrotaryengines.com

Cerveró, J. M., Álvarez, J. R., & Luque, S. (2014). Novozym 435-catalyzed synthesis of fatty acid ethyl esters from soybean oil for biodiesel production. *Biomass and Bioenergy*, *61*, 131–137. doi:10.1016/j.biombioe.2013.12.005

Chacartegui, R., Torres, M., Sánchez, D., Jiménez, F., Munoz, A., & Sánchez, T. (2011). Analysis of main gaseous emissions of heavy duty gas turbines burning several syngas fuels. *Fuel Processing Technology*, *92*(2), 213–220. doi:10.1016/j.fuproc.2010.03.014

Chandler, H. F. G., & Brown, G. I. (1992). Effect of cold flow additives on low temperature operability of diesel fuel. SAE Paper 9291186.

Chandler, M. F., Teng, Y., & Koylu, U. O. (2007). Diesel engine particulate emissions: A comparison of mobility and microscopy size measurements. *Proceedings of the Combustion Institute*, *31*(2), 2971–2979. doi:10.1016/j.proci.2006.07.200

Chandrasekaran, V., Arthanarisamy, M., Nachiappan, P., Dhanakotti, S., & Moorthy, B. (2016). The role of nano additives for biodiesel and diesel blended transportation fuels. *Transportation Research Part D, Transport and Environment*, *46*, 145–156. doi:10.1016/j.trd.2016.03.015

Charitha, V., Thirumalini, S., Prasad, M., & Srihari, S. (2018). Investigation on performance and emissions of rcci dual fuel combustion on diesel - bio diesel in a light duty engine. *Renewable Energy*. doi:10.1016/j.renene.2018.09.048

Chastek, T. Q. (2011). Improving cold flow properties of canola-based biodiesel. *Biomass and Bioenergy*, *35*(1), 600–607. doi:10.1016/j.biombioe.2010.10.024

Chaudhari, S., Bej, S., Bakhshi, N., & Dalai, A. (2001). Steam gasification of biomass-derived char for the production of carbon monoxide-rich synthesis gas. *Energy & Fuels*, *15*(3), 736–742. doi:10.1021/ef000278c

Chauhan, B. S., Kumar, N., & Cho, H. M. (2012). A study on the performance and emission of a diesel engine fueled with jatropha biodiesel oil and its blends. *Energy*, *37*(1), 616–622. doi:10.1016/j.energy.2011.10.043

Chaurasiya, P. K., Singh, S. K., Dwivedi, R., & Choudri, R. V. (2019). Combustion and emission characteristics of diesel fuel blended with raw jatropha, soybean and waste cooking oils. *Heliyon (London)*, *5*(5). doi:10.1016/j.heliyon.2019.e01564 PMID:31193108

Chee Loong, T., & Idris, A. (2017). One step transesterification of biodiesel production using simultaneous cooling and microwave heating. *Journal of Cleaner Production*, *146*, 57–62. doi:10.1016/j.jclepro.2016.03.155

Chen, A. F., Adzmi, M. A., Adam, A., Othman, M. F., Kamaruzzaman, M. K., & Mrwan, A. G. (2018). Combustion characteristics, engine performances and emissions of a diesel engine using nanoparticle-diesel fuel blends with aluminium oxide, carbon nanotubes and silicon oxide. *Energy Conversion and Management*, *171*, 461–477. doi:10.1016/j.enconman.2018.06.004

Cheng, J., Yu, T., Li, T., Zhou, J., & Cen, K. (2013). Using wet microalgae for direct biodiesel production via microwave irradiation. *Bioresource Technology*, *131*, 531–535. doi:10.1016/j.biortech.2013.01.045 PMID:23403061

Chen, H., Wang, J., Shuai, S., & Chen, W. (2008). Study of oxygenated biomass fuel blends on a diesel engine. *Fuel*, *87*(15-16), 3462–3468. doi:10.1016/j.fuel.2008.04.034

Chen, K. S., Lin, Y. C., Hsu, K. H., & Wang, H. K. (2012). Improving biodiesel yields from waste cooking oil by using sodium methoxide and a microwave heating system. *Energy*, *38*(1), 151–156. doi:10.1016/j.energy.2011.12.020

Cheruiyot, N. K., Hou, W.-C., Wang, L.-C., & Chen, C.-Y. (2019). The impact of low to high waste cooking oil-based biodiesel blends on toxic organic pollutant emissions from heavy-duty diesel engines. *Chemosphere*, *235*, 726–733. doi:10.1016/j.chemosphere.2019.06.233 PMID:31279123

Chiaramonti, D., Bonini, M., Fratini, E., Tondi, G., Gartner, K., Bridgwater, A. V., ... Baglioni, P. (2003a). Development of emulsions from biomass pyrolysis liquid and diesel and their use in engines—Part 1: Emulsion production. *Biomass and Bioenergy*, *25*(1), 85–99. doi:10.1016/S0961-9534(02)00183-6

Chiaramonti, D., Bonini, M., Fratini, E., Tondi, G., Gartner, K., Bridgwater, A. V., ... Baglioni, P. (2003b). Development of emulsions from biomass pyrolysis liquid and diesel and their use in engines—Part 2: Tests in diesel engines. *Biomass and Bioenergy*, *25*(1), 101–111. doi:10.1016/S0961-9534(02)00184-8

Choedkiatsakul, I., Ngaosuwan, K., Assabumrungrat, S., Mantegna, S., & Cravotto, G. (2015). Biodiesel production in a novel continuous flow microwave reactor. *Renewable Energy*, *83*, 25–29. doi:10.1016/j.renene.2015.04.012

Choi, B., Jiang, X., Kim, Y. K., Jung, G., Lee, C., Choi, I., & Song, C. S. (2015). Effect of diesel fuel blend with n-butanol on the emission of a turbocharged common rail direct injection diesel engine. *Applied Energy*, *146*, 20–28. doi:10.1016/j.apenergy.2015.02.061

Chotwichien, A., Luengnaruemitchai, A., & Jai-In, S. (2009). Utilization of palm oil alkyl esters as an additive in ethanol-diesel and butanol-diesel blends. *Fuel*, *88*(9), 1618–1624. doi:10.1016/j.fuel.2009.02.047

Christensen, M., & Johansson, B. (2000). Supercharged Homogeneous Charge Compression Ignition (HCCI) with Exhaust Gas Recirculation and Pilot Fuel. *SAE Technical Papers*, (724), SAE 2000-01-1835. doi:10.4271/2000-01-1835

Christensen, E., & McCormick, R. L. (2014). Long-term storage stability of biodiesel and biodiesel blends. *Fuel Processing Technology*, *128*, 339–348. doi:10.1016/j.fuproc.2014.07.045

Chrysikou, L. P., Dagonikou, V., Dimitriadis, A., & Bezergianni, S. (2019). Waste cooking oils exploitation targeting EU 2020 diesel fuel production: Environmental and economic benefits. *Journal of Cleaner Production*, *219*, 566–575. doi:10.1016/j.jclepro.2019.01.211

Chung, K. H., Kim, J., & Lee, K. Y. (2009). Biodiesel production by transesterification of duck tallow with methanol on alkali catalysts. *Biomass and Bioenergy*, *33*(1), 155–158. doi:10.1016/j.biombioe.2008.04.014

Chynoweth, D. P., Owens, J. M., & Legrand, R. (2000). Renewable methane from anaerobic digestion of biomass. *Renewable Energy*, *22*(1–3), 1–8. doi:10.1016/S0960-1481(00)00019-7

Cinar, C., Ozer, C., Fatih, S., & Serdar, Y. H. (2010). Effects of premixed diethyl ether (DEE) on combustion and exhaust emissions in a HCCI-DI diesel engine. *Applied Thermal Engineering*, *30*(4), 360–365. doi:10.1016/j.applthermaleng.2009.09.016

Cole, C. V., Duxbury, J., Freney, J., Heinemeyer, O., Minami, K., Mosier, A., ... Zhao, Q. (1997). Global estimates of potential mitigation of greenhouse gas emissions by agriculture. *Nutrient Cycling in Agroecosystems*, *49*(1), 221–228. doi:10.1023/A:1009731711346

Colucci, J. A., Borrero, E. E., & Alape, F. (2005). Biodiesel from an alkaline transesterification reaction of soybean oil using ultrasonic mixing. *JAOCS. Journal of the American Oil Chemists' Society*, *82*(7), 525–530. doi:10.100711746-005-1104-3

Corma, A., de la Torre, O., Renz, M., & Villandier, N. (2011). Production of High-Quality Diesel from Biomass Waste Products. *Angewandte Chemie International Edition*, *50*(10), 2375–2378. doi:10.1002/anie.201007508 PMID:21351358

Costa, R. C., & Sodré, J. R. (2011). Compression ratio effects on an ethanol/gasoline fuelled engine performance. *Applied Thermal Engineering*, *31*(2), 278–283. doi:10.1016/j.applthermaleng.2010.09.007

Dayong, J., & Yun, B. B. (2012). Performances on engine-out emissions and combustion of ethylene glycol monobutyl ether palm oil monoester as a new type biodiesel. *Journal of Renewable and Sustainable Energy*, *4*, 1–10.

Demirbas, A. (2000). A direct route to the calculation of heating values of liquid fuels by using their density and viscosity measurements. *Energy Conversion and Management*, *41*(15), 1609–1614. doi:10.1016/S0196-8904(00)00012-1

Demirbas, A. (2004). Combustion characteristics of different biomass fuels. *Progress in Energy and Combustion Science*, *30*(2), 219–230. doi:10.1016/j.pecs.2003.10.004

Demirbas, A. (2005). Biodiesel production from vegetable oils by supercritical methanol. *Journal of Scientific and Industrial Research*, *64*(11), 858–865.

Demirbas, A. (2007). Progress and recent trends in biofuels. *Progress in Energy and Combustion Science*, *33*(1), 1–18. doi:10.1016/j.pecs.2006.06.001

Demirbas, A. (2008). Studies on cottonseed oil biodiesel prepared in non-catalytic SCF conditions. *Bioresource Technology*, *99*(5), 1125–1130. doi:10.1016/j.biortech.2007.02.024 PMID:17433669

Demirbas, A. (2009). Biodiesel from waste cooking oil via base-catalytic and supercritical methanol transesterification. *Energy Conversion and Management*, *50*(4), 923–927. doi:10.1016/j.enconman.2008.12.023

Demirbas, A. (2011). Competitive liquid biofuels from biomass. *Applied Energy*, *88*(1), 17–28. doi:10.1016/j.apenergy.2010.07.016

Demirbas, A., & Demirbas, I. (2007). Importance of rural bioenergy for developing countries. *Energy Conversion and Management*, *48*(8), 2386–2398. doi:10.1016/j.enconman.2007.03.005

Demirbas, M. F. (2006). Current technologies for biomass conversion into chemicals and fuels. *Energy Sources. Part A, Recovery, Utilization, and Environmental Effects*, *28*(13), 1181–1188. doi:10.1080/00908310500434556

Demirbas, M., Balat, M., & Balat, H. (2011). Biowastes-to-biofuels. *Energy Conversion and Management*, *52*(4), 1815–1828. doi:10.1016/j.enconman.2010.10.041

Dept. of Animal Husbandry Dairying & Fisheries. (2012). *19th Livestock Census-2012: All India Report. Ministry of Agriculture*. Retrieved from http://dahd.nic.in/sites/default/files/Livestock 5.pdf

Devarajan, Y., Munuswamy, D. B., & Mahalingam, A. (2017). Performance, combustion and emission analysis on the effect of ferrofluid on neat biodiesel. *Process Safety and Environmental Protection*, *111*, 283–291. doi:10.1016/j.psep.2017.07.021

Dhamodaran, G., & Esakkimuthu, G. (2018). A Comparison Study on Emission Characteristics of Using Higher Alcohol Oxygenates with Gasoline in a Multipoint Fuel Injection Spark-Ignition Engine. *Journal of Testing and Evaluation*, *48*(1), 1–13.

Dhanasekaran, R., Krishnamoorthy, V., Rana, D., Saravanan, S., Nagendran, A., & Kumar, B. R. (2017). A sustainable and eco-friendly fueling approach for direct-injection diesel engines using restaurant yellow grease and n-pentanol in blends with diesel fuel. *Fuel*, *193*, 419–431. doi:10.1016/j.fuel.2016.12.030

Dhar, A., Kevin, R., & Agarwal, A. K. (2012). Production of biodiesel from high-FFA neem oil and its performance, emission and combustion characterization in a single cylinder DICI engine. *Fuel Processing Technology*, *97*, 118–129. doi:10.1016/j.fuproc.2012.01.012

Di Serio, M., Ledda, M., Cozzolino, M., Minutillo, G., Tesser, R., & Santacesaria, E. (2006). Transesterification of soybean oil to biodiesel by using heterogeneous basic catalysts. *Industrial & Engineering Chemistry Research*, *45*(9), 3009–3014. doi:10.1021/ie051402o

Diebold, J. P., & Czernik, S. (1997). Additives To Lower and Stabilize the Viscosity of Pyrolysis Oils during Storage. *Energy & Fuels*, *11*(5), 1081–1091. doi:10.1021/ef9700339

Dinesha, P., Kumar, S., & Rosen, M. A. (2019). Combined effects of water emulsion and diethyl ether additive on combustion performance and emission of a compression ignition engine using biodiesel blends. *Energy*, *179*, 928–937. doi:10.1016/j.energy.2019.05.071

Dinesh, K., Tamilvanan, A., Vaishnavi, S., Gopinath, M., & Mohan, K. R. (2019). Biodiesel production using Calophyllum inophyllum (Tamanu) seed oil and its compatibility test in a CI engine. *Biofuels*, *10*(3), 347–353. doi:10.1080/17597269.2016.1187543

Ding, Y., Chen, H., Wang, L., & Yang, C. (2007).. . *Heat Transfer Intensification Using Nanofluids*, *25*(25), 23–38.

Diórdinis, T., Lanzanova, M., Dalla, M., Eduardo, M., Martins, S., Romeu, P., ... Maria, S. (2019). The effects of residual gas trapping on part load performance and emissions of a spark ignition direct injection engine fuelled with wet ethanol. *Applied Energy*, *253*(January), 113508. doi:10.1016/j.apenergy.2019.113508

Di, Y., Cheung, C., & Huang, Z. (2009). Experimental study on particulate emission of a diesel engine fuelled with blended ethanol-dodecanol-diesel. *Journal of Aerosol Science*, *40*(2), 101–112. doi:10.1016/j.jaerosci.2008.09.004

Dogan, O. (2011). The influence of n-butanol/diesel fuel blends utilization on a small diesel engine performance and emissions. *Fuel*, *90*(7), 2467–2472. doi:10.1016/j.fuel.2011.02.033

Domkundwar, V. M. (2009). *A course in internal combustion engines*. Dhanpat Rai & Co.

Douvartzides, S. L., Charisiou, N. D., Papageridis, K. N., & Goula, M. A. (2019). Green Diesel: Biomass Feedstocks, Production Technologies, Catalytic Research, Fuel Properties and Performance in Compression Ignition Internal Combustion Engines. *Energies*, *12*(5), 809. doi:10.3390/en12050809

Dubé, M. A., Tremblay, A. Y., & Liu, J. (2007). Biodiesel production using a membrane reactor. *Bioresource Technology*, *98*(3), 639–647. doi:10.1016/j.biortech.2006.02.019 PMID:16580197

Dubey, A. (2000). Wet scrubbing of carbon dioxide. *Annual Report of CIAE, Bhopal (India)*. Retrieved from https://scholar.google.co.in/scholar?q=Wet+scrubbing+of+carbon+dioxide+dubey&btnG=&hl=en&as_sdt=0%2C5#0

Duc, P. M., & Wattanavichien, K. (2007). Study on biogas premixed charge diesel dual fuelled engine. *Energy Conversion and Management*, *48*(8), 2286–2308. doi:10.1016/j.enconman.2007.03.020

Duraisamy, G., Rangasamy, M., & Govindan, N. (2020). A comparative study on methanol/diesel and methanol/PODE dual fuel RCCI combustion in an automotive diesel engine. *Renewable Energy*, *145*, 542–556. doi:10.1016/j.renene.2019.06.044

Du, W., Xu, Y. Y., Liu, D. H., & Li, Z. B. (2005). Study on acyl migration in immobilized lipozyme TL-catalyzed transesterification of soybean oil for biodiesel production. *Journal of Molecular Catalysis. B, Enzymatic*, *37*(1–6), 68–71. doi:10.1016/j.molcatb.2005.09.008

Duz, M. Z., Saydut, A., & Ozturk, G. (2011). Alkali catalyzed transesterification of safflower seed oil assisted by microwave irradiation. *Fuel Processing Technology*, *92*(3), 308–313. doi:10.1016/j.fuproc.2010.09.020

Dwivedi, G., Jain, S., & Sharma, M. P. (2013). Diesel engine performance and emission analysis using biodiesel from various oil sources. *J Mater Environ Sci*, *4*, 434–447.

Eevera, T., Rajendran, K., & Saradha, S. (2009). Biodiesel production process optimization and characterization to assess the suitability of the product for varied environmental conditions. *Renewable Energy*, *34*(3), 762–765. doi:10.1016/j.renene.2008.04.006

EIA, U. (2015). Total Petroleum and Other Liquids Production 2014.

Elfasakhany, A. (2015). Investigations on the effects of ethanol–methanol–gasoline blends in a spark-ignition engine: Performance and emissions analysis. *Engineering Science and Technology, an International Journal, 18*(4), 713-719.

Elfasakhany, A. (2016). Performance and emissions of spark-ignition engine using ethanol–methanol–gasoline, n-butanol–iso-butanol–gasoline and iso-butanol–ethanol–gasoline blends: A comparative study. *Engineering Science and Technology, an International Journal, 19*(4), 2053-2059.

Elfasakhany, A., & Mahrous, A.-F. (2016). Performance and emissions assessment of n-butanol–methanol–gasoline blends as a fuel in spark-ignition engines. *Alexandria Engineering Journal*, *55*(3), 3015–3024. doi:10.1016/j.aej.2016.05.016

El-Galad, M. I., El-Khatib, K. M., & Zaher, F. A. (2015). Economic feasibility study of biodiesel production by direct esterification of fatty acids from the oil and soap industrial sector. *Egyptian Journal of Petroleum*, *24*(4), 455–460. doi:10.1016/j.ejpe.2015.06.002

El-Gendy, N. Sh., Deriase, S. F., Hamdy, A., & Abdallah, R. I. (2015). Statistical optimization of biodiesel production from sunflower waste cooking oil using basic heterogeneous biocatalyst prepared from egg shells. *Egyptian Journal of Petroleum*, *24*(1), 37–48. doi:10.1016/j.ejpe.2015.02.004

Elliott, D. C. (2007). Historical Developments in Hydroprocessing Bio-oils. *Energy & Fuels*, *21*(3), 1792–1815. doi:10.1021/ef070044u

El-Seessy, A. I. (2018). Atiat AMA and El-Batsh HM. The effect of aluminum oxide nanoparticles addition with jojoba methyl ester –diesel fuel blends on a diesel engine performance, combustion, and emission characteristics. *Fuel*, *224*, 147–166. doi:10.1016/j.fuel.2018.03.076

EL-Seesy. A. I., Abdel-Rahman, A. K., Hassan, H., Ookawara, S., & Hawi, M. (2017). *Investigation of the performance of a diesel engine fueled by biodiesel-diesel fuel mixture with addition of nanoparticles.* Paper presented at the ASME 2017 Power Conference Joint With ICOPE-17 collocated with the ASME 2017 11th International Conference on Energy Sustainability, the ASME 2017 15th International Conference on Fuel Cell Science, Engineering and Technology, and the ASME 2017 Nuclear Forum.

Encinar, J. M., Gonzalez, J. F., & Rodriguez-Reinares, A. (2007). Ethanolysis of used frying oil. Biodiesel preparation and characterization. *Fuel Processing Technology*, *88*(5), 513–522. doi:10.1016/j.fuproc.2007.01.002

Energy Outlook, B. P. 2030, London, (2012, January). Retrieved from https://www.bp.com/ content/dam/bp/en/corporate/pdf/energy-economics/energy-outlook/bp-energy-outlook-2012.pdf

Enweremadu, C. C., & Rutto, H. L. (2010). Combustion, emission and engine performance characteristics of used cooking oil biodiesel-A review. *Renewable & Sustainable Energy Reviews*, *14*(9), 2863–2873. doi:10.1016/j.rser.2010.07.036

EPA. (2002). A comprehensive analysis of biodiesel impacts on exhaust emissions. Draft Technical Report EPA420-P-02-001. Retrieved from http://www.epa.gov

Ezeji, T. C., Qureshi, N., & Blaschek, H. P. (2007). Bioproduction of butanol from biomass: From genes to bioreactors. *Current Opinion in Biotechnology*, *18*(3), 220–227. doi:10.1016/j.copbio.2007.04.002 PMID:17462877

Fadhil, A. B., & Ali, L. H. (2013). Alkaline-catalyzed transesterification of Silurus triostegus Heckel fish oil: Optimization of transesterification parameters. *Renewable Energy*, *60*, 481–488. doi:10.1016/j.renene.2013.04.018

Fahd, M. E. A., Wenming, Y., Lee, P. S., Chou, S. K., & Yap, C. R. (2013). Experimental investigation of the performance and emission characteristics of the direct injection diesel engine by water emulsion diesel under varying load condition. *Applied Energy*, *102*, 1042–1049. doi:10.1016/j.apenergy.2012.06.041

Fang, Q., Fang, J., Zhuang, J., & Huang, Z. (2013). Effects of ethanol-diesel-biodiesel blends on combustion and emissions in premixed low temperature combustion. *Applied Thermal Engineering*, *54*(2), 541–548. doi:10.1016/j.applthermaleng.2013.01.042

Fangsuwannarak, K., & Triratanasirichai, K. (2013). Effect of metalloid compound and bio-solution additives on biodiesel engine performance and exhaust emissions. *American Journal of Applied Sciences*, *10*(10), 1201–1213. doi:10.3844/ajassp.2013.1201.1213

Farag, H. A., El-Maghraby, A., & Taha, N. A. (2011). Optimization of factors affecting esterification of mixed oil with high percentage of free fatty acid. *Fuel Processing Technology*, *92*(3), 507–510. doi:10.1016/j.fuproc.2010.11.004

Farooq, M., Ramli, A., Naeem, A., Mahmood, T., Ahmad, S., Humayun, M., & Islam, M. G. U. (2018). Biodiesel production from date seed oil (Phoenix dactylifera L.) via eggshell derived heterogeneous catalyst. *Chemical Engineering Research & Design*, *132*, 644–651. doi:10.1016/j.cherd.2018.02.002

Farooq, M., Ramli, A., & Subbarao, D. (2013). Biodiesel production from waste cooking oil using bifunctional heterogeneous solid catalysts. *Journal of Cleaner Production*, *59*, 131–140. doi:10.1016/j.jclepro.2013.06.015

Farrauto, R. J., Dorazio, L., & Bartholomew, C. H. (2015). *Introduction to Catalysis and Industrial Catalytic Processes*. John Wiley & Sons.

Fatih Demirbas, M. (2009). Biorefineries for biofuel upgrading: A critical review. *Applied Energy*, *86*, S151–S161. doi:10.1016/j.apenergy.2009.04.043

Feroskhan, M., & Ismail, S. (2016). Investigation of the effects of biogas composition on the performance of a biogas – diesel dual fuel CI engine. *Biofuels*, *7*(6), 593–601. doi:10.1080/17597269.2016.1168025

Feroskhan, M., & Ismail, S. (2017). A review on the purification and use of biogas in compression ignition engines. *International Journal of Automotive and Mechanical Engineering*, *14*(3), 4383–4400. doi:10.15282/ijame.14.3.2017.1.0348

Freedman, B., Butterfield, R. O., & Pryde, E. H. (1986). Transesterification kinetics of soybean oil 1. *Journal of the American Oil Chemists' Society*, *63*(10), 1375–1380. doi:10.1007/BF02679606

FSSAI. (2019). Retrieved from https://fssai.gov.in/upload/advisories/2019/06/5cf8fe9068fc1Order_Used_Cooking_Oil_06_06_2019.pdf

Fuskele, V., & Sarviya, R. (2017). Recent developments in nanoparticles synthesis, preparation and stability of nanofluids. *Materials Today: Proceedings*, *4*(2), 4049–4060.

Gad, M. S., El-Araby, R., Abed, K. A., El-Ibiari, N. N., El Morsi, A. K., & El-Diwani, G. I. (2018). Performance and emission characteristics of C.I. engine fueled with palm oil/palm oil methyl ester blended with diesel fuel. *Egyptian Journal of Petroleum*, *27*(2), 215–219. doi:10.1016/j.ejpe.2017.05.009

Gairing, M., Marriott, J., Reders, K., & Reglitzky, A. (1995). The effects of modern additive technology on diesel fuel performance. SAE Technical Paper.

Ganesh, D., & Gowrishankar, G. (2011). *Effect of nano-fuel additive on emission reduction in a biodiesel fuelled CI engine*. Paper presented at the 2011 International conference on electrical and control engineering. 10.1109/ICECENG.2011.6058240

Gao, M., & Gao, Q. G. (2013b). Base-Catalyzed Transesterification of Soybean Oil by Ultrasonication. *Advanced Materials Research*, *724–725*, 356–359. doi:10.4028/www.scientific.net/AMR.724-725.356

Gao, M., Guo, C. B., & Lin, X. (2013c). Alkali-Catalyzed Transesterification of Waste Cooking Oil to Prepare Biodiesel by Ultrasonication. *Advanced Materials Research*, *803*, 17–20. doi:10.4028/www.scientific.net/AMR.803.17

Gao, M., Yan, Y. Q., & Lin, X. (2013). Alkali-Catalyzed Transesterification of Palm Oil by Ultrasonication. *Advanced Materials Research*, *803*, 13–16. doi:10.4028/www.scientific.net/AMR.803.13

García-Martín, J. F., Alés-Álvarez, F. J., López-Barrera, M. C., Martín-Domínguez, I., & Álvarez-Mateos, P. (2019). Cetane number prediction of waste cooking oil-derived biodiesel prior to transesterification reaction using near infrared spectroscopy. *Fuel*, *240*, 10–15. doi:10.1016/j.fuel.2018.11.142

Garnier, C., Bilcan, A., Le Corre, O., & Rahmouni, C. (2005). Characterisation of a syngas-diesel fuelled CI engine: SAE Technical Paper.

Gassner, M., & Maréchal, F. (2012). Thermo-economic optimisation of the polygeneration of synthetic natural gas (SNG), power and heat from lignocellulosic biomass by gasification and methanation. *Energy & Environmental Science*, *5*(2), 5768–5789. doi:10.1039/c1ee02867g

Ghogare, P., & Kale, N. (2016). Experimental investigation on single cylinder diesel engine fuelled with soya bean biodiesel blends with nano additives. *International Journal of Pure and Applied Research in Engineering and Technology*, *4*(9), 247–257.

Ghojel, J. I., & Tran, X. (2010). Ignition characteristics of diesel-water emulsion sprays in a constant-volume vessel: Effect of injection pressure and water content. *Energy & Fuels*, *24*(7), 3860–3866. doi:10.1021/ef1001875

Go, A. W., Sutanto, S., Nguyenthi, B. T., Cabatingan, L. K., Ismadji, S., & Ju, Y. H. (2014). Transesterification of soybean oil with methanol and acetic acid at lower reaction severity under subcritical conditions. *Energy Conversion and Management*, *88*, 1159–1166. doi:10.1016/j.enconman.2014.03.014

Gobbato, P., Masi, M., & Benetti, M. (2015). Performance analysis of a producer gas-fuelled heavy-duty SI engine at full-load operation. *Energy Procedia*, *82*, 149–155. doi:10.1016/j.egypro.2015.12.007

Golas, P. J., & Needham, J. (1999). Science and Civilisation in China, Cambridge University Press, pp. 186-91.

Gopal, K. N., & Karupparaj, R. T. (2015). Effect of pongamia biodiesel on emission and combustion characteristics of DI compression ignition engine. *Ain Shams Engineering Journal*, *6*(1), 297–305. doi:10.1016/j.asej.2014.10.001

Gowtham, M., Mohan, C. G., & Prakash, R. (2019). Effect of n-butanol fumigation on the regulated and unregulated emission characteristics of a diesel engine. *Fuel*, *242*(2), 84–95. doi:10.1016/j.fuel.2019.01.019

Gramajo de Doz, M. B., Bonatti, C. M., & Sólimo, H. N. (2004). Water Tolerance and Ethanol Concentration in Ethanol-Gasoline Fuels at Three Temperatures. *Energy & Fuels*, *18*(2), 334–337. doi:10.1021/ef034040a

Guan, G., Kusakabe, K., Sakurai, N., & Moriyama, K. (2009). Transesterification of vegetable oil to biodiesel fuel using acid catalysts in the presence of dimethyl ether. *Fuel*, *88*(1), 81–86. doi:10.1016/j.fuel.2008.07.021

Guha, A., Majumdar, S., & Sirkar, K. (1991). A larger-scale study of gas separation by hollow-fiber-contained liquid membrane permeator. *Journal of Membrane Science*, *62*(3), 293–307. doi:10.1016/0376-7388(91)80044-7

Gui, M. M., Lee, K. T., & Bhatia, S. (2009). Supercritical ethanol technology for the production of biodiesel: Process optimization studies. *The Journal of Supercritical Fluids*, *49*(2), 286–292. doi:10.1016/j.supflu.2008.12.014

Gumus, S., Ozcan, H., Ozbey, M., & Topaloglu, B. (2016). Aluminum oxide and copper oxide nanodiesel fuel properties and usage in a compression ignition engine. *Fuel*, *163*, 80–87. doi:10.1016/j.fuel.2015.09.048

Gunaseelan, V. N. (1997). Anaerobic digestion of biomass for methane production: A review. *Biomass and Bioenergy*, *13*(1–2), 83–114. doi:10.1016/S0961-9534(97)00020-2

Guoxin, H., & Hao, H. (2009). Hydrogen rich fuel gas production by gasification of wet biomass using a CO2 sorbent. *Biomass and Bioenergy*, *33*(5), 899–906. doi:10.1016/j.biombioe.2009.02.006

Gupta, A. R., & Rathod, V. K. (2018). Waste cooking oil and waste chicken eggshells derived solid base catalyst for the biodiesel production: Optimization and kinetics. *Waste Management (New York, N.Y.)*, *79*, 169–178. doi:10.1016/j.wasman.2018.07.022 PMID:30343743

Gupta, H., Agrawal, G., & Mathur, J. (2012). An overview of Nanofluids: A new media towards green environment. *International Journal of Environmental Sciences (China)*, *3*(1), 433–440.

Hajjari, M., Tabatabaei, M., Aghbashlo, M., & Ghanavati, H. (2017). A review on the prospects of sustainable biodiesel production: A global scenario with an emphasis on waste-oil biodiesel utilization. *Renewable & Sustainable Energy Reviews*, *72*, 445–464. doi:10.1016/j.rser.2017.01.034

Hamelinck, C. N., Hooijdonk, G., & Faaij, A. P. C. (2005). Ethanol from lignocellulosic biomass: Techno-economic performance in short-, middle- and long-term. *Biomass and Bioenergy*, *28*(4), 384–410. doi:10.1016/j.biombioe.2004.09.002

Handayani, P. A., Abdullah, A., & Hadiyanto, H. (2017). Biodiesel production from Nyamplung (Calophyllum inophyllum) oil using ionic liquid as a catalyst and microwave heating system. *Bulletin of Chemical Reaction Engineering & Catalysis*, *12*(2), 293–298. doi:10.9767/bcrec.12.2.807.293-298

Hannah, R., & Max, R. (2018). Fossil Fuels. *Published online at OurWorldInData.org,* Retrieved from https://ourworldindata.org/fossil-fuels.

Happonen, M., Heikkila, J., Saksa, P. A., Murtonen, T., Lehto, K., Rostedt, A., ... Virtanen, A. (2013). Diesel exhaust emissions and particle hygroscopicity with HVO fuel-oxygenate blend. *Fuel*, *103*, 380–386. doi:10.1016/j.fuel.2012.09.006

Harrington, K. J. (1986). Chemical and physical properties of vegetable oil esters and their effect on diesel fuel performance. *Biomass*, *9*(1), 1–17. doi:10.1016/0144-4565(86)90008-9

Hassan, S. N., Sani, Y. M., Abdul Aziz, A. R., Sulaiman, N. M. N., & Daud, W. M. A. W. (2015). Biogasoline: An out-of-the-box solution to the food-for-fuel and land-use competitions. *Energy Conversion and Management*, *89*, 349–367. doi:10.1016/j.enconman.2014.09.050

Hawi, M., Elwardany, A., Ismail, M., & Ahmed, M. (2019). Experimental investigation on performance of a compression ignition engine fuelled with waste cooking oil biodiesel-diesel blend enhanced with Iron-doped cerium oxide nanoparticles. *Energies*, *12*(5), 798. doi:10.3390/en12050798

Hazar, H. (2009). Effects of biodiesel on a low heat loss diesel engine. *Renewable Energy*, *34*(6), 1533–1537. doi:10.1016/j.renene.2008.11.008

He, B.-Q., Jian-Xin, W., Hao, J.-M., Yan, X.-G., & Xiao, J.-H. (2003). A study on emission characteristics of an EFI engine with ethanol blended gasoline fuels. *Atmospheric Environment*, *37*(7), 949–957. doi:10.1016/S1352-2310(02)00973-1

He, B.-Q., Liu, M.-B., Yuan, J., & Zhao, H. (2013). Combustion and emission characteristics of a HCCI engine fuelled with n-butanol-gasoline blends. *Fuel*, *108*, 668–674. doi:10.1016/j.fuel.2013.02.026

Henham, A., & Makkar, M. (1998). Combustion of simulated biogas in a dual-fuel diesel engine. *Energy Conversion and Management*, *39*(16–18), 2001–2009. doi:10.1016/S0196-8904(98)00071-5

Heywood, J. (2011). *Internal Combustion Engine Fundamentals*. New Delhi, India: Tata McGraw Hill.

Hileman, J. I., & Stratton, R. W. (2014). Alternative jet fuel feasibility. *Transport Policy*, *34*, 52–62. doi:10.1016/j.tranpol.2014.02.018

Hirner, F. S., Hwang, J., Bae, C., Patel, C., Gupta, T., & Agarwal, A. K. (2019). Performance and emission evaluation of a small-bore biodiesel compression-ignition engine. *Energy*, *183*, 971–982. doi:10.1016/j.energy.2019.07.015

Hokeman, S. K., & Robbins, C. (2012). Review of the effects of biodiesel on NO_x emissions. *Fuel Processing Technology*, *96*, 237–249. doi:10.1016/j.fuproc.2011.12.036

Holman, J. P. (2007). Experimental Techniques for Engineers (7th Ed). New Delhi, India: Tata McGraw Hill.

Holman, J. P. (2011). Experimental methods for engineers, Eighth edition, McGraw-Hill: Company, New York.

Hossain, A. K., Ouadi, M., Siddiqui, S. U., Yang, Y., Brammer, J., Hornung, A., ... Davies, P. A. (2013). Experimental investigation of performance, emission and combustion characteristics of an indirect injection multi-cylinder CI engine fuelled by blends of de-inking sludge pyrolysis oil with biodiesel. *Fuel*, *105*, 135–142. doi:10.1016/j.fuel.2012.05.007

Hosseini, S. H., Taghizadeh-Alisaraei, A., Ghobadian, B., & Abbaszadeh-Mayvan, A. (2017). Effect of added alumina as nano-catalyst to diesel-biodiesel blends on performance and emission characteristics of CI engine. *Energy*, *124*, 543–552. doi:10.1016/j.energy.2017.02.109

Hosseinzadeh-Bandbafha, H., Tabatabaei, M., Aghbashlo, M., Khanali, M., & Demirbas, A. (2018). A comprehensive review on the environmental impacts of diesel/biodiesel additives. *Energy Conversion and Management*, *174*, 579–614. doi:10.1016/j.enconman.2018.08.050

Hsieh, W.-D., Chen, R.-H., Wu, T.-L., & Lin, T.-H. (2002). Engine performance and pollutant emission of an SI engine using ethanol–gasoline blended fuels. *Atmospheric Environment*, *36*(3), 403–410. doi:10.1016/S1352-2310(01)00508-8

Huang, J., Wang, Y., Li, S., Roskilly, A. P., Yu, H., & Li, H. (2009). Experimental investigation on the performance and emissions of a diesel engine fuelled with ethanol-diesel blends. *Applied Thermal Engineering*, *29*(11-12), 484–490. doi:10.1016/j.applthermaleng.2008.12.016

Huber, G. W., Iborra, S., & Corma, A. (2006). Synthesis of Transportation Fuels from Biomass: Chemistry, Catalysts, and Engineering. *Chemical Reviews*, *106*(9), 4044–4098. doi:10.1021/cr068360d PMID:16967928

Hullu, D. (2008). Biogas Upgrading: Comparing Different Techniques, Eindhoven, The Netherlands. Retrieved from students.chem.tue.nl/ifp24/Finalpresentation.pps%3E accessed 08.03.2013

Hulwan, D. B., & Joshi, S. V. (2011). Performance, emission and combustion characteristic of a multicylinder DI diesel engine running on diesel–ethanol–biodiesel blends of high ethanol content. *Applied Energy*, *88*(12), 5042–5055. doi:10.1016/j.apenergy.2011.07.008

Huo, M., Lin, S., Liu, H., & Lee, C. F. (2014). Study on the spray and combustion characteristics of water–emulsified diesel. *Fuel*, *123*, 218–229. doi:10.1016/j.fuel.2013.12.035

Husam Al-Mashhadani, & Fernando, S. (2017). Properties, performance, and applications of biofuel blends: a review. *AIMS Energy, 5*(4), 735-767.

Hwang, J., Bae, C., & Gupta, T. (2016). Application of waste cooking oil (WCO) biodiesel in a compression ignition engine. *Fuel*, *176*, 20–31. doi:10.1016/j.fuel.2016.02.058

Hyvönen, J., Haraldsson, G., & Johansson, B. (2003). Operating range in a Multi Cylinder HCCI engine using Variable Compression Ratio. *SAE Technical Paper 2003-01-1829*. doi:10.4271/2003-01-1829

Ibrahim, M. M., Varuna Narasimhan, J., & Ramesh, A. (2015). Comparison of the predominantly premixed charge compression ignition and the dual fuel modes of operation with biogas and diesel as fuels. *Energy*, *89*, 990–1000. doi:10.1016/j.energy.2015.06.033

IEA, C. (2017). Emissions from fossil fuel combustion. Highlights: OECD/IEA.

iea. (2018). Retrieved from www.iea.org/renewables2018/transport/

Ikura, M., Stanciulescu, M., & Hogan, E. (2003). Emulsification of pyrolysis derived bio-oil in diesel fuel. *Biomass and Bioenergy*, *24*(3), 221–232. doi:10.1016/S0961-9534(02)00131-9

Ilham, Z., & Saka, S. (2010). Two-step supercritical dimethyl carbonate method for biodiesel production from Jatropha curcas oil. *Bioresource Technology*, *101*(8), 2735–2740. doi:10.1016/j.biortech.2009.10.053 PMID:19932022

Imran, A., Varman, M., Masjuki, H., & Kalam, M. (2013). Review on alcohol fumigation on diesel engine: A viable alternative dual fuel technology for satisfactory engine performance and reduction of environment concerning emission. *Renewable & Sustainable Energy Reviews*, *26*, 739–751. doi:10.1016/j.rser.2013.05.070

Imtenan, S., Masjuki, H. H., Varman, M., Arbab, M. I., Sajjad, H., Fattah, I. M. R., ... Hasib, A. S. M. (2014). Emission and performance improvements analysis of biodiesel-diesel blends with additives. *Procedia Engineering*, *90*, 472–477. doi:10.1016/j.proeng.2014.11.759

Imtenan, S., Masjuki, H. H., Varman, M., Rizwanul Fattah, I. M., Sajjad, H., & Arbab, M. (2015). Effect of n-butanol and diethyl ether as oxygenated additives on combustion–emission-performance characteristics of a multiple cylinder diesel engine fuelled with diesel–jatropha biodiesel blend. *Energy Conversion and Management*, *94*, 84–94. doi:10.1016/j.enconman.2015.01.047

Indrawan, N., Thapa, S., Bhoi, P. R., Huhnke, R. L., & Kumar, A. (2017). Engine power generation and emission performance of syngas generated from low-density biomass. *Energy Conversion and Management*, *148*, 593–603. doi:10.1016/j.enconman.2017.05.066

Isayama, Y., & Saka, S. (2008). Biodiesel production by supercritical process with crude bio-methanol prepared by wood gasification. *Bioresource Technology*, *99*(11), 4775–4779. doi:10.1016/j.biortech.2007.09.056 PMID:17977720

Jain, S., & Sharma, M. P. (2010). Stability of biodiesel and its blends: A review. *Renewable & Sustainable Energy Reviews*, *14*(2), 667–678. doi:10.1016/j.rser.2009.10.011

Jayanthi, P., & Rao, M. S. (2016). Effects of nanoparticles additives on performance and emissions characteristics of a DI diesel engine fuelled with biodiesel. *International Journal of Advances in Engineering and Technology*, *9*(6), 689.

Jehad, A. A. (2019). Effect of Some Oxygenates on the Opacity Level of a DI Diesel Engine with and without DPF. *Modern Applied Science*, *13*(3), 35–41. doi:10.5539/mas.v13n3p35

Jelles, S., Krul, R., Makkee, M., & Moulijn, J. (1999). The influence of NO_x on the oxidation of metal activated diesel soot. *Catalysis Today*, *53*(4), 623–630. doi:10.1016/S0920-5861(99)00150-9

Ji, J., Wang, J., Li, Y., Yu, Y., & Xu, Z. (2006). Preparation of biodiesel with the help of ultrasonic and hydrodynamic cavitation. *Ultrasonics*, *44*(SUPPL.), e411–e414. doi:10.1016/j.ultras.2006.05.020 PMID:16797656

Jin, C., Yao, M., Liu, H., Lee, C. F., & Ji, J. (2011). Progress in the production and application of n-butanol as a biofuel. *Renewable & Sustainable Energy Reviews*, *15*(8), 4080–4106. doi:10.1016/j.rser.2011.06.001

Jones, M., Li, C. H., Afjeh, A., & Peterson, G. (2011). Experimental study of combustion characteristics of nanoscale metal and metal oxide additives in biofuel (ethanol). *Nanoscale Research Letters*, *6*(1), 246. doi:10.1186/1556-276X-6-246 PMID:21711760

Jun, D., Ishii, K., & Iida, N. (2003). Combustion Analysis of Natural Gas in a Four Stroke HCCI Engine Using Experiment and Elementary Reactions Calculation. *SAE Technical Papers, 2003*(724). doi:10.4271/2003-01-1089

Jung, H., Kittelson, D. B., & Zachariah, M. R. (2005). The influence of a cerium additive on ultrafine diesel particle emissions and kinetics of oxidation. *Combustion and Flame*, *142*(3), 276–288. doi:10.1016/j.combustflame.2004.11.015

Kalligeros, S., Zannikos, F., Stournas, S., & Teas, C. (2003). An investigation of using biodiesel/marine biodiesel blends on the performance of a stationary diesel engine. *Biomass and Bioenergy*, *24*(2), 141–149. doi:10.1016/S0961-9534(02)00092-2

Kalsi, S. S., & Subramanian, K. A. (2017). Effect of simulated biogas on performance, combustion and emissions characteristics of a bio-diesel fueled diesel engine. *Renewable Energy*, *106*, 78–90. doi:10.1016/j.renene.2017.01.006

Kannan, G., Karvembu, R., & Anand, R. (2011). Effect of metal-based additive on performance emission and combustion characteristics of diesel engine fuelled with biodiesel. *Applied Energy*, *88*(11), 3694–3703. doi:10.1016/j.apenergy.2011.04.043

Karabektas, M. (2009). The effects of turbocharger on the performance and exhaust emissions of a diesel engine fuelled with biodiesel. *Renewable Energy*, *34*(4), 989–993. doi:10.1016/j.renene.2008.08.010

Karabetkas, M., & Hosoz, M. (2009). Performance and emission characteristics of a diesel engine using isobutanol-diesel fuel blends. *Renewable Energy*, *34*(6), 1554–1559. doi:10.1016/j.renene.2008.11.003

Kara, K., Ouanji, F., & El, M. (2018). Biodiesel production from waste fish oil with high free fatty acid content from Morroccan fish processing industries. *Egyptian Journal of Petroleum*, *27*(2), 249–255. doi:10.1016/j.ejpe.2017.07.010

Karavalakis, G., Hilari, D., Givalou, L., Karonis, D., & Stournas, S. (2011). Storage stability and ageing effect of biodiesel blends treated with different antioxidants. *Energy*, *36*(1), 369–374. doi:10.1016/j.energy.2010.10.029

Karthickeyan, V., & Balamurugan, P. (2017). Effect of thermal barrier coating with various blends of pumpkin seed oil methyl ester in DI diesel engine. *Heat and Mass Transfer/Waerme- Und Stoffuebertragung*, *53*(10), 3141–3154. doi:10.100700231-017-2058-8

Karthickeyan, V. (2019a). Data set for effect of cetane enhancer on ceramic coated diesel engine fuelled with neat Moringa oleifera methyl ester. *Data in Brief*, *24*. doi:10.1016/j.dib.2019.103932 PMID:31193703

Karthickeyan, V. (2019a). Effect of cetane enhancer on Moringa oleifera biodiesel in a thermal coated direct injection diesel engine. *Fuel*, *235*, 538–550. doi:10.1016/j.fuel.2018.08.030

Karthickeyan, V. (2019b). Effect of combustion chamber bowl geometry modification on engine performance, combustion and emission characteristics of biodiesel fuelled diesel engine with its energy and exergy analysis. *Energy*, *176*, 830–852. doi:10.1016/j.energy.2019.04.012

Karthickeyan, V., Ashok, B., Nanthagopal, K., Thiyagarajan, S., & Geo, V. E. (2019). Investigation of novel Pistacia khinjuk biodiesel in DI diesel engine with post combustion capture system. *Applied Thermal Engineering*, *159*, 113969. doi:10.1016/j.applthermaleng.2019.113969

Karthickeyan, V., Balamurugan, P., & Senthil, R. (2017). Investigation on environmental effects of thermal barrier coating with waste cooking palm oil methyl ester blends in diesel engine. *Biofuels*, 1–14. doi:10.1080/17597269.2017.1316142

Karthickeyan, V., Thiyagarajan, S., Geo, V. E., Ashok, B., Nanthagopal, K., Hwai, O., & Vignesh, R. (2019). Simultaneous reduction of NOx and smoke emissions with low viscous biofuel in low heat rejection engine using selective catalytic reduction technique. *Fuel*, *255*(January), 115854. doi:10.1016/j.fuel.2019.115854

Katahira, S., Mizuike, A., Fukuda, H., & Kondo, A. (2006). Ethanol fermentation from lignocellulosic hydrolysate by a recombinant xylose- and cellooligosaccharide-assimilating yeast strain. *Applied Microbiology and Biotechnology*, *72*(6), 1136–1143. doi:10.100700253-006-0402-x PMID:16575564

Keera, S. T., El Sabagh, S. M., & Taman, A. R. (2011). Transesterification of vegetable oil to biodiesel fuel using alkaline catalyst. *Fuel*, *90*(1), 42–47. doi:10.1016/j.fuel.2010.07.046

Keskin, A., Guru, M., & Altiparmak, D. (2007). Biodiesel production from tall oil with synthesized Mn and Ni based additives: Effects of the additives on fuel consumption and emissions. *Fuel*, *86*(7-8), 1139–1143. doi:10.1016/j.fuel.2006.10.021

Keskin, A., Gürü, M., & Altıparmak, D. (2008). Influence of tall oil biodiesel with Mg and Mo based fuel additives on diesel engine performance and emission. *Bioresource Technology*, *99*(14), 6434–6438. doi:10.1016/j.biortech.2007.11.051 PMID:18164614

Khalife, E., Tabatabaei, M., Demirbas, A., & Aghbashlo, M. (2017). Impacts of additives on performance and emission characteristics of diesel engines during steady state operation. *Progress in Energy and Combustion Science*, *59*, 32–78. doi:10.1016/j.pecs.2016.10.001

Khan, M. Y., Abdul Karim, Z. A., Hagos, F. Y., Aziz, A. R. A., & Tan, I. M. (2014). Current trends in the water-in-diesel emulsion as a fuel. *The Scientific World Journal*, 1–15. PMID:24563631

Khan, S., Dewang, Y., Raghuwanshi, J., & Sharma, V. (2019). Nanoparticles exceptional properties: Applications in internal combustion engines. In *AIP Conference Proceedings*. 10.1063/1.5098700

Khapre, U. (1989). Studies on biogas utilization for domestic cooking. *XXV Annual Convention of ISAE, Held at CTAE, Udaipur*. Retrieved from https://scholar.google.co.in/scholar?q=Studies+on+biogas+utilization+for+domestic+cooking&btnG=&hl=en&as_sdt=0%2C5#0

Khoiyangbam, R. S., Gupta, N., & Kumar, S. (2011). *Biogas Technology Towards Sustainable Development*. Teri Publications.

Kim, H. G., & Cho, S. H. (2013). The characteristics of biodiesel and oxygenated additives in a compression ignition engine. *Journal of Mechanical Science and Technology*, *27*(5), 1539–1543. doi:10.100712206-013-0334-1

Kim, H. J., Suh, H. K., Park, S. H., & Lee, C. S. (2008). An Experimental and Numerical Investigation of Atomization Characteristics of Biodiesel, Dimethyl Ether, and Biodiesel–Ethanol Blended Fuel. *Energy & Fuels*, *22*(3), 2091–2098. doi:10.1021/ef700692w

Kim, H., & Choi, B. (2010). The effect of biodiesel and bioethanol blended diesel fuel on nanoparticles and exhaust emissions from CRDI diesel engine. *Renewable Energy*, *35*(1), 157–163. doi:10.1016/j.renene.2009.04.008

Kim, Y. D., Yang, C. W., Kim, B. J., Kim, K. S., Lee, J. W., Moon, J. H., ... Do Lee, U. (2013). Air-blown gasification of woody biomass in a bubbling fluidized bed gasifier. *Applied Energy*, *112*, 414–420. doi:10.1016/j.apenergy.2013.03.072

Kirillov, V. A., Sobyanin, V. A., Kuzin, N. A., Brizitski, O. F., & Terentiev, V. Y. (2012). Synthesis gas generation onboard a vehicle: Development and results of testing. *International Journal of Hydrogen Energy*, *37*(21), 16359–16366. doi:10.1016/j.ijhydene.2012.04.106

Kirubakaran, K., & Arul Mozhi Selvan, V. (2018). Eggshell as heterogeneous catalyst for synthesis of biodiesel from high free fatty acid chicken fat and its working characteristics on a CI engine. *Journal of Environmental Chemical Engineering*, *6*(4), 4490–4503. doi:10.1016/j.jece.2018.06.027

Knothe, G. (2006). Analyzing biodiesel: Standards and other methods. *Journal of the American Oil Chemists' Society*, *83*(10), 823–833. doi:10.100711746-006-5033-y

Knothe, G., & Steidley, K. R. (2009). A comparison of used cooking oils; A very heterogeneous feedstock for biodiesel. *Bioresource Technology*, *100*(23), 5796–5801. doi:10.1016/j.biortech.2008.11.064 PMID:19592239

Koc, A. B., & Abdullah, M. (2013). Performance and NOX emissions of a diesel engine fueled with biodiesel-diesel-water nanoemulsions. *Fuel Processing Technology*, *109*, 70–77. doi:10.1016/j.fuproc.2012.09.039

Kojima, Y., & Takai, S. (2019). Transesterification of vegetable oil with methanol using solid base catalyst of calcium oxide under ultrasonication. *Chemical Engineering and Processing - Process Intensification*, 101–106. doi:10.1016/j.cep.2018.12.007

Königsson, F., Stalhammar, P., & Angstrom, H.-E. (2011). Characterization and Potential of Dual Fuel Combustion in a Modern Diesel Engine. Technical report, SAE Technical Paper 2011–01–2223. doi:10.4271/2011-01-2223

Koutsouki, A. A., Tegou, E., Kontakos, S., Kontominas, M. G., Pomonis, P. J., & Manos, G. (2015). In situ transesterification of Cynara cardunculus L. seed oil via direct ultrasonication for the production of biodiesel. *Fuel Processing Technology*, *134*, 122–129. doi:10.1016/j.fuproc.2015.01.024

Kozlov, V. E., Chechet, I. V., Matveev, S. G., Titova, N. S., & Starik, A. M. (2016). Modeling study of combustion and pollutant formation in HCCI engine operating on hydrogen rich fuel blends. *International Journal of Hydrogen Energy*, *41*(5), 3689–3700. doi:10.1016/j.ijhydene.2015.12.078

Kulkarni, M. G., Gopinath, R., Meher, L. C., & Dalai, A. K. (2006). Solid acid catalyzed biodiesel production by simultaneous esterification and transesterification. *Green Chemistry*, *8*(12), 1056–1062. doi:10.1039/b605713f

Kumaran, P., Mazlini, N., Hussein, I., Nazrain, M., & Khairul, M. (2011). Technical feasibility studies for Langkawi WCO (waste cooking oil) derived – biodiesel. *Energy*, *36*(3), 1386–1393. doi:10.1016/j.energy.2011.02.002

Kumar, D., Kumar, G., Poonam, & Singh, C. P. (2010a). Ultrasonic-assisted transesterification of Jatropha curcus oil using solid catalyst, Na/SiO2. *Ultrasonics Sonochemistry*, *17*(5), 839–844. doi:10.1016/j.ultsonch.2010.03.001 PMID:20359932

Kumar, D., Kumar, G., Poonam, & Singh, C. P. (2010b). Fast, easy ethanolysis of coconut oil for biodiesel production assisted by ultrasonication. *Ultrasonics Sonochemistry*, *17*(3), 555–559. doi:10.1016/j.ultsonch.2009.10.018 PMID:19945331

Kumar, N., Varun, S. R., & Chauhan, S. R. (2013). Performance and emission characteristics of biodiesel from different origins: A review. *Renewable & Sustainable Energy Reviews*, *21*, 633–658. doi:10.1016/j.rser.2013.01.006

Kumar, P., Sharma, M. P., & Dwivedi, G. (2016). Impact of ternary blends of biodiesel on diesel engine performance. *Egyptian Journal of Petroleum*, *25*(2), 255–261. doi:10.1016/j.ejpe.2015.06.010

Kumar, S., Dinesha, P., & Bran, I. (2017). Experimental investigation of the effects of nanoparticles as an additive in diesel and biodiesel fuelled engines: A review. *Biofuels*, 1–8. doi:10.1080/17597269.2017.1358948

Kumar, V., Nanda, M., Joshi, H. C., Singh, A., Sharma, S., & Verma, M. (2018). Production of biodiesel and bioethanol using algal biomass harvested from freshwater river. *Renewable Energy*, *116*, 606–612. doi:10.1016/j.renene.2017.10.016

Kusworo, T. D., Johari, S., & Widiasa, I. N. (2012). The Uses of Carbon Nanotubes Mixed Matrix Membranes (MMM) for Biogas Purification. [IJWR]. *International Journal of Waste Resources*, *2*(1), 5–10. doi:10.12777/ijwr.2.1.2012.5-10

Lambe, S., & Watson, H. (1992). Low polluting, energy efficient CI hydrogen engine. *International Journal of Hydrogen Energy*, *17*(7), 513–525. doi:10.1016/0360-3199(92)90151-L

Lapuerta, M. N., Garcia-Contreras, R., Campos-Femandez, J., & Dorado, M. P. (2010). Stability lubricity viscosity and cold flow properties of alcohol-diesel blends. *Energy & Fuels*, *24*(8), 4497–4502. doi:10.1021/ef100498u

Lapuerta, M., Rodriguez-Fernandez, J., & Armas, O. (2010). Correlation for the estimation of the density of fatty acid esters fuels and its implications. A proposed Biodiesel Cetane Index. *Chemistry and Physics of Lipids*, *163*(7), 720–727. doi:10.1016/j.chemphyslip.2010.06.004 PMID:20599853

Lee, J. S., & Saka, S. (2010). Biodiesel production by heterogeneous catalysts and supercritical technologies. *Bioresource Technology*, *101*(19), 7191–7200. doi:10.1016/j.biortech.2010.04.071 PMID:20570143

Lee, J., & Castaldi, M. J. (2010). A Study on Performance and Emissions of a 4-stroke IC Engine Operated on Landfill Gas with Syngas Addition. In *Proceedings 18th Annual North American Waste-to-Energy Conference*. 10.1115/NAWTEC18-3565

Lele, A., Soni, K., Narayanaswamy, K., & Krishnasamy, A. (2019). Experimental and Modeling Investigation of NO formation mechanism for biodiesel and its blends with methanol. SAE International, Technical Paper 2019-01-0217, 13, Doi:10.4271/2019-01-0217

Lenin, M., Swaminathan, M., & Kumaresan, G. (2013). Performance and emission characteristics of a DI diesel engine with a nanofuel additive. *Fuel*, *109*, 362–365. doi:10.1016/j.fuel.2013.03.042

Leung, D. Y. C., Wu, X., & Leung, M. K. H. (2010). A review on biodiesel production using catalyzed transesterification. *Applied Energy*, *87*(4), 1083–1095. doi:10.1016/j.apenergy.2009.10.006

Li, H., Karim, G. A., & Sohrabi, A. (2003). Knock and combustion characteristics of CH4, CO, H2 and their binary mixtures: SAE Technical Paper.

Li, C., Zhao, X., Wang, A., Huber, G. W., & Zhang, T. (2015). Catalytic Transformation of Lignin for the Production of Chemicals and Fuels. *Chemical Reviews*, *115*(21), 11559–11624. doi:10.1021/acs.chemrev.5b00155 PMID:26479313

Li, K., & Teo, W. K. (1993). Use of an internally staged permeator in the enrichment of methane from biogas. *Journal of Membrane Science*, *78*(1–2), 181–190. doi:10.1016/0376-7388(93)85259-Y

Li, L., Du, W., Liu, D., Wang, L., & Li, Z. (2006). Lipase-catalyzed transesterification of rapeseed oils for biodiesel production with a novel organic solvent as the reaction medium. *Journal of Molecular Catalysis. B, Enzymatic*, *43*(1–4), 58–62. doi:10.1016/j.molcatb.2006.06.012

Lin, Y. C., Chen, S. C., Wu, T. Y., Yang, P. M., Jhang, S. R., & Lin, J. F. (2015). Energy-saving and rapid transesterification of jatropha oil using a microwave heating system with ionic liquid catalyst. *Journal of the Taiwan Institute of Chemical Engineers*, *49*, 72–78. doi:10.1016/j.jtice.2014.11.014

Liu, G., Yan, B., & Chen, G. (2013). Technical review on jet fuel production. *Renewable & Sustainable Energy Reviews*, *25*, 59–70. doi:10.1016/j.rser.2013.03.025

Liu, H., Cooper, V. R., Dai, S., & Jiang, D. (2012). Windowed Carbon Nanotubes for Efficient CO 2 Removal from Natural Gas. *The Journal of Physical Chemistry Letters*, *3*(22), 3343–3347. doi:10.1021/jz301576s

Lounici, M. S., Loubar, K., Tazerout, M., Balistrou, M., & Tarabet, L. (2014). Experimental Investigation on the Performance and Exhaust Emission of Biogas-Diesel Dual-Fuel Combustion in a CI Engine. *Technical report, SAE Technical Paper 2014–01–2689*. doi:10.4271/2014-01-2689

Luijten, C. C. M., & Kerkhof, E. (2011). Jatropha oil and biogas in a dual fuel CI engine for rural electrification. *Energy Conversion and Management*, *52*(2), 1426–1438. doi:10.1016/j.enconman.2010.10.005

Lu, Q., Yang, X., & Zhu, X. (2008). Analysis on chemical and physical properties of bio-oil pyrolyzed from rice husk. *Journal of Analytical and Applied Pyrolysis*, *82*(2), 191–198. doi:10.1016/j.jaap.2008.03.003

Luque, R., Herrero-Davila, L., Campelo, J. M., Clark, J. H., Hidalgo, J. M., Luna, D., ... Romero, A. A. (2008). Biofuels: A technological perspective. *Energy & Environmental Science*, *1*(5), 542–564. doi:10.1039/b807094f

Lu, X. C., Yang, J. G., Zhang, W. G., & Huang, Z. (2005). Improving the combustion and emissions of direct injection compression ignition engines using oxygenated fuel additives combined with a cetane number improver. *Energy & Fuels*, *19*(5), 1879–1888. doi:10.1021/ef0500179

Lu, X., Ma, J., Ji, L., & Huang, Z. (2008). Simultaneous reduction of NO_x emission and smoke opacity of biodiesel-fueled engines by port injection of ethanol. *Fuel*, *87*(7), 1289–1296. doi:10.1016/j.fuel.2007.07.006

Maceiras, R., Vega, M., Costa, C., Ramos, P., & Marquez, M. C. (2009). Effect of methanol content on enzymatic production of biodiesel from waste frying oil. *Fuel*, *88*(11), 2130–2134. doi:10.1016/j.fuel.2009.05.007

Ma, F., Clements, L. D., & Hanna, M. A. (1998). The effects of catalyst, free fatty acids, and water on transesterification of beef tallow. *Transactions of the ASAE. American Society of Agricultural Engineers*, *41*(5), 1261–1264. doi:10.13031/2013.17292

Ma, F., & Hanna, M. A. (1999). Biodiesel production: A review. *Bioresource Technology*, *70*(1), 1–15. doi:10.1016/S0960-8524(99)00025-5

Mahgoub, B., Sulaiman, S., Karim, Z., & Hagos, F. (2015). Experimental study on the effect of varying syngas composition on the emissions of dual fuel CI engine operating at various engine speeds. In *Proceedings IOP Conference Series: Materials Science and Engineering*. 10.1088/1757-899X/100/1/012006

Mansir, N., Hwa Teo, S., Lokman Ibrahim, M., & Yun Hin, T. Y. (2017). Synthesis and application of waste eggshell derived CaO supported W-Mo mixed oxide catalysts for FAME production from waste cooking oil: Effect of stoichiometry. *Energy Conversion and Management*, *151*(August), 216–226. doi:10.1016/j.enconman.2017.08.069

Marulanda, V. F. (2012). Biodiesel production by supercritical methanol transesterification: Process simulation and potential environmental impact assessment. *Journal of Cleaner Production*, *33*, 109–116. doi:10.1016/j.jclepro.2012.04.022

Mat Yasin, M. H., Yusaf, T., Mamat, R., & Fitri Yusop, A. (2014). Characterization of a diesel engine operating with a small proportion of methanol as a fuel additive in biodiesel blend. *Applied Energy*, *114*, 865–873. doi:10.1016/j.apenergy.2013.06.012

Mattson, J., Burnete, N. V., Depcik, C., Moldovanu, D., & Burnete, N. (2019). Second law analysis of waste cooking oil biodiesel versus ULSD during operation of a CI engine. *Fuel*, *255*(June), 115753. doi:10.1016/j.fuel.2019.115753

Ma, Y., Zhu, M., & Zhang, D. (2013). The effect of a homogeneous combustion catalyst on exhaust emissions from a single cylinder diesel engine. *Applied Energy*, *102*, 556–562. doi:10.1016/j.apenergy.2012.08.028

McKendry, P. (2002). Energy production from biomass (part 1): Overview of biomass. *Bioresource Technology*, *83*(1), 37–46. doi:10.1016/S0960-8524(01)00118-3 PMID:12058829

McMillan, J. D. (1997). Bioethanol production: Status and prospects. *Renewable Energy*, *10*(2), 295–302. doi:10.1016/0960-1481(96)00081-X

Meher, L. C., Dharmagadda, V. S. S., & Naik, S. N. (2006). Optimization of alkali-catalyzed transesterification of Pongamia pinnata oil for production of biodiesel. *Bioresource Technology*, *97*(12), 1392–1397. doi:10.1016/j.biortech.2005.07.003 PMID:16359862

Meher, L. C., Vidya Sagar, D., & Naik, S. N. (2006). Technical aspects of biodiesel production by transesterification—A review. *Renewable & Sustainable Energy Reviews*, *10*(3), 248–268. doi:10.1016/j.rser.2004.09.002

Mehta, R. N., Chakaraborty, M., Mahanta, P., & Parikh, P. A. (2010). Evaluation of fuel properties of butanol-biodiesel-diesel blends and their impact on engine performance and emissions. *Industrial & Engineering Chemistry Research*, *49*(16), 7660–7665. doi:10.1021/ie1006257

Mehta, R. N., Chakraborty, M., & Parikh, P. A. (2014). Nanofuels: Combustion, engine performance and emissions. *Fuel*, *120*, 91–97. doi:10.1016/j.fuel.2013.12.008

Mendonça, I. M., Paes, O. A. R. L., Maia, P. J. S., Souza, M. P., Almeida, R. A., Silva, C. C., ... de Freitas, F. A. (2019). New heterogeneous catalyst for biodiesel production from waste tucumã peels (Astrocaryum aculeatum Meyer): Parameters optimization study. *Renewable Energy*, *130*, 103–110. doi:10.1016/j.renene.2018.06.059

Miao, X., Li, R., & Yao, H. (2009). Effective acid-catalyzed transesterification for biodiesel production. *Energy Conversion and Management*, *50*(10), 2680–2684. doi:10.1016/j.enconman.2009.06.021

Moffat, R. J. (1988). Describing the uncertainties in experimental results. *Experimental Thermal and Fluid Science*, *1*(1), 3–17. doi:10.1016/0894-1777(88)90043-X

Mofijur, M., Masjuki, H. H., Kalam, M. A., Atabani, A. E., Arbab, M. I., Cheng, S. F., & Gouk, S. W. (2014). Properties and use of Moringa oleifera biodiesel and diesel fuel blends in a multi-cylinder diesel engine. *Energy Conversion and Management*, *82*, 169–176. doi:10.1016/j.enconman.2014.02.073

Mofijur, M., Masjuki, H., Kalam, M., & Shahabuddin, M. (2012). Experimental study of additive added palm biodiesel in a compression ignition engine. *Energy Education Science and Technology, Part A. Energy Science and Research, 30*, 737–748.

Mohamed Musthafa, M., Ashok Kumar, T., Mohanraj, T., & Chandramouli, R. (2018). A comparative study on performance, combustion and emission characteristics of a diesel engine fuelled by biodiesel blends with and without an additive. *Fuel, 225*, 343–348. doi:10.1016/j.fuel.2018.03.147

Mootabadi, H., Salamatinia, B., Bhatia, S., & Abdullah, A. Z. (2010). Ultrasonic-assisted biodiesel production process from palm oil using alkaline earth metal oxides as the heterogeneous catalysts. *Fuel, 89*(8), 1818–1825. doi:10.1016/j.fuel.2009.12.023

Morais, S., Mata, T. M., Martins, A. A., Pinto, G. A., & Costa, C. A. V. (2010). Simulation and life cycle assessment of process design alternatives for biodiesel production from waste vegetable oils. *Journal of Cleaner Production, 18*(13), 1251–1259. doi:10.1016/j.jclepro.2010.04.014

Morozumi, Y., & Saito, Y. (2010). Effect of physical properties on micro explosion occurrence in water-in-oil emulsion droplets. *Energy & Fuels, 24*(3), 1854–1859. doi:10.1021/ef9014026

Mosier, N., Wyman, C., Dale, B., Elander, R., Lee, Y. Y., Holtzapple, M., & Ladisch, M. (2005). Features of promising technologies for pretreatment of lignocellulosic biomass. *Bioresource Technology, 96*(6), 673–686. doi:10.1016/j.biortech.2004.06.025 PMID:15588770

Motasemi, F., & Ani, F. N. (2012). A review on microwave-assisted production of biodiesel. *Renewable & Sustainable Energy Reviews, 16*(7), 4719–4733. doi:10.1016/j.rser.2012.03.069

Mura, E., Josset, C., Loubar, K., Huchet, G., & Bellettre, J. (2010). Effect of dispersed water droplet size in micro explosion phenomenon for water in oil emulsion. *Atomization and sprays, 20*(9), 791–799.

Mura, E., Massoli, P., Josset, C., Loubar, K., & Bellettre, J. (2012). Study of the micro-explosion temperature of water in oil emulsion droplets during the leidenfrost effect. *Experimental Thermal and Fluid Science, 43*, 63–70. doi:10.1016/j.expthermflusci.2012.03.027

Murugesan, A., Subramaniam, D., & Avinash, A. (2015). An empirical and statistical analysis of biodiesel production by transesterification process. *Biofuels, 6*(1–2), 79–86. doi:10.1080/17597269.2015.1050643

Mustafi, N. N., Raine, R. R., & Verhelst, S. (2013). Combustion and emissions characteristics of a dual fuel engine operated on alternative gaseous fuels. *Fuel, 109*, 669–678. doi:10.1016/j.fuel.2013.03.007

Nabi, M. N., Najmul Hoque, S. M., & Akhter, M. S. (2009). Karanja (Pongamia Pinnata) biodiesel production in Bangladesh, characterization of Karanja biodiesel and its effect on diesel emissions. *Fuel Processing Technology, 90*(9), 1080–1086. doi:10.1016/j.fuproc.2009.04.014

Nabi, M. N., Rahman, M. M., & Akhter, M. S. (2009). Biodiesel from cotton seed oil and its effects on engine performance and exhaust emissions. *Applied Thermal Engineering, 29*(11-12), 2265–2270. doi:10.1016/j.applthermaleng.2008.11.009

Nair, P., Singh, B., Upadhyay, S. N., & Sharma, Y. C. (2012). Synthesis of biodiesel from low FFA waste frying oil using calcium oxide derived from Mereterix mereterix as a heterogeneous catalyst. *Journal of Cleaner Production, 29-30*, 82–90. doi:10.1016/j.jclepro.2012.01.039

Nakagawa, Y., Tamura, M., & Tomishige, K. (2019). Recent development of production technology of diesel- and jet-fuel-range hydrocarbons from inedible biomass. *Fuel Processing Technology, 193*, 404–422. doi:10.1016/j.fuproc.2019.05.028

Nanda, S., Azargohar, R., Dalai, A. K., & Kozinski, J. A. (2015). An assessment on the sustainability of lignocellulosic biomass for biorefining. *Renewable & Sustainable Energy Reviews*, *50*, 925–941. doi:10.1016/j.rser.2015.05.058

Nanda, S., Rana, R., Sarangi, P. K., Dalai, A. K., & Kozinski, J. A. (2018). *A Broad Introduction to First-, Second-, and Third-Generation Biofuels, Recent Advancements in Biofuels and Bioenergy Utilization* (pp. 1–25). Springer.

Nanthagopal, K., Ashok, B., Tamilarasu, A., Johny, A., & Mohan, A. (2017). Influence on the effect of zinc oxide and titanium dioxide as an additive with calophylluminophyllum methyl ester in a CI engine. *Energy Conversion and Management*, *146*, 8–19. doi:10.1016/j.enconman.2017.05.021

Nathan, S. S., & Mallikarjuna, J. M. (2010). An experimental study of the biogas-diesel HCCI mode of engine operation. *Energy Conversion and Management*, *51*(7), 1347–1353. doi:10.1016/j.enconman.2009.09.008

Nathan, S. S., Mallikarjuna, J. M., & Ramesh, A. (2007). Effect of mixture preparation in a diesel HCCI engine using early in-cylinder injection during the suction stroke. *International Journal of Automotive Technology*, *8*(5), 543–553. Retrieved from http://www.scopus.com/scopus/inward/record.url?eid=2-s2.0-36849090822&partnerID=40&rel=R8.2.0

Nathan, S. S., Mallikrajuna, J. M., & Ramesh, A. (2009). Homogeneous charge compression ignition versus dual fuelling for utilizing biogas in compression ignition engines. *Proceedings of the Institution of Mechanical Engineers. Part D, Journal of Automobile Engineering*, *223*(3), 413–422. doi:10.1243/09544070JAUTO970

NBB. (2007). Biodiesel production and quality. Fuel fact sheets, USA: National Biodiesel Board, Retrieved from http://biodiesel.org/what-is-biodiesel/biodiesel-fact-sheets.

Ndaba, B., Chiyanzu, I., & Marx, S. (2015). n-Butanol derived from biochemical and chemical routes: A review. *Biotechnology Reports (Amsterdam, Netherlands)*, *8*, 1–9. doi:10.1016/j.btre.2015.08.001 PMID:28352567

Nejat, P., Jomehzadeh, F., Taheri, M. M., Gohari, M., & Abd. Majid, M. Z. (2015). A global review of energy consumption, CO2 emissions and policy in the residential sector (with an overview of the top ten CO2 emitting countries). *Renewable & Sustainable Energy Reviews*, *43*, 843–862. doi:10.1016/j.rser.2014.11.066

Nigam, P. S., & Singh, A. (2011). Production of liquid biofuels from renewable resources. *Progress in Energy and Combustion Science*, *37*(1), 52–68. doi:10.1016/j.pecs.2010.01.003

Nijaguna, B. (2006). Biogas technology. New Age International Retrieved from https://scholar.google.co.in/scholar?q=nijaguna&btnG=&hl=en&as_sdt=0%2C5#2

Nirmala, N., Dawn, S. S., & Harindra, C. (2019). Analysis of performance and emission characteristics of Waste cooking oil and Chlorella variabilis MK039712.1 biodiesel blends in a single cylinder, four strokes diesel engine. Renewable Energy, doi: j.renene.2019.08.133.

Noge, H., Kidoguchi, Y., Yahya, W. J., Imai, Y., & Tajima, K. (2015). An investigation into the relationship between the formation of thermal cracked components and PM reduction during diesel combustion using water emulsified fuel. *J. Therm. Sci. Technol*, *10*(2), 1–11. doi:10.1299/jtst.2015jtst0024

Noor El-Din, M. R., & Sabrnal, H., El-Hamouly, M. H. M., Mishrif, M. R., & Ragab, A. M. (2013). Water-in-diesel fuel nanoemulsions: Preparation, stability and physical properties Egyptian. *Journal of Petrology*, *22*(4), 517–530.

Nour, M., EL-Seesy, A. I., Abdel-Rahman, A. K., & Bady, M. (2018). Influence of adding aluminum oxide nanoparticles to diesterol blends on the combustion and exhaust emission characteristics of a diesel Engine. *Experimental Thermal and Fluid Science*, *98*, 634–644. doi:10.1016/j.expthermflusci.2018.07.009

Obert, R., & Dave, B. C. (1999). Enzymatic Conversion of Carbon Dioxide to Methanol: Enhanced Methanol Production in Silica Sol–Gel Matrices. *Journal of the American Chemical Society*, *121*(51), 12192–12193. doi:10.1021/ja991899r

Ochoterena, R., Lif, A., Nyden, M., Anderson, S., & Denbratt, I. (2010). Optical studies of spray development and combustion of water in- diesel emulsion and microemulsion fuels. *Fuel*, *89*(1), 122–132. doi:10.1016/j.fuel.2009.06.039

Ojolo, S. J. O. B. S., Adelaja, A. O., & Ogbonnaya, M. (2011). Study of an effective technique for the production of biodiesel from jatropha oil. *Journal of Emerging Trends in Engineering and Applied Sciences*, *2*, 79–86.

Okoro, O. V., Sun, Z., & Birch, J. (2018). Catalyst-free biodiesel production methods: A comparative technical and environmental evaluation. *Sustainability (Switzerland)*, *10*(1). doi:10.3390u10010127

Ospina, G., Selim, M. Y. E., Al Omari, S. A. B., Hassan Ali, M. I., & Hussien, A. M. M. (2019). Engine roughness and exhaust emissions of a diesel engine fueled with three biofuels. *Renewable Energy*, *134*, 1465 1472. doi:10.1016/j.renene.2018.09.046

Oxygenates Fact Book. (2003). *A compilation of information on the benefits of oxygenates in gasoline, clean fuels development coalition*. CFDC.

Özçimen, D., & Yücel, S. (2011). *Novel Methods in Biodiesel Production. Biofuel's Engineering Process Technology Marco Aurélio dos Santos Bernardes*. Intech Open; doi:10.5772/18750

Özgür, T., Özcanli, M., & Aydin, K. (2015). Investigation of nanoparticle additives to biodiesel for improvement of the performance and exhaust emissions in a compression ignition engine. *International Journal of Green Energy*, *12*(1), 51–56. doi:10.1080/15435075.2014.889011

Ozsezen, A. N., Canakci, M., Turkcan, A., & Sayin, C. (2009). Performance and combustion characteristics of a DI diesel engine fueled with waste palm oil and canola oil methyl esters. *Fuel*, *88*(4), 629–636. doi:10.1016/j.fuel.2008.09.023

Pande, D. R., & Fabiani, C. (1989). Feasibility studies on the use of a naturally occurring molecular sieve for methane enrichment from biogas. *Gas Separation & Purification*, *3*(3), 143–147. doi:10.1016/0950-4214(89)80026-X

Pandiaraj, S., Subbaiyan, D., Ayyasamy, T., & Nagarajan, S. (2019). Design, Development, and Analysis of Mullite Catalytic Converter for CI Engines: SAE Technical Paper.

Pandit, P. R., & Fulekar, M. H. (2017). Eggshell waste as heterogeneous nanocatalyst for biodiesel production: Optimized by response surface methodology. *Journal of Environmental Management*, *198*, 319–329. doi:10.1016/j.jenvman.2017.04.100 PMID:28494420

Pandit, P. R., & Fulekar, M. H. (2019). Biodiesel production from microalgal biomass using CaO catalyst synthesized from natural waste material. *Renewable Energy*, *136*, 837–845. doi:10.1016/j.renene.2019.01.047

Papagiannakis, R., Rakopoulos, C., Hountalas, D., & Giakoumis, E. (2007). Study of the performance and exhaust emissions of a spark-ignited engine operating on syngas fuel. *International Journal of Alternative Propulsion*, *1*(2/3), 190–215. doi:10.1504/IJAP.2007.013022

Paramasivam, B., Kasimani, R., & Rajamohan, S. (2019). Experimental assessment and multi-response optimization of diesel engine performance and emission characteristics fuelled with Aegle marmelos seed cake pyrolysis oil-diesel blends using Grey relational analysis coupled principal component analysis. *Environmental Science and Pollution Research International*, *26*(7), 6980–7004. doi:10.100711356-019-04164-8

Paramasivam, B., K, R., & R, S. (2019). Investigation and improvement on storage stability of pyrolysis oil obtained from Aegle marmelos de-oiled seed cake. *Energy Sources. Part A, Recovery, Utilization, and Environmental Effects*, 1–15. doi:10.1080/15567036.2019.1632989

Park, S. H., & Lee, C. S. (2013). Combustion performance and emission reduction characteristics of automotive DME engine system. *Progress in Energy and Combustion Science*, *39*(1), 147–168. doi:10.1016/j.pecs.2012.10.002

Park, S. H., & Yoon, S. H. (2016). Effect of dual-fuel combustion strategies on combustion and emission characteristics in reactivity-controlled compression ignition (RCCI) engine. *Fuel, 181*, 310–318. doi:10.1016/j.fuel.2016.04.118

Park, S. H., Yoon, S. H., Cha, J., & Lee, C. S. (2014). Mixing effects of biogas and dimethyl ether (DME) on combustion andemission characteristics of DME fueled high-speed diesel engine. *Energy, 66*, 413–422. doi:10.1016/j.energy.2014.02.007

Patel, N. K., & Shah, S. N. (2015). Biodiesel from Plant Oils. In Food, Energy, and Water: The Chemistry Connection (pp. 277–307). doi:10.1016/B978-0-12-800211-7.00011-9

Payne, C. M., Knott, B. C., Mayes, H. B., Hansson, H., Himmel, M. E., Sandgren, M., ... Beckham, G. T. (2015). Fungal Cellulases. *Chemical Reviews, 115*(3), 1308–1448. doi:10.1021/cr500351c PMID:25629559

Perumal, V., & Ilangkumaran, M. (2018). Water emulsified hybrid pongamia biodiesel as a modified fuel for the experimental analysis of performance, combustion and emission characteristics of a direct injection diesel engine. *Renewable Energy, 121*, 623–631. doi:10.1016/j.renene.2018.01.060

Phoon, L., Mustaffa, A., Hashim, H., Mat, R., Manan, Z., & Yunus, N. (2017). Performance and emission characteristics of green diesel blends containing diethyl-succinate and 1-octanol. *Journal of Cleaner Production, 161*, 1192–1202. doi:10.1016/j.jclepro.2017.06.219

Plamondon, E., & Seers, P. (2019). Parametric study of pilot –main injection strategies on the performance of a light-duty diesel engine fueled with diesel or a WCO biodiesel–diesel blend. *Fuel, 236*, 1273–1281. doi:10.1016/j.fuel.2018.09.111

Polk, A. C., Gibson, C. M., Shoemaker, N. T., Srinivasan, K. K., & Krishnan, S. R. (2013). Analysis of Ignition Behavior in a Turbocharged Direct Injection Dual Fuel Engine Using Propane and Methane as Primary Fuels. *Journal of Energy Resources Technology, 135*(3). doi:10.1115/1.4023482

Prabakaran, B., & Udhoji, A. (2016). Experimental investigation into effects of addition of zinc oxide on performance, combustion and emission characteristics of diesel-biodiesel-ethanol blends in CI engine. *Alexandria Engineering Journal, 55*(4), 3355-3362.

Prabakaran, B., & Vijayabalan, P. (2018). Influence of zinc oxide nano particles on performance, combustion and emission characteristics of butanol-diesel-ethanol blends in DI CI engine. In *Proceedings IOP Conference Series: Materials Science and Engineering*. 10.1088/1757-899X/377/1/012069

Prabhu, A. (2018). Nanoparticles as additive in biodiesel on the working characteristics of a diesel engine. *Ain Shams Engineering Journal, 9*(4), 2343–2349. doi:10.1016/j.asej.2017.04.004

Prabu, A., & Anand R. B. (2015). Emission control strategy by adding alumina and cerium oxide nanoparticle in biodiesel. *Journal of Energy Institute,* 1-7.

Prabu, A. (2017). *Nanoparticles as additive in biodiesel on the working characteristics of a DI diesel engine*. Ain Shams Engineering Journal.

Prabu, A., & Anand, R. B. (2015). Influence of oxygenate additives on the performance and emission characteristics of Jatropha fuelled direct injection diesel engine. *Biofuels*, 667–672.

Prabu, A., & Anand, R. B. (2016). Emission control strategy by adding alumina and cerium oxide nano particle in biodiesel. *Journal of the Energy Institute, 89*(3), 366–372. doi:10.1016/j.joei.2015.03.003

Prabu, A., & Anand, R. B. (2018). Influence of Antioxidant Addition in Jatropha Biodiesel on the Performance, Combustion and Emission Characteristics of a DI Diesel Engine. *J. Inst. Eng. India Ser. C, 99*(2), 207–216. doi:10.100740032-017-0350-5

Pradhan, D., Bendu, H., Singh, R. K., & Murugan, S. (2017). Mahua seed pyrolysis oil blends as an alternative fuel for light-duty diesel engines. *Energy, 118*, 600–612. doi:10.1016/j.energy.2016.10.091

Prakash, R., Singh, R. K., & Murugan, S. (2013). Experimental investigation on a diesel engine fueled with bio-oil derived from waste wood–biodiesel emulsions. *Energy, 55*, 610–618. doi:10.1016/j.energy.2013.03.085

Praveen, A., Rao, G. L. N., & Balakrishna, B. (2018). Performance and emission characteristics of a diesel engine using Calophyllum Inophyllum biodiesel blends with TiO_2 nanoadditives and EGR. *Egyptian Journal of Petroleum, 27*(4), 731–738. doi:10.1016/j.ejpe.2017.10.008

Puhan, S., Nagarajan, G., Vedaraman, N., & Ramabramhmam, B. V. (2007). Mahua oil derivatives as a renewable fuel for diesel engine systems in India, a performance and emissions comparative study. *International Journal of Green Energy, 4*(1), 89–104. doi:10.1080/15325000601015627

Puhan, S., Vedaraman, N., Sankaranarayanan, G., & Ram, B. V. B. (2005). Performance and emission study of Mahuca oil (madhucaindica oil) ethyl ester in a four stroke natural aspirated direct injection diesel engine. *Renewable Energy, 30*(8), 1269–1278. doi:10.1016/j.renene.2004.09.010

Qibai, W., Xialin, X., Yaodong, W., & Roskilly, T. (2018). Effect of carbon coated aluminium nanoparticles as additive to biodiesel-diesel blends on performance and emission characteristics of diesel engine. *Applied Energy, 221*, 597–604. doi:10.1016/j.apenergy.2018.03.157

Qi, D. H., Chen, H., Geng, L. M., & Bian, Y. Z. (2010). Experimental studies on the combustion characteristics and performance of a direct injection engine fuelled with biodiesel/diesel blends. *Energy Conversion and Management, 51*(12), 2985–2992. doi:10.1016/j.enconman.2010.06.042

Qi, D. H., Chen, H., Geng, L. M., & Bian, Y. Z. (2011). Effect of diethyl ether and ethanol additives on the combustion and emission characteristics of biodiesel-diesel blended fuel engine. *Renewable Energy, 36*(4), 1252–1258. doi:10.1016/j.renene.2010.09.021

Ragauskas, A. J., Williams, C. K., Davison, B. H., Britovsek, G., Cairney, J., Eckert, C. A., ... Tschaplinski, T. (2006). The Path Forward for Biofuels and Biomaterials. *Science, 311*(5760), 484–489. doi:10.1126cience.1114736 PMID:16439654

Rahees, K., & Meera, V. (2014). Production of Biodiesel from Dairy Waste Scum. *International Journal of Scientific & Engineering Research, 5*(7), 194–199. doi:10.1016/j.psep.2019.03.021

Rahman, A. K., & Ramesh, A. (2017). Effect of reducing the methane concentration on the combustion and performance of a biogas diesel predominantly premixed charge compression ignition engine. *Fuel, 206*, 117–132. doi:10.1016/j.fuel.2017.05.100

Rahman, M. M., Hassan, M. H., Kalam, M. A., Atabani, A. E., Memon, L. A., & Rahman, S. M. A. (2014). Performance and emission analysis of jatropha curcas and Motinga oleifers methyl ester fuel blends in a multi-cylinder diesel engine. *Journal of Cleaner Production, 65*, 304–310. doi:10.1016/j.jclepro.2013.08.034

Rahman, S. M. A., Masjuki, H. H., Kalam, M. A., Abedin, M. J., Sanjid, A., & Imtenan, S. (2014). Effect of idling on fuel consumption and emissions of a diesel engine fueled by jatropha biodiesel blends. *Journal of Cleaner Production, 69*, 208–215. doi:10.1016/j.jclepro.2014.01.048

Rahman, S. M. A., Masjuki, H. H., Kalam, M. A., & Sajjad, H. (2013). Impact of idling on fuel consumption and exhaust emission and available idle reduction technologies for diesel vehicles – a review. *Energy Conversion and Management, 74*, 171–182. doi:10.1016/j.enconman.2013.05.019

Rahmat, N., Abdullah, A. Z., & Mohamed, A. R. (2010). Recent progress on innovative and potential technologies for glycerol transformation into fuel additives: A critical review. *Renewable & Sustainable Energy Reviews*, *14*(3), 987–1000. doi:10.1016/j.rser.2009.11.010

Rajan, S., & Saniee, F. F. (1983). Water—ethanol—gasoline blends as spark ignition engine fuels. *Fuel*, *62*(1), 117–121. doi:10.1016/0016-2361(83)90265-X

Rajasekar, E., & Selvi, S. (2014). Review of combustion characteristics of CI engines fuelled with biodiesel. *Renewable & Sustainable Energy Reviews*, *35*, 390–399. doi:10.1016/j.rser.2014.04.006

Raju, V. D. (2019). V., Kishore, P. S., Kumar, M. H., & Reddy, S. R. Experimental investigation of alumina nanoparticles effects on the performance and emission characteristics of the tamarind seed biodiesel fuelled diesel engine, Materials today. *Proceedings*, *18*(03), 1229–1242. doi:10.1016/j.matpr.2019.06.585

Raju, V. D., Kishore, P. S., & Yamini, K. (2018). Experimental studies on four stroke diesel engine fuelled with tamarind seed oil as potential alternative fuel sustainable green environment. *European Journal of Sustainable Development Research*, *2*. doi:10.20897/ejosdr/78489

Rakopoulos, D. C. (2012). Heat release analysis of combustion in heavy-duty turbocharged diesel engine operating on blends of diesel fuel with cottonseed or sunflower oils and their bio-diesel. *Fuel*, *96*, 524–534. doi:10.1016/j.fuel.2011.12.063

Rakopoulos, D. C., Rakopoulos, C. D., Giakoumis, E. G., Dimaratos, A. M., & Kyritsis, D. C. (2010). Effects of butanol-diesel fuel blends on the performance and emissions of a high-speed DI diesel engine. *Energy Conversion and Management*, *51*(10), 1989–1997. doi:10.1016/j.enconman.2010.02.032

Rakopoulos, D. C., Rakopoulos, C. D., Kakaras, E. C., & Giakoumis, E. G. (2008). Effects of ethanol-diesel fuel blends on the performance and exhaust emissions of heavy duty DI diesel engine. *Energy Conversion and Management*, *49*(11), 3155–3162. doi:10.1016/j.enconman.2008.05.023

Ramadhas, A. S., Jayaraj, S., & Muraleedharan, C. (2004). Use of vegetable oils as I.C engine fuels- A review. *Renewable Energy*, *29*(5), 727–742. doi:10.1016/j.renene.2003.09.008

Ramadhas, A. S., Muraleedharan, C., & Jayaraj, S. (2005). Performance and emission evaluation of a diesel engine fueled with methyl esters of rubber seed oil. *Renewable Energy*, *30*(12), 1789–1800. doi:10.1016/j.renene.2005.01.009

Ramakrishnan. M, Thansekhar, M. R., & Viswanathan, K. (n.d.). Comparative studies on the performance and emissions of a direct injection diesel engine fuelled with neem oil and pumpkin seed oil biodiesel with and without fuel preheater, Environmental Science and Pollution Control, 25(13), 1-11. Doi:10.100711356-017-0838-9

Ramesh, D. K., Dhanajaya Kumar, J. L., Hemanth Kumar, S. G., Namith, V., & Parashuram, B. J. (2018). And Sharath S. Study on effects of alumina nanoparticles as additive with poultry litter biodiesel on performance, combustion and emission characteristics of diesel engine. Materials today. *Proceedings*, *5*, 1114–1120. doi:10.1016/j.matpr.2017.11.190

Ranjan, A., Dawn, S. S., Jayaprabakar, J., Nirmala, N., Saikiran, K., & Sai Sriram, S. (2018). Experimental investigation on effect of MgO nanoparticles on cold flow properties, performance, emission, and combustion characteristics of waste cooking oil biodiesel. *Fuel*, *220*(November 2017), 780–791. doi:10.1016/j.fuel.2018.02.057

Rao, P. V., Ramesh, S., & Kumar, S. A. (2019). Effects of Oxygenated Additives with Diesel on the Performance of D I Diesel Engine. *Journal of Energy Research and Reviews*, *2*(3), 1–9. doi:10.9734/jenrr/2019/v2i330076

Rashdul, H. K., Masjuki, H. H., Kalam, M. A., & Ashraful, A. M. (2014). Theeffect of additives on properties, performance and emission of biodiesel fuelled compression ignition engine. *Energy Conversion and Management*, *88*, 348–364. doi:10.1016/j.enconman.2014.08.034

Rathinam, S., Justin Abraham Baby, S., Devarajan, Y., & T, A. (2018). Influence of water on exhaust emissions on unmodified diesel engine propelled with biodiesel. *Energy Sources. Part A, Recovery, Utilization, and Environmental Effects*, *40*(21), 2511–2517. doi:10.1080/15567036.2018.1503756

Ray, N. H. S. (2013). Biogas as Alternate Fuel in Diesel Engines: A Literature Review. *IOSR Journal of Mechanical and Civil Engineering*, *9*(1), 23–28. doi:10.9790/1684-0912328

Refaat, A. A., El Sheltawy, S. T., & Sadek, K. U. (2008). Optimum reaction time, performance and exhaust emissions of biodiesel produced by microwave irradiation. *International Journal of Environmental Science and Technology*, *5*(3), 315–322. doi:10.1007/BF03326026

Ren, Y. Z. H., Huang, D., Jiang, M., Li, W., Liu, B., & Wang, X. B. (2008). Effects of the addition of ethanol and cetane number improver on the combustion and emission characteristics of a compression ignition engine. *Proceedings of the Institution of Mechanical Engineers. Part D, Journal of Automobile Engineering*, *222*(6), 1077–1087. doi:10.1243/09544070JAUTO516

Reyman, D., Saiz Bermejo, A., Ramirez Uceda, I., & Rodriguez Gamero, M. (2014). A new FTIR method to monitor transesterification in biodiesel production by ultrasonication. *Environmental Chemistry Letters*, *12*(1), 235–240. doi:10.100710311-013-0440-4

Ribeiro, N. M., Pinto, A. C., Quintella, C. M., da Rocha, G. O., & Teixeria, L. S. (2007). The role of additives for diesel and diesel blended (ethanol or biodiesel) fuels: A review. *Energy & Fuels*, *21*(4), 2433–2455. doi:10.1021/ef070060r

Risso, R., Ferraz, P., Meireles, S., Fonseca, I., & Vital, J. (2018). Highly active Cao catalysts from waste shells of egg, oyster and clam for biodiesel production. *Applied Catalysis A, General*, *567*, 56–64. doi:10.1016/j.apcata.2018.09.003

Rizwanul Fattah, I. M., Masjuki, H. H., Kalam, M. A., Mofijur, M., & Abedin, M. J. (2014). Effect of antioxidant on the performance and emission characteristics of a diesel engine fueled with palm biodiesel blends. *Energy Conversion and Management*, *79*, 265–272. doi:10.1016/j.enconman.2013.12.024

Roy, M. M., Tomita, E., Kawahara, N., Harada, Y., & Sakane, A. (2009). Performance and emission comparison of a supercharged dual-fuel engine fueled by producer gases with varying hydrogen content. *International Journal of Hydrogen Energy*, *34*(18), 7811–7822. doi:10.1016/j.ijhydene.2009.07.056

Ruddy, D. A., Hensley, J. E., Nash, C. P., Tan, E. C. D., Christensen, E., Farberow, C. A., ... Schaidle, J. A. (2019). Methanol to high-octane gasoline within a market-responsive biorefinery concept enabled by catalysis. *Nature Catalysis*, *2*(7), 632–640. doi:10.103841929-019-0319-2

Ruth, L. (2008). Bio or bust? *EMBO Reports*, *9*(2), 130–133. doi:10.1038j.embor.2008.6 PMID:18246105

Ryckebosch, E., Drouillon, M., & Vervaeren, H. (2011). Techniques for transformation of biogas to biomethane. *Biomass and Bioenergy*, *35*(5), 1633–1645. doi:10.1016/j.biombioe.2011.02.033

Sadiq Al-Baghdadi, M. A.-R. (2000). Performance study of a four-stroke spark ignition engine working with both of hydrogen and ethyl alcohol as supplementary fuel. *International Journal of Hydrogen Energy*, *25*(10), 1005–1009. doi:10.1016/S0360-3199(00)00012-4

Sadiq Al-Baghdadi, M. A.-R. (2001). The safe operation zone of the spark ignition engine working with dual renewable supplemented fuels (hydrogen+ethyl alcohol). *Renewable Energy*, *22*(4), 579–583. doi:10.1016/S0960-1481(00)00109-9

Saha, A. (2010). *Vaporization characteristics of pure and blended biofuel droplet injected into hot stream of air. (Master in Science)*. Orlando, FL: University of Central Florida.

Sahin, Z., Tuti, M., & Durgun, O. (2014). Experimental investigation of the effects of water adding to the intake air on the engine performance and exhaust emissions in a DI automotive diesel engine. *Fuel*, *115*, 884–895. doi:10.1016/j.fuel.2012.10.080

Sahoo, B., Sahoo, N., & Saha, U. (2009). Effect of engine parameters and type of gaseous fuel on the performance of dual-fuel gas diesel engines—A critical review. *Renewable & Sustainable Energy Reviews*, *13*(6-7), 1151–1184. doi:10.1016/j.rser.2008.08.003

Sahoo, P. K., Das, L. M., Babu, M. K. G., Arora, P., & Sing, V. P. (2009). Comparative evaluation of performance and emission characteristics of jatropha, karanja, and polanga-based biodiesel as fuel in a tractor engine. *Fuel*, *88*(9), 1698–1707. doi:10.1016/j.fuel.2009.02.015

Saibabu, K., Raju, V. D., Rao, K. A., Reddy, S. R., & Sai, P. T. Experimental studies on the influence of antioxidant additives with waste tamarind biodiesel on the diverse characteristics of diesel engine, International Journal of Ambient Energy, Published online July 9, 2019. . doi:10.1080/01430750.2019.1636871

Saifuddin, N., & Chua, K. H. (2004). Production of Ethyl Ester (Biodiesel) From Used Frying Oil: Optimization of Transesterification Process using Microwave Irradiation. *Malaysian Journal of Chemistry*, *6*(1), 77–82.

Sajeevan, A. C., & Sajith, V. (2016). Synthesis of stable cerium zirconium oxide nanoparticle - Diesel suspension and investigation of its effects on diesel properties and smoke. *Fuel*, *183*, 155–163. doi:10.1016/j.fuel.2016.06.048

Sajith, V., Sobhan, C. B., & Peterson, G. P. (2010). Experimental investigations on the effects of cerium oxide nanoparticle fuel additives on biodiesel. *Advances in Mechanical Engineering*, 1–6.

Saka, S., & Kusdiana, D. (2001). Biodiesel fuel from rapeseed oil as prepared in supercritical methanol. *Fuel*, *80*(2), 225–231. doi:10.1016/S0016-2361(00)00083-1

Sakthivel, R., Ramesh, K., Mohamed Shameer, P., & Purnachandran, R. (2018a). Experimental investigation on improvement of storage stability of bio-oil derived from intermediate pyrolysis of Calophyllum inophyllum seed cake. *Journal of the Energy Institute*. doi:10.1016/j.joei.2018.02.006

Sakthivel, R., & Ramesh, K. (2017). Influence of temperature on yield, composition and properties of the sub-fractions derived from slow pyrolysis of Calophyllum inophyllum de-oiled cake. *Journal of Analytical and Applied Pyrolysis*, *127*, 159–169. doi:10.1016/j.jaap.2017.08.012

Sakthivel, R., Ramesh, K., Joseph John Marshal, S., & Sadasivuni, K. K. (2019). Prediction of performance and emission characteristics of diesel engine fuelled with waste biomass pyrolysis oil using response surface methodology. *Renewable Energy*, *136*, 91–103. doi:10.1016/j.renene.2018.12.109

Sakthivel, R., Ramesh, K., Purnachandran, R., & Shameer, P. M. (2017). A review on the properties, performance and emission aspects of the third-generation biodiesels. *Renewable & Sustainable Energy Reviews*.

Sakthivel, R., Ramesh, K., Purnachandran, R., & Shameer, P. M. (2018). A review on the properties, performance and emission aspects of the third generation biodiesels. *Renewable & Sustainable Energy Reviews*, *82*, 2970–2992. doi:10.1016/j.rser.2017.10.037

Sakthivel, R., Ramesh, K., Shameer, P. M., & Purnachandran, R. (2018b). A Complete Analytical Characterization of Products Obtained from Pyrolysis of Wood Barks of Calophyllum inophyllum. *Waste and Biomass Valorization*. doi:10.100712649-018-0236-7

Salamatinia, B., Mootabadi, H., Bhatia, S., & Abdullah, A. Z. (2010). Optimization of ultrasonic-assisted heterogeneous biodiesel production from palm oil: A response surface methodology approach. *Fuel Processing Technology*, *91*(5), 441–448. doi:10.1016/j.fuproc.2009.12.002

Samiran, N. A., Ng, J.-H., Jaafar, M. N. M., Valera-Medina, A., & Chong, C. T. (2016). H2-rich syngas strategy to reduce NOx and CO emissions and improve stability limits under premixed swirl combustion mode. *International Journal of Hydrogen Energy*, *41*(42), 19243–19255. doi:10.1016/j.ijhydene.2016.08.095

Sandouqa, A., & Al-Hamamre, Z. (2019). Energy analysis of biodiesel production from jojoba seed oil. *Renewable Energy*, *130*, 831–842. doi:10.1016/j.renene.2018.07.015

Sanjid, A., Masjuki, H. H., Kalam, M. A., Rahman, A. S. M., Abedin, M. J., & Palash, S. M. (2014). Production of palm and jatropha based biodiesel and investigation of palm-jatropha combined blend properties, performance, exhaust emission and noise in an unmodified diesel engine. *Journal of Cleaner Production*, *65*, 295–303. doi:10.1016/j.jclepro.2013.09.026

Sankaralingam, R. K., & Venugopal, (2016). Current trends on combustion control methodsusing fuel reactivities. *International Journal of Automotive & Mechanical Engineering, 13*(3), 3788–3811. Retrieved from http://10.0.59.178/ijame.13.3.2016.20.0310%0Ahttp://ezlibrary.technion.ac.il/login?url=http://search.ebscohost.com/login.aspx?direct=true&db=aci&AN=122893253&lang=he&site=eds-live&scope=site

Santhanamuthu, M., Chittibabu, S., Tamizharasan, T., & Mani, T. (2014). Evaluation of CI engine performance fuelled by diesel-polanga oil blends doped with iron oxide nano particles. *International Journal of Chemtech Research*, *6*(2), 1299–1308.

Saraee, H. S., Jafarmadar, S., Taghavifar, H., & Ashrafi, S. (2015). Reduction of emissions and fuel consumption in a compression ignition engine using nanoparticles. *International Journal of Environmental Science and Technology*, *12*(7), 2245–2252. doi:10.100713762-015-0759-4

Sathiyamoorthi, R., & Sankaranarayanan, G. (2016). Effect of antioxidant additives on the performance and emission characteristics of a DICI engine using neat lemongrass oil-diesel blend. *Fuel*, *174*, 89–96. doi:10.1016/j.fuel.2016.01.076

Savery, C., & Cruzan, D. (1972). Methane recovery from chicken manure digestion. *Journal (Water Pollution Control Federation), 14*(2), 2349–2354. Retrieved from http://www.jstor.org/stable/25037689

Saxena, V., Kumar, N., & Saxena, V. K. (2017). A comprehensive review on combustion and stability aspects of metal nanoparticles and its additive effect on diesel and biodiesel fuelled CI engine. *Renewable & Sustainable Energy Reviews*, *70*, 563–588. doi:10.1016/j.rser.2016.11.067

Sebbar, N., Habisreuther, P., Bockhorn, H., Auzmendi-Murua, I., & Bozzelli, J. W. (2017). Di-tertiary-butyl peroxide decomposition and combustion with air: Reaction mechanism, Ignition, flame structure, and laminar flame velocities. *Energy & Fuels*, *31*(3), 2660–2273. doi:10.1021/acs.energyfuels.6b02201

Selim, M. Y., & Ghannam, M. T. (2009). Combustion study of stabilized water-in-diesel fuel emulsion. *Energy Sources, Part A: Recovery, Utilization, and Environmental Effects, 32*(3), 256-274.

Selim, M. Y. E., & Ghannam, M. (2010). Combustion study of stabilized water-in-diesel fuel emulsion. Energy Sources A Recov. *Utiliz. Environ. Effects*, *32*(3), 256–274. doi:10.1080/15567030802467621

Selvan, V. A. M., Anand, R. B., & Udayakumar, M. (2009). Effects of cerium oxide nanoparticle addition in diesel and diesel-biodiesel-ethanol blends on the performance and emission characteristics of a CI engine. *Journal of Engineering and Applied Sciences (Asian Research Publishing Network)*, *4*, 1819–1823.

Selvan, V. A. M., Anand, R. B., & Udayakumar, M. (2009). Effects of cerium oxide nanoparticle addition in diesel and diesel-biodiesel-ethanol blends on the performance and emission characteristics of a CI engine. *Journal of Engineering and Applied Sciences (Asian Research Publishing Network)*, *4*(7), 1819–6608.

Selvan, V. A. M., Anand, R. B., & Udayakumar, M. (2014). Effect of Cerium Oxide Nanoparticles and Carbon Nanotubes as fuel-borne additives in Diesterol blends on the performance, combustion and emission characteristics of a variable compression ratio engine. *Fuel*, *130*, 160–167. doi:10.1016/j.fuel.2014.04.034

Semelsberger, T. A., Borup, R. L., & Howard, L. (2006). Dimethyl ether (DME) as an alternative fuel. *Journal of Power Sources*, *156*(2), 497–511. doi:10.1016/j.jpowsour.2005.05.082

Semwal, S., Arora, A. K., Badoni, R. P., & Tuli, D. K. (2011). Biodiesel production using heterogeneous catalysts. *Bioresource Technology*, *102*(3), 2151–2161. doi:10.1016/j.biortech.2010.10.080 PMID:21106371

Serrano-Ruiz, J. C., Ramos-Fernández, E. V., & Sepúlveda-Escribano, A. (2012). From biodiesel and bioethanol to liquid hydrocarbon fuels: New hydrotreating and advanced microbial technologies. *Energy & Environmental Science*, *5*(2), 5638–5652. doi:10.1039/C1EE02418C

Shaafi, T., & Velraj, R. (2015). Influence of alumina nanoparticles, ethanol and isopropanol blend as additive with diesel–soybean biodiesel blend fuel: Combustion, engine performance and emissions. *Renewable Energy*, *80*, 655–663. doi:10.1016/j.renene.2015.02.042

Shabih, F., Amir, H., Muhammad, N., Temoor, A., Muhammad, S., Mohsin, T., ... Romana, T. (2018). Lignocellulosic Biomass: A Sustainable Bioenergy Source for the Future. *Protein and Peptide Letters*, *25*(2), 148–163. doi:10.2174/0929866525666180122144504 PMID:29359659

Shahabuddin, M., Kalam, M., Masjuki, H., & Mofijur, M. (2012). An experimental investigation into biodiesel stability by means of oxidation and property determination. *Energy*, *44*(1), 616–622. doi:10.1016/j.energy.2012.05.032

Shahabuddin, M., Liaquat, A., Masjuki, H., Kalam, M., & Mofijur, M. (2013). Ignition delay, combustion and emission characteristics of diesel engine fuelled with biodiesel. *Renewable & Sustainable Energy Reviews*, *21*, 623–632. doi:10.1016/j.rser.2013.01.019

Shahid, E. M., & Jamal, Y. (2008). A review of biodiesel as vehicular fuel. *Renewable & Sustainable Energy Reviews*, *12*(9), 2484–2494. doi:10.1016/j.rser.2007.06.001

Shahir, S., Masjuki, H., Kalam, M., Imran, A., & Ashraful, A. (2015). Performance and emission assessment of diesel-biodiesel-ethanol/bioethanol blend as a fuel in diesel engines: A review. *Renewable & Sustainable Energy Reviews*, *48*, 62–78. doi:10.1016/j.rser.2015.03.049

Shah, S., & Gupta, M. N. (2007). Lipase catalyzed preparation of biodiesel from Jatropha oil in a solvent free system. *Process Biochemistry*, *42*(3), 409–414. doi:10.1016/j.procbio.2006.09.024

Shamsul, N. S., Kamarudin, S. K., & Rahman, N. A. (2017). Conversion of bio-oil to bio gasoline via pyrolysis and hydrothermal: A review. *Renewable & Sustainable Energy Reviews*, *80*, 538–549. doi:10.1016/j.rser.2017.05.245

Sharon, H., Jai Shiva Ram, P., Jenis Fernando, K., Murali, S., & Muthusamy, R. (2013). Fuelling a direct injection diesel engine with diesel-used palm oil-butanol blends – an experimental study. *Energy Conversion and Management*, *73*, 95–105. doi:10.1016/j.enconman.2013.04.027

Sharuddin, S. D. A., Abnisa, F., Daud, W. M. A. W., & Aroua, M. K. (2018). Pyrolysis of plastic waste for liquid fuel production as prospective energy resource. *IOP Conference Series. Materials Science and Engineering*, *334*, 012001. doi:10.1088/1757-899X/334/1/012001

Shaw, D., Jamal Akhtar, S., Priyam, A., & Kumar Singh, R. (2016). Performance Study of Dual Fuel Engine Using Producer Gas as Secondary Fuel. *Carbon: Science and Technology*, *8*, 63–71.

Shivakumar., Pai, P.S., Rao, B.R.S., (2011). Artificial Neural Network based prediction of performance and emission characteristics of a variable compression ratio CI engine using WCO as a biodiesel at different injection timings. *Applied Energy 88, 2344–2354.*

Singh, A. K., Fernando, S. D., & Hernandez, R. (2007). Base-catalyzed fast transesterification of soybean oil using ultrasonication. *Energy & Fuels*, *21*(2), 1161–1164. doi:10.1021/ef060507g

Singh, D., Sharma, D., Soni, S. L., Sharma, S., & Kumari, D. (2019). Chemical compositions, properties, and standards for different generation biodiesels: A review. *Fuel*, *253*, 60–71. doi:10.1016/j.fuel.2019.04.174

Singh, P. J., Khurma, J., & Singh, A. (2010). Preparation, characterization, engine performance and emission characteristics of coconut oil-based hybrid fuels. *Renewable Energy*, *35*(9), 2065–2070. doi:10.1016/j.renene.2010.02.007

Singh, S. K. (2019). Solubility of lignin and chitin in ionic liquids and their biomedical applications. *International Journal of Biological Macromolecules*, *132*, 265–277. doi:10.1016/j.ijbiomac.2019.03.182 PMID:30930266

Singh, S. K., & Dhepe, P. L. (2016a). Isolation of lignin by organosolv process from different varieties of rice husk: Understanding their physical and chemical properties. *Bioresource Technology*, *221*, 310–317. doi:10.1016/j.biortech.2016.09.042 PMID:27648850

Singh, S. K., & Dhepe, P. L. (2016b). Ionic liquids catalyzed lignin liquefaction: Mechanistic studies using TPO-MS, FT-IR, RAMAN and 1D, 2D-HSQC/NOESY NMR. *Green Chemistry*, *18*(14), 4098–4108. doi:10.1039/C6GC00771F

Singh, S. K., & Dhepe, P. L. (2018). Experimental evidences for existence of varying moieties and functional groups in assorted crop waste derived organosolv lignins. *Industrial Crops and Products*, *119*, 144–151. doi:10.1016/j.indcrop.2018.04.002

Singh, S. K., Savoy, A., Yuan, Z., Luo, H., Stahl, S. S., Hegg, E. L., & Hodge, D. B. (2019). Integrated Two-Stage Alkaline-Oxidative Pretreatment of Hybrid Poplar. Part 1: Impact of Alkaline Pre-Extraction Conditions on Process Performance and Lignin Properties. *Industrial & Engineering Chemistry Research*, *58*(35), 15989–15999. doi:10.1021/acs.iecr.9b01124

Sivakumar, M., Shanmugasundaram, N., Rameshkumar, R., & Syed Thasthagir, M. (2018). Effects of aluminium oxide nanoparticles blended pongamia methyl ester on performance, combustion and emission characteristics of diesel engine. *Renewable Energy*, *116*, 518–526. doi:10.1016/j.renene.2017.10.002

Sivakumar, M., Shanmugasundram, N., & Rameshkumar, R. (2017). Effects of pongamia methyl esters and its blends on a diesel engine performance, combustion and emission characteristics. *Environmental Progress & Sustainable Energy*, *36*(1), 269–276. doi:10.1002/ep.12492

Sivalakshmi, S., & Balusamy, T. (2013). Effect of biodiesel and its blends with diethyl ether on the combustion, performance and emissions of a diesel engine. *Fuel*, *106*, 106–110. doi:10.1016/j.fuel.2012.12.033

Sobyanin, V., Sadykov, V., Kirillov, V., Kuzmin, V., Kuzin, N., Vostrikov, Z., ... Terentyev, V. (2005). Syngas as a fuel for IC and diesel engines: efficiency and harmful emissions cut-off. In *Proceedings of International Hydrogen Energy Congress and Exhibition IHEC*.

Solubility of alcohols. (2019). Retrieved from https://www.solubilityofthings.com/water/alcohols

Song, J., Wang, J., & Boehman, A. L. (2006). The role of fuel-borne catalyst in diesel particulate oxidation behaviour. *Combustion and Flame*, *146*(1-2), 73–84. doi:10.1016/j.combustflame.2006.03.012

Soni, G. S., Rathod, P. P., & Goswami, J. J. (2015). Performance and emission characteristics of CI engine using diesel and biodiesel blends with nanoparticles as additive-A review study. *International Journal of Engineering Development and Research, 3*(4), 879–884.

Sorathia, H. S., & Yadav, H. J. (2012). Energy Analyses To a Ci-Engine Using Diesel and Bio-Gas Dual Fuel- a Review Study. *International Journal of Advanced Engineering Research and Studies, Vol.*, 212–217.

Soudagar M. E. M., Ghazali, N. N., Kalam, M. A., & Badruddin, I. A. (2019).The effects of grapheme oxide nanoparticle additive stably dispersed in dairy scum oil biodiesel-diesel fuel blend on CI engine: performance, emission and combustion characteristics. Fuel, 257, ID:116015, . doi:10.1016/j.fuel.2019.116015

Soudagar, M. E. M., Nik-Ghazali, N.-N., Kalam, M. A., Badruddin, I., Banapurmath, N., & Akram, N. (2018). The effect of nano-additives in diesel-biodiesel fuel blends: A comprehensive review on stability, engine performance and emission characteristics. *Energy Conversion and Management, 178*, 146–177. doi:10.1016/j.enconman.2018.10.019

Sprules, F. J., & Price, D. (1950). Production of fatty esters. United States: NOPCO CHEM CO. Retrieved from http://www.freepatentsonline.com/ 2494366.html

Sridhar, G., Sridhar, H., Dasappa, S., Paul, P., Rajan, N., & Mukunda, H. (2005). Development of producer gas engines. *Proceedings of the Institution of Mechanical Engineers. Part D, Journal of Automobile Engineering, 219*(3), 423–438. doi:10.1243/095440705X6596

Srihari, S., & Thirumalini, S. (2017). Investigation on reduction of emission in PCCI-DI engine with biofuel blends. *Renewable Energy, 114*, 1232–1237. doi:10.1016/j.renene.2017.08.008

Srihari, S., Thirumalini, S., & Prashanth, K. (2017). An experimental study on the performance and emission characteristics of PCCI-DI engine fuelled with diethyl ether-biodiesel-diesel blends. *Renewable Energy, 107*, 440–447. doi:10.1016/j.renene.2017.01.015

Srinivasa Rao, M., & Anand, R. B. (2015). Production characterization and working characteristics in DICI engine of pongamia biodiesel. *Ecotoxicology and Environmental Safety, 121*, 16–21. doi:10.1016/j.ecoenv.2015.07.031 PMID:26242967

Steinberg, M. (1975). Electrolytic synthesis of methanol from CO2, US Pat. No. US3959094A.

Stöcker, M. (2008). Biofuels and Biomass-To-Liquid Fuels in the Biorefinery: Catalytic Conversion of Lignocellulosic Biomass using Porous Materials. *Angewandte Chemie International Edition, 47*(48), 9200–9211. doi:10.1002/anie.200801476 PMID:18937235

Straathof, A. J. J. (2014). Transformation of Biomass into Commodity Chemicals Using Enzymes or Cells. *Chemical Reviews, 114*(3), 1871–1908. doi:10.1021/cr400309c PMID:23987659

Subramaniam, D., Murugesan, A., Avinash, A., & Kumaravel, A. (2013). Bio-diesel production and its engine characteristics - An expatiate view. *Renewable & Sustainable Energy Reviews*. doi:10.1016/j.rser.2013.02.002

Subramanian, K. A. (2011). A comparison of water-diesel emulsion and timed injection of water into the intake manifold of a diesel engine for simultaneous control of NO and smoke emissions. *Energy Conversion and Management, 52*(2), 849–857. doi:10.1016/j.enconman.2010.08.010

Sudheesh, K., & Mallikarjuna, J. M. (2010). Diethyl ether as an ignition improver for biogas homogeneous charge compression ignition (HCCI) operation - An experimental investigation. *Energy, 35*(9), 3614–3622. doi:10.1016/j.energy.2010.04.052

Sukjit, E., Herreros, J. M., Dearn, K. D., Garcia-Contreras, R., & Tsolakis, A. (2012). The effect of the addition of individual methyl esters on the combustion and emissions of ethanol and butanol-diesel blends. *Energy*, *42*(1), 364–374. doi:10.1016/j.energy.2012.03.041

Supple, B., Howard-Hildige, R., Gonzalez-Gomez, E., & Leahy, J. J. (2002). The effect of steam treating waste cooking oil on the yield of methyl ester. *Journal of the American Oil Chemists' Society*, *79*(2), 175–178. doi:10.100711746-002-0454-1

Suresh, S., Venkitaraj, K., & Selvakumar, P. (2011). *Synthesis, Characterisation of Al_2O_3-Cu Nano composite powder and water based nanofluids.* Paper presented at the Advanced Materials Research.

Suresh, V., & Amirthagadeswaran, K. S. (2015). Combustion and performance characteristics of water in diesel emulsion fuel. Energy Sources A Recov. *Utiliz. Environ. Effects*, *37*(18), 2020–2028. doi:10.1080/15567036.2015.1072605

Suresh, V., & Amirthagadeswaran, K. S. (2016). The role of water-in-diesel emulsion and its additives on diesel engine performance and emission levels: A retrospective review. *Alexandria Engineering Journal*, *55*(3), 2463–2472. doi:10.1016/j.aej.2016.07.021

Suryanto, A., Sabara, Z., Artiningsih, A., & Ismail, H. (2018). Microwave Assisted Trans-esterification of Cotton Seed Oil for Biodiesel Production. *Reaktor*, *18*, 27–30. doi:10.14710/reaktor.18.1.27-30

Tabatabaei, M., Aghbashlo, M., Dehhaghi, M., Panahi, H. K. S., Mollahosseini, A., Hosseini, M., & Soufiyan, M. M. (2019). Reactor technologies for biodiesel production and processing: A review. *Progress in Energy and Combustion Science*, *74*, 239–303. doi:10.1016/j.pecs.2019.06.001

Taghizadeh-Alisaraei, A., & Rezaei-Asl, A. (2016). The effect of added ethanol to diesel fuel on performance, vibration, combustion and knocking of a CI engine. *Fuel*, *185*, 718–733. doi:10.1016/j.fuel.2016.08.041

Takase, M., Feng, W., Wang, W., Gu, X., Zhu, Y., Li, T., ... Wu, X. (2014). Silybum marianum oil as a new potential non-edible feedstock for biodiesel: A comparison of its production using conventional and ultrasonic assisted method. *Fuel Processing Technology*, *123*, 19–26. doi:10.1016/j.fuproc.2014.01.032

Tamilvanan, A., Balamurugan, K., Ponappa, K., & Madhan Kumar, B. (2016). Using response surface methodology in synthesis of ultrafine copper nanoparticles by electrolysis. *International Journal of Nanoscience, 15*(01n02).

Tamilvanan, A., Balamurugan, K., Ponappa, K., & Kumar, B. M. (2014). Copper nanoparticles: Synthetic strategies, properties and multifunctional application. *International Journal of Nanoscience*, *13*(02). doi:10.1142/S0219581X14300016

Tamilvanan, A., Balamurugan, K., & Vijayakumar, M. (2019). Effects of nano-copper additive on performance, combustion and emission characteristics of Calophyllum inophyllum biodiesel in CI engine. *Journal of Thermal Analysis and Calorimetry*, *136*(1), 317–330. doi:10.100710973-018-7743-4

Tan, Y. H., Abdullah, M. O., Nolasco-Hipolito, C., & Ahmad Zauzi, N. S. (2017). Application of RSM and Taguchi methods for optimizing the transesterification of waste cooking oil catalyzed by solid ostrich and chicken-eggshell derived CaO. *Renewable Energy, 114*(PB), 437–447. doi:10.1016/j.renene.2017.07.024

Tan, K. T., Lee, K. T., & Mohamed, A. R. (2010). A glycerol-free process to produce biodiesel by supercritical methyl acetate technology: An optimization study via Response Surface Methodology. *Bioresource Technology*, *101*(3), 965–969. doi:10.1016/j.biortech.2009.09.004 PMID:19773156

Tan, Y. H., Abdullah, M. O., & Nolasco-Hipolito, C. (2015). The potential of waste cooking oil-based biodiesel using heterogeneous catalyst derived from various calcined eggshells coupled with an emulsification technique: A review on the emission reduction and engine performance. *Renewable & Sustainable Energy Reviews*, *47*, 589–603. doi:10.1016/j.rser.2015.03.048

Taylor, M., Atri, S. S., & Minhas, S. (2005). Evalueserve Analysis: Developments in Microwave Chemistry. Retrieved from www.rsc.org/images/ evalueserve_ tcm18-16758.pdf

Tenenbaum, D. J. (2008). Food vs. Fuel: Diversion of Crops Could Cause More Hunger. *Environmental Health Perspectives*, *116*(6), 254–257. doi:10.1289/ehp.116-a254 PMID:18560500

Teoh, Y. H., How, H. G., Masjuki, H. H., Kalam, M. A., & Alabdulkarem, A. (2019). Investigation on Particulate Emissions and Combustion Characteristics of a Common-Rail Diesel Engine Fueled with Moringa Oleifera Biodiesel-Diesel Blends. *Renewable Energy*, *136*, 521–534. doi:10.1016/j.renene.2018.12.110

Tesfa, B., Mishra, R., Gu, F., & Ball, A. (2013). Combustion and performance characteristics of CI engine running with biodiesel. *Energy*, *51*(1), 101–115. doi:10.1016/j.energy.2013.01.010

Thakur, A. K., Kaviti, A. K., Mehra, R., & Mer, K. K. S. (2017). Progress in performance analysis of ethanol-gasoline blends on SI engine. *Renewable & Sustainable Energy Reviews*, *69*, 324–340. doi:10.1016/j.rser.2016.11.056

Thanh, L. T., Okitsu, K., Sadanaga, Y., Takenaka, N., Maeda, Y., & Bandow, H. (2010). A two-step continuous ultrasound assisted production of biodiesel fuel from waste cooking oils: A practical and economical approach to produce high quality biodiesel fuel. *Bioresource Technology*, *101*(14), 5394–5401. doi:10.1016/j.biortech.2010.02.060 PMID:20219362

Thirugnanasambandham, K., Shine, K., Aziz, H. A., & Gimenes, M. L. (2017). Biodiesel synthesis from waste oil using novel microwave technique: Response surface modeling and optimization. *Energy Sources. Part A, Recovery, Utilization, and Environmental Effects*, *39*(7), 636–642. doi:10.1080/15567036.2016.1196270

Tilman, D., Hill, J., & Lehman, C. (2006). Carbon-Negative Biofuels from Low-Input High-Diversity Grassland Biomass. *Science*, *314*(5805), 1598–1600. doi:10.1126cience.1133306 PMID:17158327

Tippayawong, N., & Thanompongchart, P. (2010). Biogas quality upgrade by simultaneous removal of CO2 and H2S in a packed column reactor. *Energy*, *35*(12), 4531–4535. doi:10.1016/j.energy.2010.04.014

Tomar, M., & Kumar, N. (2019). Influence of nanoadditives on the performance and emission characteristics of a CI engine fuelled with diesel, biodiesel, and blends–a review. *Energy Sources. Part A, Recovery, Utilization, and Environmental Effects*, 1–18. doi:10.1080/15567036.2019.1623347

Trifirò, F. (2019). Fuels from Biomass. *TECNICA ITALIANA-Italian Journal of Engineering Science*, *63*(1), 86–93. doi:10.18280/ti-ijes.630112

Tse, H., Leung, C. W., & Cheung, C. S. (2015). Investigation on the combustion characteristics and particulate emissions from a diesel engine fueled with diesel biodiesel- ethanol blends. *Energy*, *83*, 343–350. doi:10.1016/j.energy.2015.02.030

Tshizanga, N., Aransiola, E. F., & Oyekola, O. (2017). Optimisation of biodiesel production from waste vegetable oil and eggshell ash. *South African Journal of Chemical Engineering*, *23*, 145–156. doi:10.1016/j.sajce.2017.05.003

Tsolakis, A., Rounce, P., Leung, P., & York, A. P. E. (2010). A comparison of diesel and biodiesel emissions using dimethyl carbonate as an oxygenated additive. *Energy & Fuels*, *24*(9), 4812–4819. doi:10.1021/ef100103z

Turner, J. W., Lewis, A. G., Akehurst, S., Brace, C. J., Verhelst, S., Vancoillie, J., ... Edwards, P. P. (2018). Alcohol fuels for spark-ignition engines: Performance, efficiency and emission effects at mid to high blend rates for binary mixtures and pure components. *Proceedings of the Institution of Mechanical Engineers. Part D, Journal of Automobile Engineering*, *232*(1), 36–56. doi:10.1177/0954407017752832

U.S. EIA. (2018a). International Energy Statistics. Retrieved from https://www.eia.gov/

U.S. EIA. (2018b). The Short-term Energy Outlook. Retrieved from https://www.eia.gov/outlooks/steo/

Ullman, T. L., Spreen, K. B., & Mason, R. L. (1994). Effects of cetane number, cetane improver, aromatics and oxygenates on 1994 heavy-duty diesel engine emissions. SAE Technical Paper.

Uprety, B. K., Chaiwong, W., Ewelike, C., & Rakshit, S. K. (2016). Biodiesel production using heterogeneous catalysts including wood ash and the importance of enhancing byproduct glycerol purity. *Energy Conversion and Management*, *115*, 191–199. doi:10.1016/j.enconman.2016.02.032

Utlu, Z., & Kocak, M. S. (2008). The effect of biodiesel fuel obtained from waste frying oil on direct injection diesel engine performance and exhaust emissions. *Renewable Energy*, *33*(8), 1936–1941. doi:10.1016/j.renene.2007.10.006

Uzun, B. B., Kılıç, M., Özbay, N., Pütün, A. E., & Pütün, E. (2012). Biodiesel production from waste frying oils: Optimization of reaction parameters and determination of fuel properties. *Energy*, *44*(1), 347–351. doi:10.1016/j.energy.2012.06.024

Vaclav, S. (2017). *Energy Transitions: Global and National Perspectives*. BP Statistical Review of World Energy.

Vallinayagam, R., Vedharaj, S., Yang, W. M., Saravanan, C. G., Lee, P. S., Chua, K. J. E., & Chou, S. K. (2014). Impact of ignition promoting additives on the characteristics of a diesel engine powered by pine oil-diesel blend. *Fuel*, *117*, 278–285. doi:10.1016/j.fuel.2013.09.076

Varatharajan, K., Cherlathan, M., & Velraj, R. (2011). Mitigration of NO_x emissions from a jatropha biodiesel fuelled DI diesel engine using antioxidant additives. *Fuel*, *90*(8), 2721–2725. doi:10.1016/j.fuel.2011.03.047

Varma, M. N., Deshpande, P. A., & Madras, G. (2010). Synthesis of biodiesel in supercritical alcohols and supercritical carbon dioxide. *Fuel*, *89*(7), 1641–1646. doi:10.1016/j.fuel.2009.08.012

Varun, S., Singh, P., Tiwari, S. K., Singh, R., & Kumar, N. (2017). Modification in combustion chamber geometry of CI engines for suitability of biodiesel: A review. *Renewable & Sustainable Energy Reviews*, *79*, 1016–1033. doi:10.1016/j.rser.2017.05.116

Venkatesan, S. P., & Kadiresh, P. N. (2016). Influence of an aqueous cerium oxide nanofluid fuel additive on performance and emission characteristics of a compression ignition engine. *International Journal of Ambient Energy*, *37*(1), 64–67. doi:10.1080/01430750.2014.882863

Venu, H., & Madhavan, V. (2016). Effect of nano additives (titanium and zirconium oxides) and diethyl ether on biodiesel-ethanol fuelled CI engine. *Journal of Mechanical Science Technology*, *30*(5), 2361–2368. doi:10.100712206-016-0446-5

Venu, H., & Madhavan, V. (2017). Effect of diethyl ether and Al_2O_3 nano additives in diesel-biodiesel-ethanol blends: Performance, combustion and emission characteristics. *Journal of Mechanical Science and Technology*, *31*(1), 409–420. doi:10.100712206-016-1243-x

Venu, H., Raju, V. D., & Subramani, L. (2019). Combined effect of influence of nano additives, combustion geometry and injection timing in a diesel engine fuelled with (diesel-biodiesel-ethanol) blends. *Energy*, *174*, 386–406. doi:10.1016/j.energy.2019.02.163

Venu, H., Subramani, L., & Raju, V. D. (2019). Emission reduction in a DI diesel engine using exhaust gas recirculation (EGR) of palm biodiesel blended with TiO2 nano additives. *Renewable Energy*, *140*, 245–263. doi:10.1016/j.renene.2019.03.078

Verma, S., Das, L. M., & Kaushik, S. C. (2017). Effects of varying composition of biogas on performance and emission characteristics of compression ignition engine using exergy analysis. *Energy Conversion and Management*, *138*, 346–359. doi:10.1016/j.enconman.2017.01.066

Vijay, V. K. (2007). Biogas refining for production of bio-methane and its bottling for automotive applications and holistic development. *Proc. of International Symposium on Eco Topia, 07*, 35–40. Retrieved from http://www.esi.nagoya-u.ac.jp/h/isets07/Contents/Session05/1018Vijay.pdf

Viswanathan, K., Balasubramanian, D., Subramanian, T., & Varuvel, E. G. (2019). Investigating the combined effect of thermal barrier coating and antioxidants on pine oil in DI diesel engine. *Environmental Science and Pollution Research International*, *26*(15), 15573–15599. doi:10.100711356-019-04649-6 PMID:30945081

Vohra, M., Manwar, J., Manmode, R., Padgilwar, S., & Patil, S. (2014). Bioethanol production: Feedstock and current technologies. *Journal of Environmental Chemical Engineering*, *2*(1), 573–584. doi:10.1016/j.jece.2013.10.013

Volli, V., Singh, R. K., & Murugan, S. (2014). The Use of Mustard Cake Pyrolytic Oil Blends as Fuel in a Diesel Engine. *Waste and Biomass Valorization*, *5*(4), 661–668. doi:10.100712649-013-9270-7

von Blottnitz, H., & Curran, M. A. (2007). A review of assessments conducted on bio-ethanol as a transportation fuel from a net energy, greenhouse gas, and environmental life cycle perspective. *Journal of Cleaner Production*, *15*(7), 607–619. doi:10.1016/j.jclepro.2006.03.002

Wahidin, S., Idris, A., & Shaleh, S. R. M. (2014). Rapid biodiesel production using wet microalgae via microwave irradiation. *Energy Conversion and Management*, *84*, 227–233. doi:10.1016/j.enconman.2014.04.034

Wang, T., Zhang, Q., Ding, M., Wang, C., Li, Y., Zhang, Q., & Ma, L. (2017). Bio-gasoline Production by Coupling of Biomass Catalytic Pyrolysis and Oligomerization Process. *Energy Procedia*, *105*, 858–863. doi:10.1016/j.egypro.2017.03.401

Wang, X., Qin, X., Li, D., Yang, B., & Wang, Y. (2017). One-step synthesis of high-yield biodiesel from waste cooking oils by a novel and highly methanol-tolerant immobilized lipase. *Bioresource Technology*, *235*, 18–24. doi:10.1016/j.biortech.2017.03.086 PMID:28351728

Wang, Y., Liu, H., Huang, Z., & Ke, X. (2016). Particulate Size Distribution from Diesel–Dimethyl Ether Dual Fuel Premixed Compression Ignition Combustion Engine. *Journal of Energy Engineering*, *142*(4). doi:10.1061/(ASCE)EY.1943-7897.0000351

Wang, Y., Ou, S., Liu, P., Xue, F., & Tang, S. (2006). Comparison of two different processes to synthesize biodiesel by waste cooking oil. *Journal of Molecular Catalysis A Chemical*, *252*(1–2), 107–112. doi:10.1016/j.molcata.2006.02.047

Watanabe, H., Suzuki, Y., Harada, T., Matsushita, Y., Aoki, H., & Miura, T. (2010). An experimental investigation of the breakup characteristics of secondary atomization of emulsified fuel Droplet. *Energy*, *35*(2), 806–813. doi:10.1016/j.energy.2009.08.021

Wei, C. Y., Huang, T. C., & Chen, H. H. (2013). Biodiesel production using supercritical methanol with carbon dioxide and acetic acid. *Journal of Chemistry*, *2013*, 1–6. doi:10.1155/2013/789594

Wei, H., Liu, W., Chen, X., Yang, Q., Li, J., & Chen, H. (2019). Renewable bio-jet fuel production for aviation: A review. *Fuel*, *254*. doi:10.1016/j.fuel.2019.06.007

Weisz, P. B., Haag, W. O., & Rodewald, P. G. (1979). Catalytic Production of High-Grade Fuel (Gasoline) from Biomass Compounds by Shape-Selective Catalysis. *Science*, *206*(4414), 57–58. doi:10.1126cience.206.4414.57 PMID:17812451

Wender, I. (1996). Reactions of synthesis gas. *Fuel Processing Technology*, *48*(3), 189–297. doi:10.1016/S0378-3820(96)01048-X

World Health Organization. (1989). Crude Oil. *IARC Monographs on the Evaluation of Carcinogenic Risks to Humans*, *45*, 119–158. PMID:2664246

Wu, P., Yang, Y., Colucci, J. A., & Grulke, E. A. (2007). Effect of ultrasonication on droplet size in biodiesel mixtures. *JAOCS. Journal of the American Oil Chemists' Society*, *84*(9), 877–884. doi:10.100711746-007-1114-9

Xiao, F., Sohrabi, A., Galal, M., & Karim, G. (2009). The Performance of an IDI Diesel Engine Having Low Concentrations of Hydrogen in the Intake Air: SAE Technical Paper.

Xiaolu, L., Hongyan, C., Zhiyong, Z., & Zhen, H. (2006). Study of combustion and emission characteristics of a diesel engine operated with dimethyl carbonate. *Energy Conversion and Management*, *47*(11-12), 1438–1448. doi:10.1016/j.enconman.2005.08.021

Xue, J., Grift, T. E., & Hansen, A. C. (2011). Effect of biodiesel on engine performances and emissions. *Renewable & Sustainable Energy Reviews*, *15*(2), 1098–1116. doi:10.1016/j.rser.2010.11.016

Yadav, A. K., Khan, M. E., Pal, A., & Ghosh, U. (2018). Performance and emission characteristics of a stationary diesel engine fuelled by Schleichera Oleosa Methyl Ester (SOME) produced through hydrodynamic cavitation process. *Egyptian Journal of Petroleum*, *27*(1), 89–93. doi:10.1016/j.ejpe.2017.01.007

Yamin, J. A., Sakhnini, N., Sakhrieh, A., & Hamdan, M. A. (2013). Environmental and performance study of a 4-Stroke CI engine powered with waste oil biodiesel. *Sustain. Cities Soc.*, *9*, 32–38. doi:10.1016/j.scs.2013.02.002

Yanfeng, G., Liu, S., Guo, H., Tiegang, H., & Longbao, Z. (2013). A new diesel oxygenate additive and its effects on engine combustion and emissions. *Applied Thermal Engineering*, *27*(1), 202–207. doi:10.1016/j.applthermaleng.2006.04.021

Yang, H. H, Lee, W. J., Mi, H. H, Wong, C. H., & Chen, C. B. (1998). PAH emission influenced by MN-based additive and turbocharging from a heavy-duty diesel engine. Environment inter, 24, 389-403.

Yang, B., & Wyman, C. E. (2008). Pretreatment: The key to unlocking low-cost cellulosic ethanol. *Biofuels, Bioproducts, & Biorefining*, *2*(1), 26–40. doi:10.1002/bbb.49

Yang, W. M., An, H., Chou, S. K., Chua, K. J., Mohan, V., Sivasankaralingma, V., ... Li Maghbouli, J. (2013). Impact of emulsion fuel with nano-organic additives on the performance of diesel engine. *Applied Energy*, *112*, 1206–1212. doi:10.1016/j.apenergy.2013.02.027

Yang, W., An, H., Chou, S., Vedharaji, S., Vallinagam, R., Balaji, M., ... Chua, K. J. E. (2013). Emulsion fuel with novel nano-organic additives for diesel engine application. *Fuel*, *104*, 726–731. doi:10.1016/j.fuel.2012.04.051

Yang, Z., & Xie, W. (2007). Soybean oil transesterification over zinc oxide modified with alkali earth metals. *Fuel Processing Technology*, *88*(6), 631–638. doi:10.1016/j.fuproc.2007.02.006

Yao, M., Zheng, Z., & Liu, H. (2009). Progress and recent trends in homogeneous charge compression ignition (HCCI) engines. *Progress in Energy and Combustion Science*, *35*(5), 398–437. doi:10.1016/j.pecs.2009.05.001

Yassin, F. A., El Kady, F. Y., Ahmed, H. S., Mohamed, L. K., Shaban, S. A., & Elfadaly, A. K. (2015). Highly effective ionic liquids for biodiesel production from waste vegetable oils. *Egyptian Journal of Petroleum*, *24*(1), 103–111. doi:10.1016/j.ejpe.2015.02.011

Yilmaz, I. T., & Gumus, M. (2017). Investigation of the effect of biogas on combustion and emissions of TBC diesel engine. *Fuel*, *188*, 69–78. doi:10.1016/j.fuel.2016.10.034

Yilmaz, N., & Atmanli, A. (2017). Experimental assessment of a diesel engine fuelled with diesel-biodiesel-1-pentanol blends. *Fuel*, *191*, 190–197. doi:10.1016/j.fuel.2016.11.065

Yilmaz, N., & Vigil, F. M. (2014). Potential use of a blend of diesel, biodiesel, alcohols and vegetable oil in compression ignition engines. *Fuel*, *124*, 168–172. doi:10.1016/j.fuel.2014.01.075

Ying, W., Longbao, Z., & Hewu, W. (2006). Diesel emission improvements by the use of oxygenated DME/diesel blend fuels. *Atmospheric Environment*, *40*(13), 2313–2320. doi:10.1016/j.atmosenv.2005.12.016

Yongcheng, H., Longbao, Z., Shangxue, W., & Shenghua, L. (2006). Study on the performance and emissions of a compression ignition engine fuelled with Fischer– Tropsch diesel fuel. *Proc. Institution of Mechanical Engineers Part D: Journal of Automobile Engineering*, *220*, 827–835.

Yoon, S. H., & Lee, C. S. (2011a). Effect of biofuels combustion on the nanoparticle and emission characteristics of a common-rail di diesel engine. *Fuel*, *90*(10), 3071–3077. doi:10.1016/j.fuel.2011.05.007

Yoon, S. H., & Lee, C. S. (2011b). Experimental investigation on the combustion and exhaust emission characteristics of biogas–biodiesel dual-fuel combustion in a CI engine. *Fuel Processing Technology*, *92*(5), 992–1000. doi:10.1016/j.fuproc.2010.12.021

Yuan, Z., Singh, S. K., Bals, B. D., Hodge, D. B., & Hegg, E. L. (2019). Integrated Two-Stage Alkaline-Oxidative Pretreatment of Hybrid Poplar. Part 2: Impact of Cu-Catalyzed Alkaline Hydrogen Peroxide Pretreatment Conditions on Process Performance and Economics. *Industrial & Engineering Chemistry Research*, *58*(35), 16000–16008. doi:10.1021/acs.iecr.9b00901

Yu, D., Tian, L., Wu, H., Wang, S., Wang, Y., Ma, D., & Fang, X. (2010). Ultrasonic irradiation with vibration for biodiesel production from soybean oil by Novozym435. *Process Biochemistry*, *45*(4), 519–525. doi:10.1016/j.procbio.2009.11.012

Yunus Khan, T. M., Atabani, A. E., Badruddin, I. A., Ankalgi, R. F., Mainuddin Khan, T. K., & Badarudin, A. (2015). Ceiba pentandra, Nigella sativa and their blend as prospective feedstocks for biodiesel. *Industrial Crops and Products*, *65*, 367–373. doi:10.1016/j.indcrop.2014.11.013

Yuvarajan, D., & Babu, M. D. (2018). Experimental investigation on the influence of titanium dioxide nanofluid on emission pattern of biodiesel in a diesel engine. *Atmospheric Pollution Research*, *9*(1), 47–52. doi:10.1016/j.apr.2017.06.003

Zaharin, M. S. M., Abdullah, N. R., Najafi, G., Sharudin, H., & Yusaf, T. (2017). Effects of physicochemical properties of biodiesel fuel blends with alcohol on diesel engine performance and exhaust emissions: A review. *Renewable & Sustainable Energy Reviews*, *79*, 475–493. doi:10.1016/j.rser.2017.05.035

Zeng, J., Xiao, R., Zhang, H., Chen, X., Zeng, D., & Ma, Z. (2017). Syngas production via biomass self-moisture chemical looping gasification. *Biomass and Bioenergy*, *104*, 1–7. doi:10.1016/j.biombioe.2017.03.020

Zhang, X., Reece, D., Haws, R., & Moller, G. (1998). Biodegradability of biodiesel in the aquatic environment. ASAE, 41-0(5), 1423. doi:10.13031/2013.17277

Zhang, D., Ma, Y., & Zhu, M. (2013). Nanostructure and oxidative properties of soot from a compression ignition engine: The effect of a homogeneous combustion catalyst. *Proceedings of the Combustion Institute*, *34*(1), 1869–1876. doi:10.1016/j.proci.2012.05.096

Zhang, Q., Chang, J., Wang, T., & Xu, Y. (2007). Review of biomass pyrolysis oil properties and upgrading research. *Energy Conversion and Management*, *48*(1), 87–92. doi:10.1016/j.enconman.2006.05.010

Zhang, Q., Chang, J., Wang, & Xu, Y. (2006). Upgrading Bio-oil over Different Solid Catalysts. *Energy & Fuels*, *20*(6), 2717–2720. doi:10.1021/ef060224o

Zhang, X., Yan, S., Tyagi, R. D., Drogui, P., & Surampalli, R. Y. (2016). Ultrasonication aided biodiesel production from one-step and two-step transesterification of sludge derived lipid. *Energy*, *94*, 401–408. doi:10.1016/j.energy.2015.11.016

Zhang, X., Yan, S., Tyagi, R. D., Surampalli, R. Y., & Valéro, J. R. (2014). Ultrasonication aided in-situ transesterification of microbial lipids to biodiesel. *Bioresource Technology*, *169*, 175–180. doi:10.1016/j.biortech.2014.06.108 PMID:25050978

Zhang, Y., Dubé, M. A., McLean, D. D., & Kates, M. (2003). Biodiesel production from waste cooking oil: 2. Economic assessment and sensitivity analysis. *Bioresource Technology*, *90*(3), 229–240. doi:10.1016/S0960-8524(03)00150-0 PMID:14575945

Zhang, Z., Bi, P., Jiang, P., Fan, M., Deng, S., Zhai, Q., & Li, Q. (2015). Production of gasoline fraction from bio-oil under atmospheric conditions by an integrated catalytic transformation process. *Energy*, *90*, 1922–1930. doi:10.1016/j.energy.2015.07.009

Zheng, M., Mulenga, M. C., Wang, M., & Tjong, J. (2008). Biodiesel engine performance and emissions in low temperature combustion. *Fuel*, *87*(6), 714–722. doi:10.1016/j.fuel.2007.05.039

Zhu, D., Li, X., Wang, N., Wang, X., Gao, J., & Li, H. (2009). Dispersion behavior and thermal conductivity characteristics of Al_2O_3–H_2O nanofluids. *Current Applied Physics*, *9*(1), 131–139. doi:10.1016/j.cap.2007.12.008

About the Contributors

J. Sadhik Basha obtained his B.E. degree in Mechanical Engineering with first class from the University of Madras in the year 1999. He obtained his M.E. degree in Thermal Engineering with first-class distinction from Anna University in the year 2005. He completed his Ph.D. degree at the National Institute of Technology, Tiruchirappalli, India in the year 2011. He is currently working as a Professor, Department of Process/Mechanical Engineering, International Maritime College Oman, Sohar, Sultanate of Oman. Dr. Basha is an experienced educator and researcher with more than seventeen years of experience. His current research interests include water-diesel/biodiesel emulsions as alternative fuels, combustion modeling, spray combustion, nano-additive emulsion fuels, thermal sciences, and renewable energy. He is an expert Referee, Editorial Board Member, and Advisory Board Member for more than fifty reputed International Journals in the field of Energy, Combustion, and Power Engineering. Dr. Basha is a lifetime member of EAMAE and Combustion Institute professional organizations. He has published more than fifty research articles and participated in several national and International Conferences. Further, the contents of his research work have been cited in more than a thousand times in various Technical Articles/Magazines, and International Journals.

R. B. Anand is a Professor at the Mechanical Engineering Department, National Institute of Technology, Tiruchirappalli, India where he has been a faculty member in this institution since 2007. He also worked as an Assistant Director at National Power Training Institute (NPTI), New Delhi, India from 2002 to 2007. He obtained his Ph.D. in 2002 from IIT, Delhi, and M.Tech. in 1999 from IIT Madras, India. His areas of research include Computational Fluid Dynamics, Internal combustion engines, Alternative fuels, Emission control, Renewable energy. He has published more than sixty international journals and participated in more than fifty National/International conferences. Dr. R. B. Anand has successfully guided seven Ph.D. scholars to his credit..

***

Avinash Alagumalai is working as an Assistant Professor in Department of Mechanical Engineering, GMR Institute of Technology, Rajam. His areas of interest comprise usage of bio-fuels in internal combustion engine and engine characteristics. He has authored/co-authored 30+ articles published in leading journals.Besides, he is available as reviewer in Wiley, Interscience, ACS publications, Taylor and Francis, and Elsevier.

B. Ashok received his B.E degree in Mechatronic Engineering and M.Tech degree in Automotive Engineering. He received his PhD degree from VIT University, India, in the thesis title of "Development of throttle control system for SI engine". To his credit he got a funded project from Department of science and technology (DST) under the ASEAN scheme. He has published over 80 international research papers in reputed peer reviewed journals with cumulative Thomson Reuters Impact factor (Web of Science)) of 190.143 in the field of IC engines, alternate fuels, automotive engineering and automotive control system areas. To his credit a patent has been filled in the area of automotive engineering. Presently he is guiding four full time research students.

A.Tamilvanan is currently working as an Assistant Professor in Mechanical Engineering Department, Kongu Engineering College, Erode, Tamil Nadu, India. He has 8 years of teaching experience and he published 11 journals in national and international level. His area of interest is biofuels and nano fluids especially nano fuel additive to bio-diesel in CI engines.

S. Balamurugan completed his Master of Engineering in Heat Power Engineering from Coimbatore Institute of Technology, Coimbatore, Tamil Nadu, India. He is rightly working in Sri Krishna College of Engineering and Technology, Coimbatore, as a Assistant Professor in the Department of Mechanical Engineering.

Dhinesh Balasubramanian An Optimistic and Committed Automobile Engineer at Doctorate level, with 7 years of experience (out of which nearly 3 years as fulltime researcher) in the field of renewable energy – research, industrial and teaching environs. Technically sound and good in precedency and ceasing tasks in a timely manner. Devoted to loyalty, dignity and excellence.

P. Baranitharan is currently working as post doctoral research fellow at National Institute of Technology, Trichy. He is expert in biomass pyrolysis and Engine analysis.

H. Sharon Hilary Doss received the BE degree (Mechanical) from Government College of Engineering Tirunelveli in 2009, the ME degree (Energy) from Anna University of Technology Tirunelveli in 2011 and the Ph.D. degree (Mechanical-Thermal Sciences) form the Indian Institute of Technology Madras in 2018. He served as Lecturer in National Engineering College Kovilpatti from 2011 to 2012. He has published 11 peer reviewed journal articles, 9 conference papers and 1 book chapter. He was awarded Institute Research Award and Shri. J. C. Bose Patent Award in 2017 by Indian Institute of Technology Madras for his research works quality and contribution in developing Solar Zero Discharge Toilet. His research interests are Alternate Fuels, adoption of Solar Energy for Desalination & Wastewater Treatment and conversion of Waste to Energy.

M. Feroskhan has earned his doctoral degree from Vellore Institute of Technology (VIT) Chennai, India, where he is currently working as an Assistant Professor in Mechanical Engineering. He has more than 15 research publications. His areas of interest include Advanced Internal Combustion Engine Technologies, Alternative Fuels, Nanotechnology and Heat Transfer.

K. Gomathi is working as Assistant Professor in Department of Mechatronics Engineering at Kongu Engineering College since 2001. She is expert person in power electronics and MEMS. At present she is working in engine vibration analysis using MEMS based sensors.

Saleel Ismail has completed his doctoral research at Indian Institute of Technology (IIT) Madras. He has worked as a Lead Engineer at General Electric (GE) Aviation and as an Associate Professor at Vellore Institute of Technology (VIT) Chennai. He is currently an Assistant Professor in Mechanical Engineering at National Institute of Technology (NIT) Calicut. He has more than 30 research publications. His areas of interest include Thermodynamic Analysis and Modelling, Internal Combustion Engines, Computational Fluid Dynamics, Solar Energy Systems etc.

R. Joel is a student in Karunya University doing my b tech mechanical engineering.

V. Karthickeyan obtained his BE degree in Mechanical Engineering with first class from Anna University in the year 2011. He obtained his M.E degree in Internal Combustion Engineering with first class distinction from Anna University in the year 2013. He completed his PhD degree at Anna University in the year 2018. He has published more than 30 articles in the area of IC engines, alternative fuel, engine modification and emission reduction techniques in International journals.

Joseph Marshal is currently working as Assistant Professor at Karunya Institute of Technology and Sciences. He is expert resource person in the field of biomass gasification and pyrolysis studies.

Thangamuthu Mohanraj, born in 1988, is currently working as Assistant Professor in the Department of Mechanical Engineering at Amrita School of Engineering, Coimbatore, Amrita Vishwa Vidyapeetham, India. He obtained his Ph.D. degree from Anna University, Chennai in 2018. He is specialized in Mechatronics, Sensors and Signal Processing, Sensor fusion, Energy, Optimization, Nanoparticles, and Condition Monitoring.

M. Mohamed Musthafa graduated from Madurai kamaraj University, Tamilnadu in1993 and obtained post graduate degree in Thermal Plant Engineering from SASTRA University, Thanjavur, Tamilnadu in 2002 and secured Ph.D from National Institute of Technology, Trichy, Tamilnadu in 2011. At present, he is at faculty of school of Mechanical Engineering, SASTRA Deemed University. He has a vast teaching experience of more than two decades in a various reputed Engineering college. Dr.Mohamed Musthafa has large number research publication in the high impact factor Journals. His areas of interest include fuels & combustion, LHR engines, Alternative fuels, Heat transfer, Solar, etc.

C. Prabha Working as Assistant Professor in Department of Mechanical Engineering at Karunya Institute of Technology and Sciences, Coimbatore.

A. Prabu is a Professor of Mechanical Engineering and currently pursuing his post doctoral research in Jiangsu University, China. He obtained his Ph.D. in I.C. Engines from National Institute of Technology, Tiruchirappalli, India. Previously, he worked for SRM institute of science and Technology, where his intimate knowledge towards teaching and research activities were highly appreciated by his colleagues, management and students. His area of research interest includes I.C. Engines, Renewable energy, Alternative fuels, Emission control and Automobile Engineering. He has published many high impact journal publications and presented many international and national conference papers. He is also member of various mechanical and automotive societies and active reviewer for many international journal and book publications.

V. Dhana Raju currently works Associate Professor at the Department of Mechanical Engineering, Lakireddy Bali Reddy College of Engineering, An Optimistic and Committed Mechanical Engineer at Doctorate level, with 6 years of experience in the field of renewable energy – research and teaching environs. Technically sound and good in precedence and ceasing tasks in a timely manner.

R. Sakthivel is working as Assistant Professor at Department of Mechanical Engineering in Amrita Vishwa Vidyapeetham, Coimbatore since June 2018. He is expert in biomass energy conversion and have wide range of publications in leading peer reviewed Journals. He is currently working in biomass pyrolysis and assessment of storage stability of biofuels obtained from pyrolysis.

P. Selvakumar A passionate teacher with 15 years of experience in teaching and research.My areas of expertise are Heat Transfer, Computational Fluid Dynamics, Conventional and Renewable Energy Systems.

Sandip Singh was born in 1986 in India and completed his graduation and post-graduation from Veer Bahadur Singh Purvanchal University, Jaunpur India in 2007 and in 2010 respectively. He received his Ph. D. Degree in 2017 at Catalysis and Inorganic Chemistry Division, Council of Scientific & Industrial Research- National Chemical Laboratory Pune, India in catalytic upgradation of woody biomass using ionic liquids, under the supervision of Dr. Paresh Laxmikant Dhepe. He then joined Professor Shaikh Mohamad Mobin research group's as a Postdoctoral Research Associate at Discipline of Chemistry, Indian Institute Technology Indore, India in 2017. His research was highlighted on design and synthesis of single/mixed source molecular precursors-based metal oxide nanoparticles for catalytic applications. Currently, he is working as a Research Employee with Prof. David B. Hodge's lab at Chemical and Biological Engineering, Montana State University Bozeman, MT, USA and his research focuses on alkaline-oxidative pre-treatment of woody biomass for optimal co-products production.

M. Sivakumar is currently working as Associate Professor in Department of Automobile Engineering, K.S.R. College of Engineering, Tiruchengode, Tamil Nadu, INDIA. He had received his B.E. degree in Mechanical Engineering, from Bharathiar University, Coimbatore in 1997, M.E. degree in Energy Engineering and Management from Annamalai University, Chidambaram, and he was awarded Ph.D in the faculty of Mechanical engineering in the 2018 by Anna University, Chennai. His research interest includes IC engines and alternate fuels. He has 15 years of teaching experience in reputed institutions and 5 years of industries. He has published 3 papers in Scopus indexed journals.

N. Shanmuga Sundaram received his Bachelor of Engineering Degree in Mechanical Engineering and Master of Engineering Degree in Thermal Engineering both from Bharathiar University, Coimbatore, India in 1997 and 2001 respectively. He received his Doctorate in Mechanical Engineering from Anna University, Chennai, India in 2011. He is currently working as Professor and Head of Mechanical Engineering Department at SVS College of Engineering, Coimbatore, India since May 2011. Before joining to this institute he served in PSG College of Technology, and Government College of Technology, Coimbatore. He served as a dedicated teaching faculty for the past 19 years and he has 3 years of industrial experience. He attended more than 40 FDPs / Workshops at IITs and other reputed institutions. He possesses vast experience in preparing the documents for NBA, NAAC and implementing Outcome Based Education. He established Thermal Engineering Laboratory and Heat Power Laboratory at SVS College of Engineering. He guided more than 30 UG project works, 10 PG project works and 1 Ph.D. research work. At present he is guiding 3 Ph.D. research project works. His areas of research are Friction Stir Welding, IC Engines and Heat Transfer. He published his research works in more than10 research papers in various international journals. He is a reviewer for 5 international journals. He is an active life member of Indian Welding Society and SAE Faculty Advisor, under his guidance SAE MegaATV, and SAE Efficycle are fabricated. He organized several national level workshops and 2 international conference. He is an active user of 'Virtual Lab', and he received an award for the same. He received the awards viz. 'Outstanding Faculty in Engineering', and 'Certificate of Appreciation'. and 'Innovative Technological Scientific Research & Dedicated Best Professor Award'.

S. Thiyagarajan completed his PhD in the field of 'CO2 emission reduction from CI engine' from SRMIST. He has published more than 35 research articles in peer-reviewed high impact journals. He is presently working as Assistant Professor in Department of Automobile Engineering, SRMIST. His current research work includes advanced combustion modes, hydrogen production through waste heat and waste to energy.

Index

D

E

F

G

H

I

J

L

M

N

O

P

R

S

T

U

V

W

X

Z

Ensure Quality Research is Introduced to the Academic Community

Become an IGI Global Reviewer for Authored Book Projects

The overall success of an authored book project is dependent on quality and timely reviews.

In this competitive age of scholarly publishing, constructive and timely feedback significantly expedites the turnaround time of manuscripts from submission to acceptance, allowing the publication and discovery of forward-thinking research at a much more expeditious rate. Several IGI Global authored book projects are currently seeking highly-qualified experts in the field to fill vacancies on their respective editorial review boards:

Applications and Inquiries may be sent to:
development@igi-global.com

Applicants must have a doctorate (or an equivalent degree) as well as publishing and reviewing experience. Reviewers are asked to complete the open-ended evaluation questions with as much detail as possible in a timely, collegial, and constructive manner. All reviewers' tenures run for one-year terms on the editorial review boards and are expected to complete at least three reviews per term. Upon successful completion of this term, reviewers can be considered for an additional term.

If you have a colleague that may be interested in this opportunity, we encourage you to share this information with them.

www.ingramcontent.com/pod-product-compliance
Lightning Source LLC
LaVergne TN
LVHW081252011026
849162LV00006BA/1533